Mitochondrial DNA

METHODS IN MOLECULAR BIOLOGY™

John M. Walker, SERIES EDITOR

METHODS IN MOLECULAR BIOLOGY™

Mitochondrial DNA

Methods and Protocols

Edited by

William C. Copeland

Laboratory of Molecular Genetics
National Institute of Environmental Health Sciences
National Institutes of Health, Research Triangle Park, NC

Humana Press ✳ Totowa, New Jersey

Cover illustration: Representative results from *in situ* PCR utilizing the reporter dye-quencher dye system (Taqman): Son with mtDNA deletion in marrow. *See* full caption for Fig. 2, Chapter 6 on page 122.

For additional copies, pricing for bulk purchases, and/or information about other Humana titles, contact Humana at the above address or at any of the following numbers: Tel: 973-256-1699; Fax: 973-256-8341; E-mail: humana@humanapr.com or visit our website at http://humanapress.com

Printed in the United States of America. 10 9 8 7 6 5 4 3 2 1

Library of Congress Cataloging-in-Publication Data

Mitochondrial DNA / edited by William C. Copeland.
       p. cm. -- (Methods in molecular biology ; v. 197)
    Includes bibliographical references and index.
    ISBN 0-89603-972-2 (alk. paper)
    1. Mitochondrial DNA--Laboratory manuals. I. Copeland, William C. (William Chenery) II. Methods in molecular biology (Totowa, N.J.); v. 197.

  QH603.M5 M558 2002
  572.8'6--dc21

                                                                        2002020564

Preface

Mitochondria are semiautonomous organelles that contain their own DNA and are responsible for the bulk of ATP synthesis in the eukaryotic cell. Mitochondrial functions are linked to the aging process, apoptosis, sensitivities to anti-HIV drugs, and, possibly, some cancers. Mitochondria were first visualized as discrete organelles by light microscopy in 1840. However, isolation of intact mitochondria had to wait until zonal centrifugation methods were developed in 1948. In the early 1960s it was determined that these cytoplasmic organelles contain their own DNA. The DNA sequence of human mitochondrial DNA (mtDNA) was determined in 1981 and gene products were assigned by 1985, making it the first component of the human genome to be fully sequenced. Human mtDNA is a double-stranded 16,569 bp circular genome, coding for 13 polypeptides required for oxidative phosphorylation and 22 transfer RNAs and 2 ribosomal RNAs responsible for their synthesis. One noncoding segment, the displacement loop, contains several *cis*-acting elements required for initiation of transcription and replication. Mitochondrial DNA makes up only 1% of total cellular DNA, and the mtDNA polymerase activity accounts for less than 1% of the total DNA polymerase activity in the cell. Individual cells have up to 10,000 discrete mitochondrial genomes distributed within 10 to 1000 organelles. Heteroplasmy, mitochondrial genetic diversity within a single cell, can result from point mutations or deletions in mtDNA and usually increases exponentially with age. Defects in mitochondrial function produce a wide range of human diseases and can be caused by mutations within the mtDNA. The first mutation discovered in mtDNA to be the cause of a mitochondrial disease, Leber's hereditary optic neuropathy, was first identified by Douglas Wallace and coworkers in 1988. Since that time several hundred point and deletion mutations in mtDNA have been described as the causes of mitochondrial disorders. The Mitochondrial and Metabolic Disease Center reports that more than 1 in 4000 children born in the United States each year will develop a mitochondrial disease by age 10 with a mortality rate from 10 to 50%. Over 50 million people in the United States suffer from chronic degenerative disorders, and defects in mitochondrial function have been linked to several of the most common diseases of aging. Mutations within mtDNA and nuclear genes involved in the maintenance of mtDNA are the main cause of these mitochondrial diseases.

The aim of *Mitochondrial DNA: Methods and Protocols* is to provide procedures for research scientists and clinicians to analyze mitochondrial DNA and the proteins involved in its maintenance. A review of mitochondrial function, genetics, and diseases is presented in the first chapter by Douglas Wallace. *Mitochondrial DNA: Methods and Protocols* is organized into two main parts, the first addressing protocols to analyze the DNA. Specifically, methods are presented to allow purification of mtDNA from a variety of sources and to analyze the DNA for deletions, point mutations, and damage. Part I is not limited to detection of DNA damage and mutations but includes protocols for the analysis of replication intermediates and for following the fate of mtDNA outside the mitochondria. Part II contains protocols to analyze the proteins and enzymes that maintain mtDNA. There has been a growing interest in understanding the proteins and enzymes that maintain mtDNA especially since all errors in DNA must involve these proteins before any mutation is fixed in the genome. The interest in mtDNA replication and repair proteins has expanded beyond the research scientist to the clinician seeking to identify potential causes of mitochondrial dysfunction diseases. For example, a mutation in a conserved amino acid in the human mtDNA polymerase has recently been shown to be associated with progressive external ophthalmoplegias characterized by mtDNA deletions. Clearly, it is imperative to understand how mutations in the genes for mtDNA replication and repair proteins contribute to disease. Thus, protocols are presented for the purification and analysis of these essential enzymes and proteins involved in DNA replication, repair, and transcription. These protocols will provide a starting point for investigating the cause of mitochondrial dysfunction and disease.

I thank the authors for their excellent contributions that have made this book possible and John Walker for his editorial skills and patience. I also thank Matthew Longley for his assistance with graphics and Valerie Copeland for her steadfast support.

William C. Copeland

Contents

Contributors

CHANDRAMOHAN V. AMMINI • *Department of Medicine, University of Florida College of Medicine, Gainesville, FL*

GIUSEPPE ATTARDI • *Division of Biology, California Institute of Technology, Pasadena, CA*

SAMUEL E. BENNETT • *Department of Environmental and Molecular Toxicology, Oregon State University, Corvallis, OR*

DANIEL F. BOGENHAGEN • *Department of Pharmacological Sciences, State University of New York at Stony Brook, Stony Brook, NY*

VILHELM A. BOHR • *Laboratory of Molecular Gerontology, National Institute on Aging, National Institutes of Health, Baltimore, MD*

RONALD A. BUTOW • *Department of Molecular Biology, University of Texas Southwestern Medical Center, Dallas, TX*

WILLIAM C. COPELAND • *Laboratory of Molecular Genetics, National Institute of Environmental Health Sciences, National Institutes of Health, Research Triangle Park, NC*

GINO CORTOPASSI • *Molecular Biosciences, University of California, Davis, CA*

ALLISON W. DOBSON • *Department of Cell Biology and Neuroscience, University of South Alabama College of Medicine, Mobile, AL*

MATS EKSTRAND • *Department of Medical Nutrition, Karolinska Institutet, Novum, Huddinge Hospital, Sweden*

CAROL L. FARR • *Department of Biochemistry and Molecular Biology, Michigan State University, East Lansing, MI*

FRANÇOISE FOURY • *Unité de Biochimie Physiologique, Université Catholique de Louvain, Louvain-la-Neuve, Belgium*

MARIANA GERSCHENSON • *Division of Heart and Vascular Diseases, National Heart, Lung, and Blood Institute, National Institutes of Health, Bethesda, MD*

STEVEN C. GHIVIZZANI • *Center for Molecular Orthopedics, Harvard Medical School, Boston, MA*

WILLIAM W. HAUSWIRTH • *Departments of Molecular Genetics and Ophthalmology and Powell Gene Therapy Center, University of Florida College of Medicine, Gainesville FL*

LAURIE S. KAGUNI • *Department of Biochemistry and Molecular Biology, Michigan State University, East Lansing, MI*

BRETT A. KAUFMAN • *Department of Molecular Biology, University of Texas Southwestern Medical Center, Dallas, TX*

MARK R. KELLEY • *Wells Center for Pediatric Research, Department of Pediatrics, Indiana University Medical School, Indianapolis, IN*

SILJA KUUSK • *Estonian Biocentre, Tartu, Estonia*

NILS-GÖRAN LARSSON • *Department of Medical Nutrition, Karolinska Institutet, Novum, Huddinge Hospital, Sweden*

SUSAN P. LEDOUX • *Department of Cell Biology and Neuroscience, University of South Alabama College of Medicine, Mobile, AL*

KANG LI • *Departments of Internal Medicine and Molecular Biology, University of Texas Southwestern Medical Center, Dallas, TX*

MATTHEW J. LONGLEY • *Laboratory of Molecular Genetics, National Institute of Environmental Health Sciences, National Institutes of Health, Research Triangle Park, NC*

HEATHER E. LORIMER • *Department of Biological Sciences, Youngstown State University, Youngstown, OH*

ROBERT L. LOW • *Department of Pathology, University of Colorado Health Sciences Center, Denver, CO*

CORT S. MADSEN • *Bristol Myers Squibb, Princeton, NJ*

BHASKAR S. MANDAVILLI • *Laboratory of Molecular Genetics, National Institute of Environmental Health Sciences, National Institutes of Health, Research Triangle Park, NC*

LUISA A. MARCELINO • *Center for Environmental Health Sciences, Division of Bioengineering and Environmental Health, Massachusetts Institute of Technology, Cambridge, MA*

YUICHI MICHIKAWA • *Frontier Research Center, National Institute of Radiological Sciences, Inage, Chiba, Japan*

DALE W. MOSBAUGH • *Departments of Environmental and Molecular Toxicology and of Biochemistry and Biophysics, and Environmental Health Science Center, Oregon State University, Corvallis, OR*

ALI NAINI • *Department of Neurology, Columbia University College of Physicians and Surgeons, New York, NY*

ROBERT K. NAVIAUX • *Mitochondrial and Metabolic Disease Center, Department of Medicine, University of California, San Diego, School of Medicine, San Diego, CA*

SCOTT M. NEWMAN • *Xenon Genetics, Inc., Vancouver, B.C., Canada,*

SHUJI NOMOTO • *Head and Neck Cancer Research Division, Department of Otolaryngology–Head and Neck Surgery, Johns Hopkins University School of Medicine, Baltimore, MD*

SIMON G. NYAGA • *Laboratory of Molecular Gerontology, National Institute on Aging, National Institutes of Health, Baltimore, MD*

PHILIP S. PERLMAN • *Department of Molecular Biology, University of Texas Southwestern Medical Center, Dallas, TX*

MONTSERRANT SANCHEZ-CESPEDES • *Head and Neck Cancer Research Division, Department of Otolaryngology–Head and Neck Surgery, Johns Hopkins University School of Medicine, Baltimore, MD*

JANINE H. SANTOS • *Laboratory of Molecular Genetics, National Institute of Environmental Health Sciences, National Institutes of Health, Research Triangle Park, NC*

JUHAN SEDMAN • *Estonian Biocentre, Tartu, Estonia*

TIINA SEDMAN • *Estonian Biocentre, Tartu, Estonia*

ERIC A. SCHON • *Departments of Neurology and of Genetics and Development, Columbia University College of Physicians and Surgeons, New York, NY*

SARA SHANSKE • *Department of Neurology, Columbia University College of Physicians and Surgeons, New York, NY*

MARY JANE SHROYER • *Department of Environmental and Molecular Toxicology, Oregon State University, Corvallis, OR*

DAVID SIDRANSKY • *Head and Neck Cancer Research Division, Department of Otolaryngology–Head and Neck Surgery, Johns Hopkins University School of Medicine, Baltimore, MD*

MICHELINE K. STRAND • *Laboratory of Molecular Genetics, National Institute of Environmental Health Sciences, Research Triangle Park, NC*

JUNG-SUK SUNG • *Department of Environmental and Molecular Toxicology, Oregon State University, Corvallis, OR*

WILLIAM G. THILLY • *Center for Environmental Health Sciences, Division of Bioengineering and Environmental Health, Massachusetts Institute of Technology, Cambridge, MA*

MARY K. THORSNESS • *Department of Molecular Biology, University of Wyoming, Laramie, WY*

PETER E. THORSNESS • *Department of Molecular Biology, University of Wyoming, Laramie, WY*

BENNETT VAN HOUTEN • *Laboratory of Molecular Genetics, National Institute of Environmental Health Sciences, National Institutes of Health, Research Triangle Park, NC*

DOUGLAS C. WALLACE • *Center for Molecular Medicine, Emory University School of Medicine, Atlanta, GA*

KAREN H. WHITE • *Department of Molecular Biology, University of Wyoming, Laramie, WY*

R. SANDERS WILLIAMS • *Departments of Internal Medicine and Molecular Biology, University of Texas Southwestern Medical Center, Dallas, TX (Currently: Duke University School of Medicine, Durham, NC)*

GLENN L. WILSON • *Department of Cell Biology and Neuroscience, University of South Alabama College of Physicians and Surgeons, Mobile, AL*

ALICE WONG • *Department of Molecular Biosciences, University of California, Davis, CA*

WEIMING ZHENG • *Center for Environmental Health Sciences, Division of Bioengineering and Environmental Health, Massachusetts Institute of Technology, Cambridge, MA*
STEVEN J. ZULLO • *Advanced Technology Program, National Institute of Standards and Technology, Gaithersburg, MD*

I

METHODS FOR THE ANALYSIS OF mtDNA

1

Animal Models for Mitochondrial Disease

Douglas C. Wallace

1. Introduction

Although a variety of degenerative diseases are now known to be caused by two mutations in mitochondrial genes, the pathophysiology of these diseases remains poorly understood. As a consequence, relatively little progress has been made in developing new therapies for mitochondrial diseases. What has been needed are animal models for these diseases that are amenable to detailed biochemical, physiological, and molecular analysis, and on which promising therapies can be tested. In the past 5 yr, this deficiency has begun to be addressed by the construction of a number of mouse models of mitochondrial disease. These have already revolutionized our understanding of the pathophysiology of mitochondrial disease and demonstrated the efficacy of some new antioxidant drugs.

1.1. Mitochondrial Biology and Genetics

The mitochondria generate much of the cellular energy through the process of oxidative phosphorylation (OXPHOS). As a byproduct, they produce most of the endogenous toxic reactive oxygen species (ROS). The mitochondrial are also the central regulator of apoptosis (programmed cell death), a process initiated by the activation of the mitochondrial permeability transition pore (mtPTP). These interrelated mitochondrial systems are assembled from roughly 1000 genes distributed between the two very different genetic systems of the mammalian cell: the nuclear genome and the mitochondrial genome. Hence, the complexities of mitochondrial disease reflect the intricacies of both the physiology and the genetics of the mitochondrion.

From: *Methods in Molecular Biology, vol. 197: Mitochondrial DNA: Methods and Protocols*
Edited by: W. C. Copeland © Humana Press Inc., Totowa, NJ

1.1.1. Mitochondrial Physiology

To understand the pathophysiology of mitochondrial diseases, it is necessary to understand the physiology of OXPHOS. The mitochondria oxidize hydrogen derived from carbohydrates and fats to generate water and ATP (*see* **Fig. 1**). Reducing equivalents in the form of hydrogen are recovered from carbohydrates by the tricarboxylic acid (TCA) cycle, while those recovered from fats are collected through β-oxidation. The resulting electrons are transferred to NAD^+, to give $NADH + H^+$, or to flavins located in iron–sulfur (Fe–S)-center-containing enzymes that interface with the electron transport chain (ETC). Electrons donated from $NADH + H^+$ to complex I (NADH dehydrogenase) or from succinate to complex II (succinate dehydrogenase, SDH) are passed sequentially to ubiquinone (coenzyme Q or CoQ) to give ubisemiquinone (CoQH$^\bullet$) and then ubiquinol (CoQH$_2$). Ubiquinol transfers its electrons to complex III (ubiquinol:cytochrome-c oxidoreductase) which transfers the electrons to cytochrome-c. From cytochrome-c, the electrons move to complex IV (cytochrome-c oxidase, COX) and finally to oxygen to give H_2O. The energy released by this ETC is used to pump protons out of the mitochondrial inner membrane, creating the trans-membrane, electrochemical gradient ($\Delta\mu^{H+}$),

---

Fig. 1. (*see opposite page*) Diagram showing the relationships of mitochondrial oxidative phosphorylation (OXPHOS) to (1) energy (ATP) production, (2) reactive oxygen species (ROS) production, and (3) initiation of apoptosis through the mitochondrial permeability transition pore (mtPTP). The OXPHOS complexes, designed I to V, are as follows: complex I (NADH: ubiquinone oxidoreductase) encompassing a FMN and six Fe–S centers (designated with a cube); complex II (succinate: ubiquinone oxidoreductase) involving an FAD, three Fe–S centers, and a cytochrome-*b*; complex III (ubiquinol: cytochrome-*c* oxidoreductase) encompassing cytochromes-*b* and *c*1 and the Rieske Fe–S center; complex IV (cytochrome-*c* oxidase) encompassing cytochromes $a+a_3$ and CuA and CuB; and complex V (H$^+$-translocating ATP synthase). Pyruvate from glucose enters the mitochondria via pyruvate dehydrogenase (PDH), generating acetyl CoA that enters the TCA cycle by combining with oxaloacetate (OAA). *Cis*-Aconitase converts citrate to isocitrate and contains an 4Fe–4S center. Lactate dehydrogenase (LDH) converts excess pyruvate plus NADH to lactate (*1–3*). Small molecules defuse through the outer membrane via the voltage-dependent anion channel (VDAC) or porin. The VDAC together with ANT, Bax, and the cyclophilin D (CD) protein are thought to come together at the mitochondrial inner and outer membrane contact points to create the mtPTP. The proapoptotic Bax of the mtPTP is thought to interact with the antiapoptotic Bcl2 and the benzodiazepine receptor (BD). The opening of the mtPTP is associated with the release of cytc, activating which activates Apaf-1 that then binds to and activates procaspase-9. The activated caspase-9 then initiates the proteolytic degradation of cellular proteins (*4–7*). Modified from **ref.** *8* with permission from *Science.*)

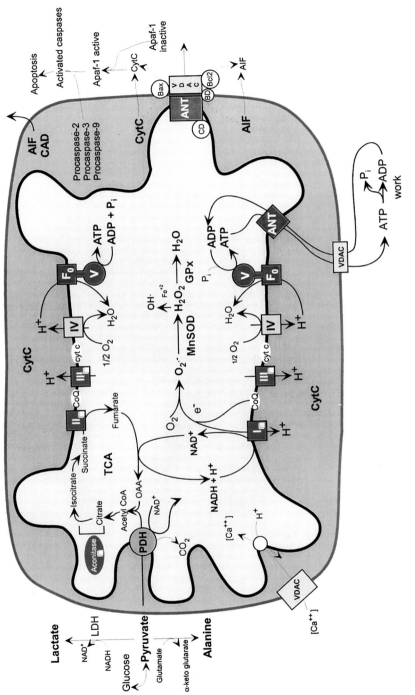

Fig. 1.

$\{\Delta\mu^{H+} = \Delta\psi + \Delta pH\}$. The potential energy stored in $\Delta\mu^{H+}$ is used to condense ADP and Pi to make ATP via complex V (ATP synthase), driven by the movement of protons back through a complex V proton channel.

Each of the ETC complexes incorporates multiple electron carriers. Complexes I, II, and III encompass several Fe–S centers, whereas complexes III and IV encompass the cytochromes. Mitochondrial aconitase also contains an Fe–S center *(8–10)*.

Matrix ATP is exchanged for cytosolic ADP by the adenine nucleotide translocator (ANT). ANT isoforms are derived from multiple genes. In humans, there are three tissue-specific isoforms *(11)*: a heart-muscle-specific isoform (ANT1) located at the chromosome 4q35 locus *(12–19)*, an inducible isoform (ANT2) located at Xq24 *(13,20–23)*, and a systemic isoform (ANT3) located in the pseudoautosomal region at Xp22.3 *(13,14,24,25)*. In mouse, there are only two ANT genes (*Ant1* and *Ant2*), homologs of the human ANT1 and ANT2 proteins *(26)*. Mouse *Ant1* maps to chromosome 8, syntenic to human 4q35 *(27,28)*, whereas mouse *Ant2* maps to regions A-D of X chromosome, syntenic to human Xq24 *(29)*.

Because the ETC is coupled to ATP synthesis through $\Delta\mu^{H+}$ mitochondrial oxygen consumption is regulated by $\Delta\mu^{H+}$ and hence the matrix concentration of ADP. In the absence of ADP, oxygen consumption is slow (state IV respiration). However, when ADP is added and transported into the matrix by the ANT, $\Delta\mu^{H+}$ falls. As the ATP synthase utilizes the proton gradient to phosphorylate the ADP back to ATP, oxygen consumption goes up as the ETC reconstitutes $\Delta\mu^{H+}$ (state III respiration). The ratio of state III and state IV respiration is called the respiratory control ratio (RCR) and the amount of molecular oxygen consumed relative to the ADP phosphorylated is the P/O ratio. Addition of uncouplers such as 2,4-dinitrophenol (DNP) and FCCP collapses $\Delta\mu^{H+}$ and permits the ETC and oxygen consumption to run at their maximum rates. Cells regulate $\Delta\mu^{H+}$ through the uncoupling proteins, Ucp. These proteins form proton channels in the mitochondrial inner membrane. Mammals have three uncoupling proteins. Uncoupler protein 1 (*Ucp1*) is primarily associated with brown adipose tissue (BAT), where it functions in thermal regulation. It is strongly induced by cold stress through a β3-adenergic response pathway *(30–33)*. Uncoupler protein 2 (*Upc2*) has 59% amino acid identity to *Ucp1* and is widely expressed in adult human tissues with mRNA levels being highest in skeletal muscle. It is also upregulated in white fat in response to an increased fat diet. In mouse, it has been linked to quantitative trait locus for hyperinsulinemia and obesity *(33)*. Uncoupler protein 3 (*Ucp3*) is 57% identical to *Ucp1* and 73% identical to *Ucp2*. *Ucp3* is widely expressed in adult tissues and at particularly high levels in skeletal muscle. Moreover, it is hormonally regulated, being induced in skeletal muscle by thyroid hormone, in white fat by β3–adrenergic agonists, and also regulated by dexamethasone, leptin, and starvation. *Ucp3* is located adjacent to *Ucp2* in human chromosome 11q13 and mouse chromosome 7 *(34–37)*.

Superoxide anion ($O_2^{\cdot-}$) is generated from OXPHOS by the transfer of one electron from the ETC to O_2 (*see* **Fig. 1**). Ubisemiquinone, localized at the CoQ binding sites of complexes I, II, and III, appears to be the primary electron donor. Because the free-radical ubisemiquinone is the probable electron donor in the ETC, conditions that maximize the levels of ubisemiquinone should also maximize mitochondrial ROS production. This would occur when the ETC is primarily but not completely reduced. This might explain why mitochondrial ROS production is further increased when uncouplers are added to Antimycin A-inhibited mitochondria *(38–41)*.

The $O_2^{\cdot-}$ is converted to H_2O_2 by Mn superoxide dismutase (MnSOD) or cytosolic Cu/ZnSOD, and the resulting H_2O_2 is reduced to water by glutathione peroxidase (GPx1) or catalase. However, H_2O_2, in the presence of reduced transition metals, can be converted to the highly reactive hydroxyl radical ($^-$OH). Major targets of ROS reactivity are the Fe–S centers of the TCA cycle and the ETC. Hence, mitochondria are particularly sensitive to oxidative stress *(8,42–45)*.

Superoxide production and H_2O_2 generation are highest when the ETC is more reduced (state IV respiration) and lowest when it is more oxidized (state III respiration) *(46–50)*. Therefore, the blocking of electron flow through the ETC by drugs such as Antimycin A, which inhibits complex III, stimulates ROS production *(38,46,48,50)*.

The mitochondria are also the major regulators of apoptosis, which is initiated though the opening of mtPTP (*see* **Fig. 1**). The mtPTP is thought to be composed of the inner membrane ANT, the outer membrane voltage-dependent anion channel (VDAC) or porin, Bax, Bcl2, cyclophilin D, and the benzodiazepine receptor *(4,51,52)*. When the mtPTP opens, $\Delta\mu^{H+}$ collapses and ions equilibrate between the matrix and cytosol, causing the mitochondria to swell. Ultimately, this disrupts the outer membrane, releasing the contents of the intermembrane space into the cytosol. The intermembrane space contains a number of cell-death-promoting factors, including cytochrome-*c*, procaspases-2, -3, and -9, apoptosis-initiating factor (AIF), as well as the caspase-activated DNase (CAD) *(5,53–56)*. On release, cytochrome-*c* interacts with the cytosolic Apaf-1 and procaspase-9 complex. This cleaves and activates procaspase-9. Caspase-9 then cleaves procaspase-3, which activates additional hydrolytic enzymes, destroying the cytoplasm. AIF and CAD are transported to the nucleus, where they degrade the chromatin *(8)*.

The mtPTP can be stimulated to open by uptake of excessive Ca^{2+}; increased oxidative stress, decreased mitochondrial $\Delta\mu^{H+}$, ADP, and ATP, and ANT ligands such as atractyloside *(4,5)*. Thus, disease states that inhibit OXPHOS and increase ROS production should also increase the propensity of cells to undergo apoptosis *(4,6,7)*.

There are two major apoptosis pathways, the "mitochondrial" or "cellular stress" pathway described earlier and the "death ligand/receptor" pathway. The "mitochondrial" pathway is initiated by cytochrome-*c* release from the mitochondrion and can be activated by multiple stress signals. These can include transfection with tBID (a caspase activated [BH3-domain-only] Bcl2 derivative) or treatment with staurosporine (a general kinase inhibitor), etoposide (topoisomerase II inhibitor), ultraviolet (UV) light, thapsigargin (inhibitor of the endoplasmic reticulum [ER] Ca^{2+} ATPase), tunicamycin (inhibitor of ER N-linked glycosylation), or brefeldin A (inhibitor of ER–Golgi transport). The "death ligand/receptor" pathway is activated by the interaction of the Fas ligand on a lymphoid effector cell with the Fas-receptor target cell. Alternatively, tumor necrosis factor (TNF)-α plus cycloheximide (CHX) can also activate the "death receptor" pathway. These signals initiate a signal transduction pathway through FADD and caspase-8, leading to the activation of caspase-3, which is central to the maturation and function of the immune system *(57,58)*.

1.1.2. Stress Response and the Mitochondria

The mitochondria interact with the cellular stress response pathways to globally regulate cellular functions, survival, and proliferation. Two such stress-response proteins are the poly(ADP-ribose) polymerase (PARP) *(59)* and the histone deacetylase SIR2.

The PARP protein is a nuclear DNA enzyme that is activated by fragments of DNA resulting from DNA damage. Utilizing NAD^+ as a substrate, it transfers 50 or more ADP-ribose moieties to nuclear proteins such as histones and PARP itself. Massive DNA damage results in excessive activation of PARP that leads to the depletion of NAD^+. The resynthesis of NAD^+ from ATP then markedly depletes cellular ATP leading to death *(60)*. Mice in which the PARP gene has been genetically inactivated show remarkable resistance to cellular stress such as cerebral ischemia (stroke) *(61,62)*. and streptozotocin-induced diabetes *(63)*.

The nuclear protein p53 is also activated by DNA damage and can initiate programmed cell death. This pathway has been shown to be mediated through mitochondrial release of cytochrome-*c*, which, in turn, activates Apaf-1 and caspase-9. The p53 initiation of mitochondrial cytochrome-*c* release requires the intervention of proapoptotic protein Bax. Hence, DNA damage activates p53, which activates Bax, which causes mitochondrial cytochrome-*c* release, which initiates apoptosis *(64)*.

Another nuclear protein, SIR2, uses NAD^+ as a cofactor to diacetylate histones. Diacetylated histones keep inactive genes, such as proto-oncogenes, silent *(65)*. Degradation of NAD^+ inactivates SIR2, permitting the histones to be acetylated and silent genes to be illegitimately expressed.

Cellular and DNA damage can be caused by ROS. NADPH oxidases reduce O_2 to generate superoxide anion in the cytosol. The best characterized of the NADPH oxidases is the macrophage "oxidative burst" complex involved in generating the $O_2^{\bullet-}$ to kill engulfed micro-organisms. However, an additional NADPH oxidase, Mox1, is a homolog of the gp91phox catalytic subunit of the phagocyte NADP oxidase. Mox1 generates $O_2^{\bullet-}$. When Mox1 is overexpressed in NIH3T3, it increases the mitotic rate, cell transformation, and tumorgenicity of cells *(66)*. This mitogenic activity of Mox1 is neutralized by overexpression of catalase, indicating that cell growth signal must be H_2O_2 *(67)*. The fact that H_2O_2 is a mitogenic signal for the cell nucleus is of great importance for the mitochondria, as H_2O_2 is the only mitochondrial ROS that it stable enough to defuse to the nucleus. Therefore, cellular H_2O_2 levels can be affected by mitochondrial H_2O2 production.

Acting together, these various enzymes and molecules form an integrated metabolic network with the mitochondria. Inhibition of the mitochondrial ETC results in increased $O_2^{\bullet-}$ production that is converted to H_2O_2 by mitochondrial MnSOD. Mitochondrial H_2O_2 can diffuse to the nucleus, where, at low concentrations, it acts as a mitogen. However, excessive mitochondrial generation of H_2O_2 can overwhelm the antioxidant defenses of the cytosol (catalase, glutathione peroxidase, etc.) and cause DNA damage. DNA damage would mutagenize proto-oncogenes, the cause of their activation. Excessive DNA damage then activates PARP, which degrades NAD^+. Depletion of NAD^+ blocks the transfer of reducing equivalents to the mitochondrial ETC, causing a depletion of ATP. Reduced NAD^+ would inactivate SIR2, causing inappropriate activation of genes, including proto-oncogenes.

1.1.3. Mitochondrial Genetics

The mitochondrial OXPHOS complexes are composed of multiple polypeptides, most encoded by the nDNA. However, 13 polypeptides are encoded by the closed circular, 16,569 base pairs (bp) mtDNA. The mtDNA also codes for the 12S and 16S rRNAs and 22 tRNAs necessary for mitochondrial protein synthesis. The 13 mtDNA polypeptides include 7 (ND1, 2, 3, 4, 4L, 5, 6) of the 43 subunits of complex I, 1 (cytb) of the 11 subunits of complex III, 3 (COI, II, III) of the 13 subunits of complex IV, and 2 (ATP6 and 8) of the 16 subunits of complex V. The mtDNA also contains an approx 1000-bp control region that encompasses the heavy (H)- and light (L)-strand promoters (P_H and P_L) and the H-strand origin of replication (O_H). The H-strand primer is generated by cleavage of the L-strand transcript by the nuclear-encoded RNase MRP at runs of G nucleotides in the conserved sequence blocks CSBIII, CSBII, and CSBI, primarily after CSBI *(68–71)*.

P_H and P_L are associated with mitochondrial transcription factor (Tfam) binding sites that are essential for the effective expression of these promoters *(72–76)*. Whereas the P_H is responsible for transcribing both of the rRNA genes and 12 of the protein coding genes, P_L transcribes the ND6 protein gene and several tRNAs and generates the primers used for initiation of H-strand replication at O_H. The L-strand origin of replication (O_L) is located two-thirds of the way around the circle from O_H *(70)*. All of the other genes necessary to assemble a mitochondrion are encoded by the nucleus *(8)*.

Each human cell contains hundreds of mitochondria and thousands of mtDNAs. The semiautonomous nature of the mitochondria has been demonstrated by showing that mitochondria and their resident mtDNAs can be transferred from one cell to another by enucleating the donor cell and fusing the mitochondria-containing cytoplast to a recipient cell. The feasibility of this cybrid transfer procedure was first demonstrated using cells harboring a mtDNA mutation that imparts resistance to the mitochondrial ribosome inhibitor chloramphenicol (CAP) *(77–79)*. This cybrid transfer process has been further refined by curing the recipient cell of its resident mtDNA by long-term growth in ethidium bromide or by treatment with the mitochondrial toxin rhodamine-6G (R6G). Cells lacking mtDNA, resulting from prolonged growth in ethidium bromide, have been designated ρ^o cells. These cells require glucose as an energy source, uridine to compensate for the block in pyrimidine biosynthesis, and pyruvate to reoxidize the NADH generated during glycolysis, a combination called GUP medium. R6G-treated cells or mtDNA-deficient ρ^o cells are ideal recipients for transmitochondrial experiments, as the resulting cybrids will not retain the recipient cells mtDNAs *(80–83)*. CAP^R was subsequently shown to result from single nucleotide substitutions in the 16S rRNA gene *(84,85)*.

The mtDNA is maternally inherited and has a very high mutation rate. When a new mtDNA mutation arises in a cell, a mixed intracellular population of mtDNAs is generated, a state termed *heteroplasmy*. As a heteroplasmic cell replicates, the mutant and normal molecules are randomly distributed into the daughter cells and the proportion of mutant mtDNAs drifts, a process called *replicative segregation*. As the percentage of mutant mtDNAs increases, the mitochondrial energetic capacity declines, ROS production increases, and the propensity for apoptosis increases. The tissues most sensitive to mitochondrial dysfunction are the brain, heart, skeletal muscle, endocrine system, and kidney *(8)*.

1.2. Mitochondrial Disease and Aging

A wide variety of neurodegenerative diseases have now been linked to mutations in mitochondrial genes located in either the mtDNA or the nDNA.

1.2.1. Mitochondrial Diseases Resulting from mtDNA Mutations

The mtDNA mutations have been associated with a variety of neuromuscular disease symptoms, including various ophthalmological symptoms, muscle degeneration, cardiovascular disease, diabetes mellitus, renal failure, movement disorders, and dementias. The mtDNA diseases can be caused by either base substitution or rearrangement mutations. Base substitution mutations can either alter proteins (missense mutations) or rRNAs and tRNAs (protein synthesis mutations). Rearrangement mutations generally delete at least one tRNA and thus cause protein synthesis defects *(86)*.

Missense mutations have been associated with myopathy, optic atrophy, dystonia, and Leigh's syndrome *(8)*. Base substitution mutations in protein synthesis genes have been associated with a wide spectrum of neuromuscular diseases, the more severe typically include mitochondrial myopathy, associated with ragged red fibers (RRFs) and subsarcolemmal aggregates of abnormal mitochondria *(8)*. Examples of maternally inherited tRNA mutations include MERRF (myoclonic epilepsy and ragged red fiber), caused by a tRNA^{Lys} np 8344 mutation *(87,88)*, and MELAS (mitochondrial encephalomyopathy, lactic acidosis, and strokelike episode), caused by a tRNA$^{Leu(UUR)}$ np 3243 mutation *(89)*. Patients with high percentages of the np 3243 mutation (>85%) can present with strokes, hypertrophic cardiomyopathy, dementia, short stature, lactic acidosis, and mitochondrial myopathy. Maternal pedigrees with low percentages (10–30%) of the np 3243 mutation may only manifest adult-onset (Type II) diabetes mellitus and deafness *(1,90–92)*. Severe tRNA mutations such as the deletion of a single base in the stem of the anticodon loop of the tRNA$^{Leu(UUR)}$ gene at np 3271, can appear "spontaneously" and result in lethal systemic disease, including short stature, deafness, seizures, cataracts, glaucoma, retinitis pigmentosa, cerebral calcifications, and death resulting from renal failure and sepsis *(93)*. The best characterized mtDNA rRNA mutation is the np 1555 base substitution in the 12S rRNA gene associated with maternally inherited sensory neural hearing loss *(94,95)*.

The mtDNA rearrangement syndromes are invariably heteroplasmic and can result in phenotypes ranging from adult-onset Type II diabetes and deafness, through ophthalmoplegia and mitochondrial myopathy, to lethal pediatric pancytopenia. Maternally inherited Type II diabetes and deafness has been linked to a trimolecular heteroplasmy encompassing normal mtDNAs, 6.1-kb insertion molecules, and reciprocal 10.8-kb deletion molecules *(96,97)*. Chronic progressive external ophthalmoplegia (CPEO) and the Kearns–Sayre syndrome (KSS) are associated with ophthalmoplegia, ptosis, and mitochondrial myopathy, together with a variety of other symptoms, including seizures, cerebellar ataxia, deafness, diabetes, heart block, and so on. *(8,98)*. CPEO and KSS patients typically develop mitochondrial myopathy with RRF that encompass COX-negative and SDH-hyperreactive muscle fiber zones where the rearranged mtDNAs are concentrated, presumably due to selective amplification *(99,100)*. Pearson's marrow/pancreas syndrome is the most severe mtDNA rearrangement syndrome. These children develop pancytopenia early in life and become transfusion dependent *(101–103)*. If they survive the pancytopenia, they progress to KSS *(104–106)*. The clinical variability of the mtDNA rearrangement syndromes appears to result from differences between insertions and deletions, the breadth of tissues that contain the rearrangement, and the percentage of the rearranged molecules in each tissues.

The OXPHOS transcript levels have been found to be upregulated in the tissues of mitochondrial disease patients, presumably as an attempt by the cells to compensate for the mitochondrial energetic defect. Analysis of the autopsy tissues of a patient with high levels of the tRNA$^{Leu(UUR)}$ np 3243 mutation who died of mitochondrial encephalomyopathy with hypertrophic cardiomyopathy and cardiac conduction defects revealed that multiple mtDNA and nDNA transcripts involved in energy metabolism were upregulated in the heart and skeletal muscle. Noteworthy among the nDNA gene transcripts were the ATP synthase β-subunit (ATPsynβ), ANT1, ANT2, muscle glycogen phosphorylase (mGP), muscle mitochondrial creatine phosphokinase (mmtCPK), and ubiquitin *(107)*. Similar results have been obtained for muscle biopsy samples from MERRF 8344, MELAS 3243, and KSS patients *(107–109)*. Muscle mtCPK is of particular interest because it is essential for muscle mitochondrial energy transfer and is a critical target for ROS inactivation *(110)*.

1.2.2. Mitochondrial Diseases Resulting from nDNA Mutations

Mitochondrial diseases have also been associated with a spectrum of different nDNA mutations *(110a,110b,110c,110d)*. Mutations in the RNA component of the mitochondrial RNAse MRP have been implicated in metaphyseal chondrodysplasia or cartilage–hair hypoplasia (CHH). CHH is an autosomal recessive disorder resulting from mutation on chromosome 9p13 that present with short stature, hypoplastic hair, ligamentous laxity, defective immunity, hypoplastic anemia, and neuronal dysplasia of the intestines, which can result in megacolon (Hirschsprung's disease) *(111)*.

The mtDNA depletion syndrome is associated with the loss of mtDNAs from various tissues during development. This results in neonatal or childhood organ failure and lethality. Pedigree analysis and somatic cell genetics have demonstrated that mtDNA depletion is the result of a nuclear gene defect *(112–114)*.

Leigh's syndrome represents the common clinical end point for mtDNA mutations in the structure or assembly of the mitochondrial OXPHOS complexes. Of Leigh's syndrome cases, about 18% involved mtDNA mutations, about 10% pyruvate dehydrogenase defects, about 19% complex I defects, about 18% complex IV, and about 35% other causes, including complex II and pyruvate carboxylase defects *(115)*.

Defects in the assembly of complex IV can result in a variety of pediatric encephalopathic disorders. Mutations in SURF1 cause Leigh's syndrome *(116,117)*, mutations in SCO1 result in encephalopathy and hepatopathy *(118)*, mutations in SCO2 cause encephalopathy with cardiomyopathy *(119,120)* and mutations in COX10 result in encephalopathy and nephropathy *(118)*.

The Mohr–Tranebjaerg syndrome manifests as early-onset deafness and dystonia and is associated with mutations in the DDP1 protein gene, a member of a family of genes involved in mitochondrial assembly and division *(121–123)*. The autosomal recessive Friedreich's ataxia is associated with cerebellar ataxia, peripheral neuropathy, hypertrophic cardiomyopathy, and diabetes, and it results from the inactivation of the frataxin gene on chromosome 9q3. Frataxin regulates free iron in the mitochondrial matrix, and its absence results in increased matrix iron that converts H_2O_2 to $^{\bullet}OH$ and inactivates the mitochondrial Fe–S center enzymes (aconitase and complexes I, II, and III) *(45,124,125)*.

Autosomal dominant progressive external ophthalmoplegia (adPEO) is associated with the accumulation of multiple mtDNA deletions in postmitotic tissues. It accounts for approx 6% of PEO cases *(126–131)*, and has been linked to two nuclear loci: one on chromosome 10q23.3–24.3 *(132,133)* and the other on chromosome 4q34-35 *(134)*. This latter locus is the ANT1 in which two missense mutations have been reported. One missense mutation changed a highly conserved alanine at codon 114 to a proline and was present in five Italian families with a common haplotype. The other mutation was found in a spontaneous case and changed the valine at codon 289 to a methionine *(134)*.

The mitochondrial neurogastrointestinal encephalomyopathy (MNGIE) syndrome is associated with mitochondrial myopathy, including RRFs and abnormal mitochondria, decreased respiratory chain activity, and multiple mtDNA deletions and mtDNA depletion. This autosomal recessive disease is the result of mutations in the nDNA thymidine phosphorylase (TP) gene, which has been hypothesized to destabilize the mtDNA, possibly through perturbing cellular thymidine pools *(135)*.

1.2.3. Mitochondrial Defects and Somatic mtDNA Mutations

Mitochondrial diseases often show a delayed onset and a progressive course. This is thought to be the result of the age-related decline in OXPHOS function in postmitotic tissues *(136–140)* associated with the progressive accumulation of somatic mtDNA rearrangement mutations *(137,141–150)* and base substitution mutations *(151–154)*.

The most likely origin of somatic mtDNA mutations is oxygen radical damage. The mtDNA has been estimated to accumulate 10 times more DNA oxidation products than the nDNA *(155,156)* and to accumulate extensive oxidative damage in a variety of tissues *(156–158)*.

In at least some postmitotic tissues, somatic mtDNA mutations are clonally amplified. The muscle of older individuals accumulate COX-negative fibers *(159,160)*, each of which harbors a different clonally expanded mutant mtDNA *(161)*. Individual human cardiomyocytes have also been found to harbor cell-specific clonally expanded mtDNA mutations *(162)*.

The age-related accumulation of mtDNA damage in mouse muscle and brain *(163)* correlates with changes in expression of mitochondrial bioenergetic genes such as the mmtCPK and a variety of stress response genes in the muscle *(164)*, as well as alteration of stress response and neurotrophic gene expression in the brain *(165)*. Caloric restriction, which is well known to extend the life-span and reduce cancer risk in laboratory rodents *(166–170)*, protects mitochondrial function from age-related decline *(44,169,171,172)*, reduces mtDNA damage *(163)*, and reverses many of the changes seen in mitochondrial gene expression *(164,165)*. Thus, the age-related decline in OXPHOS, the accumulation of oxidative damage and mtDNA mutations, and the compensatory induction of bioenergetic and stress-response gene expression are all linked in both mitochondrial diseases and in aging.

1.2.4. Mitochondrial Defects in Diabetes Mellitus

Several lines of evidence implicate mitochondrial defects as a major factor in diabetes mellitus *(1)*. Early epidemiological studies revealed that as the age of onset of diabetes in the proband increases, the probability that the mother will be the affected parent increases, ultimately reaching a ratio of 3:1. Moreover, the maternal transmission can be sustained for several generations *(173–178)*. This apparent maternal transmission of Type II diabetes is consistent with the discovery that both mtDNA rearrangement and tRNA$^{Leu(UUR)}$ np 3243 mutations can cause maternally inherited Type II diabetes mellitus *(90,92,97,179,180)*.

Diabetes mellitus has been proposed to result from defects in the glucose sensor for insulin secretion *(181–183)*, insulin resistance *(97,179,184–186)*, and from defective modulation of the β-cell K_{ATP} channels*(187)*. All three of these factors can be tied together through mitochondrial energy production (*see* **Fig. 2**).

This "glucose sensor" has been shown to involve the pancreatic β-cell glucokinase. Patients with maturity-onset diabetes of the young (MODY), Type II, have mutations in the glucokinase gene, the only hexokinase expressed in the pancreatic islet β-cells. The K_m of the islet cell glucokinase is higher than that of other cellular hexokinases and, hence, glucokinase is only active during hyperglycemia *(188–192)*. Because most of the cellular glucokinase is attached to the mitochondrial outer membrane by porin, and porin interacts with the ANT of the inner membrane *(193–195)*, it is possible that glucose sensing involves the linkage between glucokinase and OXPHOS through this trans-mitochondrial membrane macromolecular complex. Hence, mutations in either the glucokinase gene, which binds glucose during hyperglycemia, or mitochondrial OXPHOS, which provides the ATP for glucose phosphorylation, could affect the ability of the pancreas to respond to hyperglycemia *(196,197)*.

In addition to phosphorylation of glucose by β-cell glucokinase, mitochondrial ATP generation regulates the β-cell, plasma membrane, K_{ATP} channel. At low ATP/ADP ratios, the K_{ATP} channel is leaky and the plasma membrane transmembrane potential remains high. However, during active mitochondrial oxidation of glucose, cytosolic ATP/ADP ratio increases, the K_{ATP} channel closes, and the plasma membrane depolarizes. The depolarization of the β-cell plasma membrane activates the voltage-sensitive L-type Ca^{2+} channel. This causes Ca^{2+} to flow into the cytosol, which activates fusion of the insulin-containing vesicles, causing release of insulin (*see* **Fig. 2**).

The importance of the mitochondrial oxidation of NADH to generate ATP in insulin secretion has been demonstrated by the fact that elimination of mtDNAs from the rat insulinoma cell line (INS-s) resulted in the complete abolition of the insulin-secreting capacity of the β-cells *(198)*. Inhibition of the mitochondrial NADH shuttle also results in inhibition of β-cell insulin secretion *(199)*.

Based on these data, mitochondrial OXPHOS appears to play an integral role in insulin secretion: first, by keeping the ATP-binding site of glucokinase charged and primed to phosphorylate glucose and, second, by generating sufficient ATP to close the K_{ATP} channel (*see* **Fig. 2**).

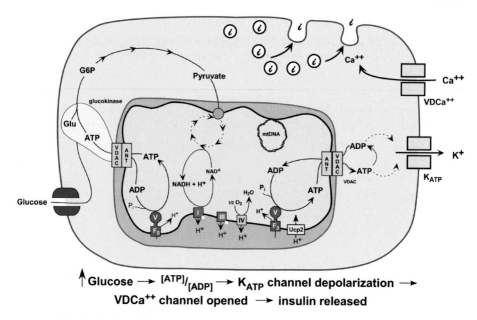

$$\uparrow \text{Glucose} \rightarrow {}^{[ATP]}/_{[ADP]} \rightarrow \text{K}_{ATP} \text{ channel depolarization} \rightarrow$$
$$\text{VDCa}^{++} \text{ channel opened} \rightarrow \text{insulin released}$$

Fig. 2. Proposed mitochondrial pathophysiology of diabetes mellitus.

In addition to mtDNA mutations, nDNA mutations in mitochondrial functions may also play an important role in diabetes. MODY has been associated with a number of nDNA mutations. MODY2 is the result of mutations in glucokinase and accounts for 10–65% of cases. MODY1 is rare and results from mutations in hepatic nuclear factor (HNF)-4α. MODY3, which accounts for 20–75% of cases, manifests as postpubertal diabetes, obesity, dyslipidema, and arterial hypertension, and results from mutations in HNF-1α. The rare MOD4 results from mutations in the insulin promoter factor (IPF)-1 HNF-4α is a member of the steroid/thyroid hormone receptor superfamily and acts as an upstream regulator of HNF-1α. HNF-1α is a transcription factor involved in the tissue-specific regulation of liver and pancreatic islet genes *(200)*. However, HNF-1α is also important in regulating nDNA-encoded mitochondrial gene expression and the expression of GLUT 2 glucose transporters *(198)*.

Type II diabetes has been associated with a Pro12A1 polymorphism in the peroxisome proliferator-activated receptor γ gene (PPARγ) *(201)*. PPARγ might play a role in the regulation of peroxisome and mitochondrial number and structure.

The insulin resistance of diabetes might also be explained by mitochondrial defects. Patients with mtDNA-based diabetes can also develop insulin resistance, which may even precede the defect in insulin secretion *(202)*. This might be explained if the systemic OXPHOS defect could inhibit the cellular utilization of the energy provided by glucose uptake. Finally, diabetic hyperglycemia is associated with a variety of secondary pathological changes affecting small vessels, arteries, and peripheral nerves. These changes are associated with (1) glucose-induced activation of protein kinase C isoforms, (2) formation of glucose-derived advanced glycation end products (AGFs); (3) increased glucose flux through the aldose reductase pathway, and (4) activation of necrosis factor kappa B (NFκB). In cultured vascular endothelial cells, all of these processes can be blocked by inhibition of complex II (SDH) by thenoyltrifluoroacetone (TTFA), uncoupling OXPHOS with carbonyl cyanide *m*-chlorophenylhydrazone (CCCP), induction of uncoupling protein 1 (*Ucp1*), or induction of mitochondrial MnSOD. Hence, all of the pathological effects of hyperglycemia are mediated through mitochondrial ROS production. Because NFκB is involved in the expression of stress-response genes such as MnSOD, mitochondrial regulation of NFκB activation may have broad effects on cellular metabolism *(203)*.

2. Animal Models of Mitochondrial Disease

Over the past 5 yr, mouse models for mitochondrial diseases have been developed for both mtDNA mutations and nDNA mutations.

2.1. Mouse Models Generated with mtDNA Mutations

Several approaches have been tried for introducing genetically distinct mtDNAs into the mouse female germline. To date, two basic procedures have been successful: (1) fusion of enucleated cell cytoplasts bearing mutant mtDNA to undifferentiated mouse female stem cells, injection of the stem cell cybrids into mouse blastocysts, and implantation of the chimeric embryos into a foster mother, and (2) fusion of cytoplasts from mutant cells directly to mouse single-cell embryos and implantation of the embryos into the oviduct of pseudopregnant females. The former method has permitted the creation of mouse strains bearing deleterious base substitution mutations *(204)*, whereas the latter has been used to create mouse strains harboring mtDNA deletions *(205)*.

2.1.1. ES Cells and Base Substitution Mutations

The first attempt to utilize the cybrid technique to introduce mutant mtDNAs into mouse stem cells involved the fusion of the cytoplasts from CAPR B16 melanoma cells to the teratocarcinoma cell line OTT6050. The resulting teratocarcinoma cybrids were injected into blastocysts and five chimeric animals were generated with 10–15% chimerism in various organs. However, no direct evidence was obtained that the CAPR mtDNA was present in the transgenic mice *(206)*.

More recent efforts have focused on mouse female embryonic stem (ES) cell lines. In two independent experiments, CAP^R mouse cell lines were enucleated and the cytoplasts fused to female ES cells. CAP^R ES cell cybrids were isolated, injected into blastocysts, and chimeric mice generated with tissues that have low percentages of chimerism and detectable levels of CAP^R mtDNAs *(207,208)*. In one of these studies, the CAP^S mtDNAs in the ES cells were removed prior to cytoplast fusion by treatment with R6G *(203)*. This greatly enriched for the CAP^R mtDNAs in the ES cell cybrids, as detected using the *Mae*III and *Tai*I restriction site polymorphism generated by the CAP^R T to C transition at np 2433 in the 16S rRNA gene *(84)*.

These studies were extended by identifying a female ES cell that would produce fertile oocytes. The successful ES cell line CC9.3.1 was then used to recover the mtDNAs from the brain of New Zealand Black (NZB) mice and introduce them into the female germline of mice that formerly harbored only the "common haplotype" mtDNA. Most inbred strains of mice from North America had the same founding female, and thus have the same mtDNA haplotype. By contrast, NZB mice were inbred in New Zealand and their mtDNAs differ from this "common haplotype" by 108 nucleotide substitutions *(209)*, one of which creates a *Bam*HI restriction site polymorphism. To transfer the mtDNAs of the NZB mice into cultured cells, the brain of an NZB mouse was homogenized, and the synaptic boutons with their resident mitochondria were isolated by Percoll gradient as synaptosomes. These synaptosomes were fused to the mouse mtDNA-deficient (ρ^0) cell line LMEB4 *(154)*. Synaptosome cybrids were recovered having the LMEB4 nucleus and the NZB mtDNA, designated the LMEB4(mtNZB) cybrids *(154)*. Next, the LMEB4(mtNZB) cybrids were enucleated and the cytoplasts fused to R6G-treated CC9.3.1 cells. This generated the CC9.3.1(mtNZB) cybrids that were injected into C57Bl/6 (B6) embryos, and mice with a high degree of chimerism generated. One female chimeric mouse, heteroplasmic for the NZB and the "common haplotype" mtDNAs, was mated with two different B6 males and the hetero-plasmic mtDNAs were transferred to all of the 7 and 10 offspring, respectively. A female of the next generation was mated to a B6 male and transmitted the heteroplasmic mtDNAs to her 7 progeny, whereas a heteroplasmic male mated to 2 B6 females did not transmit the NZB mtDNAs to any of his 16 offspring. Hence, this experiment established that exogenous mtDNA mutations could be introduced into the female mouse germline and, subsequently, be maternally transmitted through repeated generations *(204)*.

Using this same procedure, CAP^R mtDNAs from the mouse 501–1 cell line were introduced into chimeric mice. The resulting CAP^R chimeric animals developed bilateral nuclear cataracts, reduced rod and cone excitation detected by electroretinograms (ERG), and retinal hamartomatous growths emanating from the optic nerve heads. Several of the chimeric females when mated were able to transmit the CAP^R mtDNAs to their progeny in either the homoplasmic or heteroplasmic state. The resulting CAP^R progeny either died *in utero* or in the neonatal period. Mice born alive exhibited striking growth retardation, progressive myopathy with myofibril disruption and loss, dilated cardiomyopathy, and abnormal heart and muscle mitochondria morphology *(204)*. These phenotypes are remarkably similar to those seen in the patient with the single-base deletion in the anticodon stem of the tRNA$^{Leu(UUR)}$. Hence, deleterious mtDNA protein synthesis mutations can cause mitochondrial disease in the mouse with a severity and nature analogous to those seen in humans.

2.1.2. Single-Cell Embryos and Rearrangement Mutations

The alternative successful approach for introducing mutant mtDNAs into the mouse female germline has involved introduction of variant mtDNAs directly into mouse single-cell embryos, either by microinjection of mitochondria or fusion of cytoplasts. Microinjection of *Mus spretus* mitochondria into *Mus musculus domesticus* embryos has resulted in chimeric embryos, but the mutant mtDNAs appear to have been lost by replicative segregation early in preimplantation development *(210,211)*. Fusion of cytoplasts from mouse oocytes harboring one mtDNA type (NZB/BINJ) with single-cell embryos harboring a different mtDNA type (C57BL/6 or BALB/c) has resulted in heteroplasmic mice. These mice permitted the analysis of mitochondrial replicative segregation through the germline and have revealed that heteroplasmic mixtures of the NZB and "common haplotype" mtDNAs undergo directional replicative segregation in different adult tissues. However, these animals have not been found to have an abnormal phenotype *(212–214)*. Heteroplasmic animals have also been generated by fusing membrane-bound karyoplasts containing a zygote nucleus and a portion of the oocyte cytoplasm with enucleated eggs *(214,215)*.

These studies have been extended to include the fusion of cultured cell cytoplasts to single-cell embryos. Synaptosome cybrids, heteroplasmic for a 4696-bp deletion, were enucleated and the cytoplasts fused to pronucleus-stage embryos, which were then implanted into the oviducts of pseudopregnant females. The 4696-bp deletion removed six tRNAs and seven structural genes. This procedure resulted in 24 animals having 6–42% deleted mtDNAs in their muscle. Females with 6–13% deleted mtDNA were mated and the rearranged mtDNAs were transmitted through three successive generations, with the percentage of deleted mtDNAs increasing with successive generations to a maximum of 90% deletion in the muscle of some animals. Although mtDNA duplications were not observed in the original synaptosome cybrid cells, they were found in the postmitotic tissues of the animals. This raises the possibility that the maternal transmission of the rearranged mtDNA was through a duplicated mtDNA intermediate, as proposed for the human maternally inherited mtDNA rearrangement pedigree presenting with diabetes mellitus and deafness *(96,97)*. Although RRFs were not observed in these animals, fibers with greater than 85% mutant mtDNAs were COX-negative, and many fibers had aggregates of subsarcolemmal mitochondria. The heart tissue of heteroplasmic animals was also a mosaic of COX-positive and COX-negative fibers, and the amount of lactic acid in peripheral blood was proportional to the amount of mutant mtDNA in the muscle tissues. Mice with predominantly mutant mtDNAs in their muscle tissue died within 200 d with systemic ischemia and enlarged kidneys with granulated surfaces and dilation of the proximal and distal renal tubules. These animals also developed high concentrations of blood urea and creatinine *(205)*. Hence, mtDNA deletion mutations can also cause disease in mice, but the phenotypes and inheritance patterns are somewhat different from those seen in most human mtDNA rearrangement patients.

The generation of mouse strains harboring mtDNA base substitution or large deletion mutations now provide mouse models for a range of mtDNA diseases. However, the severity of the phenotypes that were observed in the two mouse mtDNA mutant strains prepared to date were the converse of those traditionally seen in humans. In humans, many base substitution mutations in protein synthesis genes are maternally inherited and usually are compatible with maturation to at least late childhood, whereas most deletion mutations are spontaneous and patients with a high-percentage mutant are severely affected in childhood. The converse was seen for the mice. The CAP[R] base substitution resulted in mice that were neonatal lethal, and the "deletion" mutation was maternally inherited and gave viable animals with up to 90% deleted mtDNAs. These aberrant findings could be explained in two ways. One possibility is that the mtDNA mutations introduced into the mouse are qualitatively different from those generally encountered in human families, resulting in differences that are more apparent than real. For example, no clinically relevant 16S rRNA mutation has been reported in humans, so they may be as lethal in man as they are in mouse. Also, mtDNA duplications can be maternally inherited in humans and the mouse rearrangement may be a duplication. Hence, the model would be more analogous to the human maternally inherited diabetes and deafness than to Pearson's marrow/pancreas syndrome. Still, this latter possibility does not explain the high levels of rearranged mtDNAs that accumulated in the mouse or the low level of duplicated molecules that were reported. The alternative possibility is that the mouse may be more tolerant of mitochondrial defects than humans. This alternative is supported by the mouse's greater tolerance of ANT1 deficiency *(27)* than is seen in human *(134)*. Many different mtDNA mutations will need to be introduced into the mouse before these alternatives can be distinguished.

2.2. Mouse Models of nDNA Mitochondrial Mutations

Four different classes of nDNA-encoded mitochondrial gene mutations have now been reported for the mouse: (1) mutations in the biosynthetic apparatus gene *Tfam*; (2) mutations in the mitochondrial bioenergetic genes *Ant1* and *Unc1–3*; (3) mutations in the mitochondrial antioxidant genes *GPx1* and *Sod2* (MnSOD); and (4) mutations in the mitochondrial apoptosis genes cytochrome-*c* (*cytc*), *Bax*, *Bak*, *Apaf1*, and caspases 9 and 3.

2.2.1. Mutations in the Mitochondrial Biosynthetic Gene Tfam

Genetic inactivation of the nuclear-encoded mitochondrial transcription factor, Tfam, may provide a model for the mtDNA depletion syndrome and possibly CHH. This follows from the importance of Tfam-directed transcription from the P_L promoter for the initiation of mtDNA H-strand replication.

2.2.1.1. SYSTEMIC TFAM DEFICIENCY RESULTS IN EMBRYONIC LETHALITY

The *Tfam* gene was inactivated in tissues by bracketing the terminal two exons of the gene with loxP sites, designated *Tfam^{loxP}*. The *Tfam* gene was then inactivated by crossing +/*Tfam^{loxP}* animals with animals bearing the *Cre* recombinase driven by the β-actin promoter. The resulting heterozygous +/*Tfam⁻* animals were viable and reproductively competent, whereas the homozygous *Tfam* –/– animals were embryonic lethals *(216)*. The *Tfam* heterozygous animals had a 50% reduction in *Tfam* transcripts and protein, a 34% reduction in mtDNA copy number, a 22% reduction in mitochondrial transcripts, and a partial reduction in the COI protein in the heart, but not the liver. The homozygous *Tfam* –/– mutant animals died between embryonic d E8.5 and E10.5, with a complete absence of Tfam protein, and either a severely reduced or a complete absence of mtDNA. The mitochondria in the *Tfam* –/– animals were enlarged with abnormal cristae and were deficient in COX but not SDH *(216)*.

2.2.1.2. HEART-MUSCLE TFAM DEFICIENCY RESULTS IN CARDIOMYOPATHY

To determine the effect of mtDNA depletion in heart and skeletal muscle, the homozygous *Tfam^{loxP}* allele was combined with the *Cre* recombinase gene driven by the mmtCPK promoter, resulting in the selective destruction of the *Tfam* genes in those tissues. Although the hearts of 18.5-d embryos had reduced levels of Tfam, they appeared to be otherwise morphologically and biochemically normal. However, after birth, the mutant animals proved to be postnatal lethals, dying at a mean age of 20 d of dilated cardiomyopathy. Under anesthesia, the animals developed cardiac conduction defects with a prolongation of the PQ interval and intermittent atrioventricular block. This was associated with a reduction in Tfam protein and mtDNA transcript levels in heart and muscle, a reduction in heart mtDNA to 26%, a reduction in skeletal muscle mtDNA to 60%, and a reduction of respiratory complexes I and IV but not complex II. Histochemical analysis of the hearts revealed a mosaic staining pattern with some cardiomyocytes being COX-negative and SDH-hyperreactive *(217)*.

2.2.1.3. PANCREATIC B-CELLS TFAM DEFICIENCY RESULTS IN DIABETES MELLITUS

To examine the importance of mtDNA depletion in diabetes, the homozygous *Tfam*^{loxP} allele was combined with a rat insulin-promoter-driven *Cre* recombinase (*RIP-Cre*). This resulted in the deletion of the *Tfam* gene in most of the β-cells of the pancreas by 7 d of age. The Tfam-depleted β-cells were found to have greatly reduced COX staining, with normal SDH staining, and to contain highly abnormal giant mitochondria. The mutant mice developed diabetes with increased blood glucose in both fasting and nonfasting states, starting at about 5 wk. They subsequently showed a progressive decline in β-cell mass, reaching a minimum at 14 wk, and a decreased ratio of endocrine to exocrine pancreatic tissue. Thus, mitochondrial diabetes progresses through two stages. The younger animals were diabetic because their β-cells could not secrete insulin, but the older animals had lost many of their β-cells. The secondary loss of the β-cells did not seem to be the product of apoptosis, however, because the number of TUNEL positive cells were not increased in the mutant animals. The mitochondria of the mutant islets showed decreased hyperpolarization and the intracellular Ca^{2+} oscillations were severely dampened in response to glucose, but not to K^+-induced Ca^{2+} modulation *(218)*. These data support a central role for the mitochondria in the β-cell signal transduction pathway leading to insulin release.

2.2.2. Mutations in Mitochondrial Bioenergetic Genes, Ant and Ucp1–3

Mouse mutants have been developed in which the mitochondrial inner membrane transport proteins *Ant1*, *Ucp1*, *Ucp2*, and *Ucp3* have been inactivates. As expected, the *Ant1* mutant reduced heart and muscle energy capacity and the *Ucp* mutants reduced proton leak and increased $\Delta\mu^{H+}$. Unexpectedly, however, all of the mutants increased mitochondrial ROS production resulting a variety of phenotypic effects.

2.2.2.1. ANT1-DEFICIENT MICE DEVELOP MYOPATHY, CARDIOMYOPATHY, AND MULTIPLE MTDNA DELETIONS

The genetic inactivation of the mouse nDNA *Ant1* gene may provide a model for the mtDNA multiple deletion syndrome, as both result from the inactivation of the human ANT1 gene *(27,134)*. Analysis of the *Ant1 –/–* mouse has also provided important insights into the significance of depleting cellular ATP, inhibiting the ETC, and increasing mitochondrial ROS production in the pathophysiology of mitochondrial disease.

ANT1-deficient [*Ant1*^{tm2Mgr} *(–/–)*] mice are viable, although they develop classical mitochondrial myopathy and hypertrophic cardiomyopathy. They also develop elevated serum lactate, alanine, succinate, and citrate, consistent with the inhibition of the ETC and the TCA cycle *(27)*.

The mouse *Ant1* isoform gene is expressed at high levels in skeletal muscle and the heart and at lower levels in the brain, whereas the mouse *Ant2* gene is expressed in all tissues but skeletal muscle *(26)*. Consequently, mice mutant in *Ant1* have a complete deficiency of ANT in skeletal muscle, a partial deficiency in the heart, but normal ANT levels in the liver; an expectation supported by the relative ADP stimulation of respiration in mitochondrial isolated from these three tissues *(27)*.

The skeletal muscle of *Ant1 –/–* animals exhibit classic RRFs and increased SDH and COX staining in the Type I oxidative muscle fibers. These elevated OXPHOS enzyme activities correlate with a massive proliferation of giant mitochondria in the skeletal muscle fibers, degeneration of the contractile fibers, and a marked exercise fatigability. The hearts of the ANT1-deficient mice also exhibited a striking hypertrophic cardiomyopathy, associated with a significant proliferation of cardiomyocyte mitochondria. The proliferation of mitochondria in the *Ant1 –/–* mouse skeletal muscle is associated with the coordinate upregulation of genes involved in energy metabolism, including most mtDNA transcripts and the nDNA complex I 18 kDa and complex IV COXVa and COXVb transcripts, genes involved in apoptosis, including the muscle Bcl-2 homolog Mcl-1, and genes potentially involved in mitochondrial biogenesis, such as SKD3 *(219)*.

The inhibition of ADP/ATP exchange would deprive the ATP synthase of substrate, block proton transport through the ATP synthase membrane channel, and result in the hyperpolarization of $\Delta\psi$ inhibiting the ETC. The inhibition of the ETC would redirect the electrons to O_2 to generate $O_2^{\bullet-}$, and the increased oxidative stress should damage the mtDNA. Consistent with this expectation, the mitochondrial H_2O_2 production rate was found to be increased sixfold to eightfold in the ANT1-deficient skeletal muscle and heart mitochondria, levels comparable to those obtained for control mitochondria inhibited by Antimycin A. In skeletal muscle, where the respiratory defect was complete, the increased oxidative stress was paralleled by a sixfold increase in mitochondrial MnSOD and a threefold increase in mitochondrial GPx1. In the heart, where the respiratory defect was incomplete, GPx1 was also increased threefold, but MnSOD was not increased *(220)*. Hence, inhibition of OXPHOS was associated with increased ROS, and the increased ROS was countered by an induction of antioxidant defenses if the oxidative stress was sufficiently severe.

The increased ROS production would also be expected to increase mitochondrial macromolecular damage. This was confirmed by the analyses of the mtDNA rearrangements in the hearts. The hearts of 16- to 20-mo ANT1-deficient mice had much higher levels of mtDNA rearrangements than did age-matched controls. In fact, the level of heart mtDNA rearrangements in middle-aged *Ant1 –/–* animals was comparable to that seen in the hearts of very old (32 mo) normal mice. Surprisingly, the mtDNAs of the skeletal muscle showed substantially less mtDNA rearrangements than the heart. However, this could be the consequence of the strong induction of MnSOD in skeletal muscle, which was not the case in the heart *(220)*.

The phenotypic, biochemical, and molecular analysis of the *Ant1 –/–* mice have confirmed that they have many of the features of patients with adPEO. These include mitochondrial myopathy with fatigability and multiple mtDNA deletions. Hence, this *Ant1 –/–* mouse model may provide valuable insights into the pathophysiological basis of adPEO. There is one striking difference between these two systems, however. In humans, the ANT1 mutation is dominant, whereas in mouse, it is recessive. There are two possible explanations for this difference. The human and mouse ANT1 mutations might be functionally different. The human mutations are missense mutations, whereas the mouse mutations are nulls mutations. Because the ANTs function as dimers, an aberrant ANT1 polypeptide could bind to normal subunits and result in a nonfunctional complex. Hence, only one-quarter of all of the ANT1 complexes might be active. This would render the human biochemical defect similarly severe to that of the mouse. Alternatively, the mouse might be less sensitive to OXPHOS defects than humans. One way to distinguish these two hypotheses would be to prepare a transgenic mouse harboring the same *Ant1* gene mutations as found in the adPEO patients. These mice could then be crossed onto an *Ant1 +/–* heterozygous background and the phenotype analyzed. If the *Ant1 +/–* transgenic mice develop myopathy and multiple deletions, then the mutation must be acting as a dominant negative. If not, then the mouse must be less sensitive to mitochondrial defects.

Comparison of the human and mouse *Ant1* mutants may also provide some insight into the cause of the mtDNA rearrangements. Two alternative hypotheses have been suggested. In the first, the ANT1 deficiency has been proposed to alter the mitochondrial nucleotide precursor pools, thus perturbing replication *(134)*. This is analogous to the proposal for why the cytosolic thymidine phosphorylase-deficiency causes multiple deletions in the MINGIE syndrome *(135)*. The difficulty with this hypothesis is that in the mouse, the ANT1 deficiency in the muscle is much more sever than that in the heart, yet the heart accumulated many more mtDNA deletions than did the skeletal muscle. The alternative hypothesis is that the inhibition of the ETC caused by the ANT1 defect increases ROS production and this acts as a mutagen, leading to rearrangements of the mtDNAs. This possibility is more consistent with the data because the antioxidant defenses in the skeletal muscle are much more strongly induced than those of the heart. Hence, the heart mtDNAs would be more vulnerable to oxidative damage and prone to rearrangements.

These studies on the *Ant1* −/− mice have demonstrated the importance of ATP deficiency in skeletal muscle and heart pathology and have suggested that increased mitochondrial ROS production is also an important factor in the pathophysiology of mitochondrial disease. Because inactivation of *Ant1* resulted in increased ROS production due to the hyperpolarization of $\Delta\psi$, which secondarily inhibited the ETC, it would follow that reduction of the mitochondrial inner membrane proton leak would also increase $\Delta\psi$ and stimulate mitochondrial ROS generation.

2.2.2.2. Ucp1-DEFICIENT MICE ARE DEFECTIVE IN THERMAL REGULATION

The uncoupler proteins (Ucp) regulate the mitochondrial inner membrane permeability to protons. Mammals have three uncoupler genes *Ucp1*, *Ucp2*, and *Ucp3*. *Ucp1* is primarily associated with brown fat, where it functions in thermogenics. On exposure to cold, rodents respond by secreting noradrenaline and adrenaline. These hormones bind the β3-adenergic receptor in brown and white fat and induce the transcription of the *Ucp1* gene. This dramatically increases *Ucp1* mRNA and protein expression. *Ucp1* then introduces a proton channel in the mitochondrial inner membrane, short circuiting $\Delta\mu^{H+}$, which activates the ETC to rapidly burn brown fat to make heat *(31,32,221–223)*.

Genetic inactivation of the dopamine β-hydroxylase gene results in mice that cannot make noradrenaline or adrenaline. These animals accumulate excess fat in the brown adipose tissue (BAT) and cannot induce *Ucp1* transcription in response to cold. Interestingly, these animals develop an increased basal metabolic rate *(224)*.

Mice with knockout mutations of the *Ucp1* gene also cannot induce the *Ucp1* transcription in response to cold and are cold sensitive. These animals accumulate excess fat in BAT, yet they do not become obese. *Ucp2* mRNA is upregulated in BAT and epididymal fat, suggesting that *Ucp1* deficiency is partially compensated by *Ucp2* expression *(225)*.

2.2.2.3. *UCP3*-DEFICIENT MICE HAVE INCREASED MUSCLE ROS PRODUCTION

Ucp2 and *Ucp3* are more systemically expressed. Mice lacking either of these two proteins exhibit increased mitochondrial ROS production, consistent with increased $\Delta\mu^{H+}$ and inhibition of the ETC *(226,227)*.

Ucp3 is expressed primarily in skeletal muscle and BAT. Inactivation of the *Ucp3* gene caused the upregulation of *Ucp1* and *Ucp2* in BAT. In skeletal muscle, the mutation increased the state 3/state 4 respiration rate by reducing state 4 respiration and, hence, the nonspecific proton leakage. The mutant animals were not obese and sustained their body temperature in response to a cold challenge. Analysis of the ROS production in isolated skeletal muscle mitochondria of *Ucp3* –/– animals revealed that superoxide anion production was increased 58% and muscle mitochondrial aconitase was reduced by 20% *(227)*. These data suggest that *Ucp3* functions in muscle to regulate ROS production by partially uncoupling OXPHOS to keep the ETC oxidized, thus reducing the steady-state levels of ubisemiquinone. This would be particularly important for skeletal muscle that normally generates ATP through aerobic metabolism, but under vigorous exercise, it becomes anaerobic and generates ATP by glycolysis. During aerobic exercise, the ETC would become fully reduced, and upon reoxidation, it would generate a burst of mitochondrial $O_2^{\bullet-}$. Expression of *Ucp3* would create a proton leak that would keep the ETC running during anaerobic exercise, thus avoiding a fully reduced ETC and a resulting burst of ROS production on reoxidation. This same mechanism might be acting to partially depolarize mitochondrial $\Delta\mu^{H+}$ in the *Ant1* –/– mouse skeletal muscle and thus explain why the muscle mtDNAs of the *Ant1* –/– animal have less deleted mtDNA in their muscle than in their hearts.

2.2.2.4. *UCP2*-DEFICIENT MICE ARE RESISTANT TO INFECTION AND DIABETES

Mice in which the *Ucp2* gene has been knocked out are also not obese and they have a normal response to cold. However, they have a marked increased resistance to *Toxoplasma gondii* infection, which forms cysts in the brain. *Ucp2* +/+ mice succumb to infection between 28 and 51 d, whereas *Ucp2* –/– mice survive over 80 d. *Toxoplasma* is eliminated by macrophages by oxidative burst, and *Ucp2* –/– macrophages produced more ROS than *Ucp2* +/+ macrophages. This is associated with an elevated expression of interleukin-1B

(Il16) and MnSOD. Thus, it would appear that the *Ucp2 –/–* mutation increases mitochondrial ROS production in tissues where *Ucp2* is expressed *(226)*.

Ucp2-deficient mice also have increased pancreatic β-cell ATP levels and a marked resistance to developing Type II diabetes. Although the body weight and cold tolerance of the *Ucp2*-deficient mice is comparable to that of wild-type animals, the *Ucp2*-deficient animals have a 2.8-fold higher serum insulin and 18% lower blood glucose levels. The *Ucp2* heterozygotes had intermediate levels with a 2.0-fold increase in insulin and an 11% decrease in blood glucose. During glucose challenge, the *Ucp2*-deficient animals produced higher levels of insulin responding to a lesser glucose challenge, and they required significantly higher glucose challenges to achieve the same elevated serum glucose level. This increased capacity to respond to glucose loads with elevated insulin secretion correlated with increased ATP levels in the isolated islets from *Ucp2*-deficient mice *(228,229)*.

To determine the ability of *Ucp2*-deficiency to compensate for diabetes, the *Ucp2* knockout locus was crossed into the diabetic *ob/ob* mouse. The *ob/ob* mouse has a 33-fold increase in serum insulin, consistent with its extreme insulin resistance, which is associated with significantly elevated levels of *Ucp2* mRNA in the pancreatic islets. Combining *Ucp2* deficiency with the *ob/ob* locus resulted in a marked reduction in the *ob/ob* hyperglycemia and significantly higher serum insulin levels *(228,229)*. These data confirm that mitochondrial energy production is essential for regulating insulin secretion.

2.2.2.5. MUTANT PANCREATIC ATP-SENSITIVE K⁺ (KATP) CHANNELS CAUSE DIABETES

It has been proposed that the mitochondria regulated insulin release through the ATP-dependent K⁺-channel (K_{ATP}). Glucose uptake provides a substrate for increased mitochondrial ATP production, increasing the cellular ATP/ADP ratio. The elevated ATP/ADP ratio causes the closure of the K_{ATP}, depolarizing the β-cell plasma membrane. Depolarization of the plasma membrane opens the voltage-dependent L-type Ca^{2+} channels, permitting extracellular Ca^{2+} to flow into the cytosol. The increased cytosolic Ca^{2+} stimulates the exocytosis of the insulin-containing secretory granules. Hence, mitochondrial function appears to be a central factor in the regulation of blood insulin levels.

To determine the importance of the K_{ATP} channel in the regulation of insulin release in the β-cells of the pancreas, transgenic mice have been developed with mutations in the Kir6.2-subunit of the K_{ATP}. The K_{ATP} channel is composed of two subunits: the sulfonylurea receptor (SUR1) and the pore-forming inward rectifier K⁺ channel Kir6.2. ATP inhibits channel activity through the interaction with the Kir6.2-subunit. Patients with defects in the K_{ATP} channel have familial persistent hyperinsulinemic hyperglycemia of infancy (PHHI) and have constitutive insulin secretion despite severe fasting hypoglycemic. This and related observations have led to the hypothesis that insulin secretion is regulated through the plasma membrane K_{ATP} channels by cytosolic ATP/ADP ratio.

To test this hypothesis, transgenic mice were prepared in which the N-terminal amino acids 2–30 of the Kir6.2 gene were removed [Kir6.2(ΔN2–30)] and the transgene expressed using the pancreas-specific *RIP* promoter. The removal of these N-terminal amino acids results in an approx 10-fold reduction in ATP sensitivity. The resulting F1 transgenic mice were severely hyper-glycemic with hypoinsulinemia and exhibited significantly elevated serum D-3–hydroxybutyrate levels. Most of the newborn mice died by d 5, and those that survived to weaning had significantly reduced body weights. Although the transgenic mice had a severe reduction in serum insulin levels, the pancreatic β-cells initially looked morphologically normal. Hence, the defect appears to be in the release of insulin from the β-cells.

The K_{ATP} channels are formed as tetramers of the Kir6.2-subunits, each associated with a SUR1 subunit. Hence, mutant Kir6.2-subunits might be expected to act as dominant negative mutants. This proved to be the case, through analysis using the inside-out patch-clamp technique. The K_{ATP} channels in the β-cell membrane patches of the transgenic mice had a shallower and significantly reduced sensitivity to ATP inhibition. These data definitively demonstrate the importance of the K_{ATP} channels, regulated by the cytosolic ATP/ADP ratio, in regulating insulin release from the β-cells of the pancreas *(187)*.

2.2.2.6. INHIBITION OF THE ETC INCREASES ROS PRODUCTION

These studies on *Ucp*-deficient mice clearing indicate the importance of regulating $\Delta\psi$ for controlling mitochondrial ROS production and oxidative stress. The apparent importance of regulating ROS production clearly attests to its toxicity. Thus, it follows that mice deficient in the mitochondrial antioxidant genes *GPx1* and *Sod2* (MnSOD) should have increased oxidative stress and develop mitochondrial disease symptoms. To determine if this is true, mice deficient in GPx1 and MnSOD have been generated.

2.2.3. Mitochondrial Defects in Antioxidant Genes GPx1 and Sod2 Antioxidant Therapy

Mitochondrial ROS are removed by the sequential action of MnSOD, which converts $O_2^{\cdot-}$ to H_2O_2 and GPx1, which converts H_2O_2 to H_2O. Because H_2O_2 is more stable than $O_2^{\cdot-}$, it should be less toxic. Hence, disruption of MnSOD would be expected to be more deleterious than GPx1. This has proven to be the case.

2.2.3.1. *GPx1* Deficiency Causes Growth Retardation

To increase the level of mitochondrial H_2O_2 production, the mouse GPx1 gene has been inactivated by homologous recombination. However, the resulting phenotype of this mutation can be understood by taking into account intertissue and intracellular distribution of GPx1. The tissue and cellular distribution of *GPx1* was studied in two ways: insertion of a reporter cassette (β-galactosidase) into the *GPx1* gene driven by the endogenous *GPx1* promoter and preparation of GPx1-specific antibodies and their use in Western blot analysis. These studies revealed that GPx1 is strongly expressed in the liver, brain, and renal cortex, but very weakly expressed in the heart and skeletal muscle. Furthermore, the GPx1 protein was found in both the cytosol and the mitochondrial of liver and kidney but only found in the cytosol of the heart. Hence, the major physiological and phenotypic consequences of the *GPx1* knockout should be found in the brain, liver and kidney, but not in the heart and skeletal muscle.

GPx1 –/– mice are viable, but they experience a 20% reduction in body weight, which suggests chronic growth retardation. Consistent with the *GPx1* expression profile, liver mitochondria of GPx1-deficient animals secreted fourfold more H_2O^2 than wild-type mitochondrial, whereas mutant heart mitochondria secreted the same levels of H_2O_2 as controls. Physiological analysis revealed that the respiratory control ratio (RCR) and the power output (state III rate times P/O ratio) levels were reduced by one-third in the *GPx1* –/– liver mitochondria but were normal in heart mitochondria *(230)*. Thus, excessive mitochondrial H_2O_2 production in the brain, liver, and kidney appears to be only mildly deleterious to the animal.

2.2.3.2. Complete Sod2 Deficiency Causes Cardiomyopathy

To examine the importance of $O_2^{•-}$ production to mitochondrial and mtDNA integrity, mice with different levels of MnSOD were generated by using different number of copies of the T-associated maternal effect (Tme) locus. This locus has a deletion in the mitochondrial MnSOD locus *Sod2* and can only be transmitted through males, but not females. Hence, it is possible to breed mice with 50%, 100%, 150%, and 200% of the normal MnSOD level. When these animals are treated with the complex 1-methyl-4-phenyl-1,2,3,6-tetrahydropyridine (MPTP), a drug known to induce parkinsonism by selectively killing neurons of the substantial nigra, the mice with 50% MnSOD activity show massive basal ganglia toxicity when compared to mice with normal or elevated MnSOD levels *(231)*.

To further investigate the importance of mitochondrial $O_2^{\cdot-}$ in the pathophysiology of disease, the MnSOD gene has been insertionally inactivated ES cells. Two mouse strains lacking *Sod2* have been reported: *Sod2*^{tm1Cje} *(232)* and *Sod2* ^{tm1Leb} *(233)*. The *Sod2*^{tm1Cje} mutation was originally studied on the CD1 background and resulted in death resulting from dilated cardiomyopathy at about 8 d of age *(232,234)*. The *Sod2*^{tm1Leb} mutation was studied on the B6 background and resulted in death after about 18 d associated with injury to the neuronal mitochondria and degeneration of the large neurons, particularly in the basal ganglia and brain stem *(233)*. Although inactivation of the mitochondrial MnSOD has proven to be lethal early in life, the inactivation of either the cytosolic Cu/ZnSOD *(235)* or extracellular Cu/ZnSOD *(236)* genes were found to have little effect on the viability or fecundity of the animals. Hence, mitochondrial toxicity of $O_2^{\cdot-}$ is far more deleterious to mammals than is the toxicity of cytosolic or extracellular $O_2^{\cdot-}$. Hence, the mitochondria must be both the major source and target for $O_2^{\cdot-}$ toxicity.

Mice harboring the *Sod2*^{tm1Cje} mutation have been extensively characterized. To determine the effects of acute $O_2^{\cdot-}$ toxicity, *Sod2*^{tm1Cje} homozygous (–/–) mutant mice have been analyzed. In addition to causing neonatal death resulting from dilated cardiomyopathy *(232,234)*, these animals developed a massive lipid deposition in the liver and a marked deficiency in SDH (complex II) in the hearts, as determined by histochemical analysis 232) and direct biochemical assays *(237)*. In addition to complex II deficiency in the heart and muscle, *Sod2* –/– mice also had partial complex I and citrate synthetase defects in the heart. However, the most striking enzyme deficiency was in mitochondrial aconitase that was almost entirely inactivated in heart and brain. Thus, the increased mitochondrial $O_2^{\cdot-}$ appears to have inactivated all of the mitochondrial Fe–S-center-containing enzymes, thus blocking the TCA cycle and ETC chain *(237)*. This would inhibit mitochondrial fatty acid oxidation, causing fat to accumulate in the liver and energy starvation in the heart, leading to dilation and failure.

Respiration studies on *Sod2*^{tm1Cje} homozygous (–/–) liver mitochondria have revealed a 40% reduction in state III respiration, consistent with impaired ETC activity. Moreover, although ADP increased the respiration rate about 1.6-fold (state III), subsequent addition of an uncoupler did not increase the respiration above the state IV rate. Mitochondria from these neonatal animals also showed a marked increased tendency toward activation of the mtPTP *(238)*. These observations are interpreted as indicating that acute exposure of the mitochondria to high levels of $O_2^{\cdot-}$ has sensitized the mtPTP. Consequently, the transient reduction in $\Delta\psi$ caused by ADP-stimulated respiration activates the mtPTP, causing the release of mitochondrial matrix cofactors and the intermembrane cytochrome-*c*, thus disrupting respiration *(239)*. Similar respiratory defects have been reported for the "senescence accelerated mouse" *(240)* suggesting that this animal may also suffer from increased mitochondrial oxidative stress.

The increased mitochondrial oxidative stress of the *Sod2* –/– animals also resulted in the development of a methylglutaconic acuidurea, associated with reduced liver HMG-CoA lyase activity. These animals also had increased oxidative damage to their DNA, with the greatest extent and level of base adducts being found in the heart, followed by the brain and then the liver *(237)*. This later observation adds credence to the hypothesis that the primary cause of mtDNA rearrangement mutations in aging and the adPEO patients is oxidative damage to the mtDNA.

2.2.3.3. Antioxidant Therapy of *Sod2*-Deficient Mice Rescues the Cardiomyopathy

To confirm that the toxicity of the MnSOD deficiency was caused by the toxicity of mitochondrial $O_2^{\cdot-}$, $Sod2^{tm1Cje}$ –/– mice on the CD1 background were treated by peritoneal injection of the catalytic antioxidant SOD mimetic, MnTBAP [manganese 5,10,15,20-tetrakis (4-benzoic acid) porphyrin]. Peritoneal injection of MnTBAP into *Sod2* –/– animals rescued them from their lethal dilated cardiomyopathy, reduced the liver lipid deposition, and extended the mean life-span of the animals to about 16 d of age. However, MnTBAP does not cross the blood-brain barrier, and by 12 d of age, the MnTBAP-treated animals began to exhibit gait disturbances that progressed by 21 d of age to ataxia, dystonia, repetitive movements, tremor, and immobility. Histological analysis of the brains of these mice revealed a symmetrical spongiform encephalopathy, together with glial fibrillary acid protein deposition, in regions of the cortex and brainstem *(234)*. This suggests that the increased mitochondrial ROS production is extremely toxic to the brain, possibly causing neuronal apoptosis. Therefore, the *Sod2* –/– mouse has permitted the demonstration of the efficacy of MnTBAP as a mitochondrial antioxidant drug, and the effectiveness of MnTBAP treatment prove that the toxic entity in the *Sod2* –/– mice is the overproduction of $O_2^{\cdot-}$.

2.2.3.4. Partial *Sod2* Deficiency Increases the Age-Related Mitochondrial Decline

To determine the effects of chronic $O_2^{\cdot-}$ toxicity, $Sod2^{tm1Cje}$ heterozygotes (+/–) animals were studied. These animals had approximately 50% of the normal MnSOD protein and, thus, a partial reduction in antioxidant capacity. Studies of 3-mo-old *Sod2* –/– animals on a B6 background revealed increased oxidative damage to mitochondrial proteins and mtDNA, reduced activity of mitochondrial glutathione, aconitase and complex I, and an increased mitochondrial predilection to undergo mtPTP transition on exposure to t-butylhydroperoxide *(241)*. Studies of young (5 mo), middle-aged (10–15 mo), and old (20–25 mo) *Sod2 +/–* mice on the CD1 background revealed that $\Delta\psi$ was reduced throughout life by the chronic oxidative stress and that $\Delta\mu^{H+}$ declined in parallel in both the heterozygous mutant and normal animals with old age. State IV respiration rates were elevated in the *Sod2 +/–* animals, whereas the state III respiration was inhibited. Moreover, the state 4 levels in normal animals increased and the state III rates declined with age. These data are consistent with chronic $O_2^{\bullet-}$ exposure partially inactivating the Fe–S-center enzymes in the TCA cycle and ETC and an increasing proton leak of the inner membrane short-circuiting $\Delta\mu^{H+}$. They also indicate that normal animals develop the same mitochondrial defects as the *Sod2 +/–* animals, but at a later age. Hence, the aging phenomena are the same for the two genotypes, but the increased $O_2^{\bullet-}$ exposure increased the rate of aging in the *Sod2 +/–* animals.

Analysis of oxidative damage in *Sod2 +/–* versus *–/–* animals revealed that total cellular and mitochondrial lipid peroxidation of the *Sod2 +/–* animals peaked at high levels in the middle-aged animals, but then fell precipitously in old age. By contrast, lipid peroxidation remained low in the normal animals during middle age, but then increased toward the heterozygote levels in older animals. Analysis of the Ca^{2+} sensitivity of the mtPTPs in the *Sod2 +/–* animals revealed that the heterozygous mitochondria were much more prone to transition than the normal mitochondria. Furthermore, the Ca^{2+} sensitivity of the mtPTP transition increased in older animals for both genotypes. TUNEL staining of hepatocytes of the older animals revealed that the apoptosis rates of the older MnSOD heterozygous animals were threefold to fourfold higher that those of older controls. Moreover, the average OXPHOS-enzyme-specific activity in isolated mitochondria was higher in the mutants than the normals. All of these observations suggest that increased mitochondrial oxidative stress of the *Sod2 +/–* animals caused the premature accumulation of mitochondrial damage, inhibition the Fe–S-center enzymes, increase in the inner membrane proton leakage, damage to the mitochondrial macromolecules, and sensitization of the mtPTP. Ultimately, the most affected cells undergo mtPTP transition, killing the cells with the most mitochondrial damage. The removal of these damaged cells increases the overall average of the mitochondrial enzyme-specific activities, but it also results in the loss of functional cell, causing a decline in overall tissue function *(238)*. Thus, chronic mitochondrial oxidative damage does have a significant deleterious effect on mitochondria and, hence, must play a central role in the progression of mitochondrial diseases and aging.

2.2.3.5. ANTIOXIDANT THERAPY CAN EXTEND LIFE-SPAN

To determine if mitochondrial ROS toxicity was also an important factor in aging, the short-lived nematode worm (*C. elegans*) was treated with the SOD mimetic EUK134, which is similar in action to MnTBAP. The *mev-1* mutant of *C. elegans* has a 30% reduction in life-span as a result of a defect in the mitochondrial complex II (SDH), which greatly increases mitochondrial ROS production in *mev-1*. Treatment of *mev-1* animals with EUK134 restored their life-span to normal. Furthermore, EUK134-treatment of normal *C. elegans* increased their life-span 50% to a level comparable to the *age-1* mutant *(242)*. Thus, mitochondrial ROS toxicity also appears to be an important factor in limiting life-span, at least of *C. elegans*, and drugs that are effective in ameliorating the symptoms of pathogenic mitochondrial mutations might also be helpful in delaying the onset and progression of symptoms in degenerative diseases and aging.

2.2.4 Mutations in Apoptosis Genes, cytc, Bax, Bak, Apaf1, Casp9, and Casp3

Production of mice with mutations in the mitochondrial apoptosis genes has confirmed that apoptosis can be initiated by the interaction of BAX and BAK with the mitochondria, resulting in the release of mitochondrial cytochrome-*c* into the cytosol. Cytochrome-*c* interacts with the Apaf1 and caspase-9 complex, activating caspase-9 and initiating cell destruction through the activation of caspase-3.

2.2.4.1. CYTC DEFICIENCY BLOCKS MITOCHONDRIAL APOPTOSIS

The genetic inactivation of the *cytc* gene should have two effects: disruption of OXPHOS and inhibition of apoptosis. In OXPHOS, the loss of cytochrome-*c* would block electron transfer from complex III to complex IV. For apoptosis, the loss of cytochrome-*c* would be expected to block the "mitochondrial" stress-response pathway of apoptosis while leaving the "death ligand/receptor" pathway intact.

As expected from its central role in OXPHOS, inactivation of both of the *cytc* alleles (–/–) in the mouse results in embryonic lethality, even though heterozygous *cytc* (+/–) mice appear to be normal. The embryonic lethality of mice devoid of cytochrome-*c* is apparent by marked developmental delay by d 8.5 postcoitum (E8.5). Although the *cytc* –/– embryos developed all three germ layers, these embryos were strikingly smaller than their normal or heterozygous counterparts. By d E9.5, the embryos were encased in a ball-like structure formed by the yolk sac and amnion and formed primitive heart tube, somites, and allantois. By d E10.5, no viable *cytc* –/– embryos could be observed *(58)*. This embryonic lethality of the *cytc* –/– mice is reminiscent of the *Tfam* –/– mice, indicating the normal OXPHOS function becomes essential in early postimplantation gestation.

Although the *cytc* –/– embryos did not develop successfully, the cells of the early embryos were viable and could be explained into cell culture. As is observed for cells lacking mtDNA (ρ^0 cells) *(82)*, these *cytc* –/– and OXPHOS-deficient cells required GUP medium for growth.

The GUP-dependent, *cytc*-deficient cells were found to have a complete deficiency in the "mitochondrial" apoptotic pathway, although they retained the "death ligand/receptor" pathway. Exposure of the explanted mouse cells to "mitochondrial" apoptosis inducers, including ultraviolet (UV) radiation, staurosporine, or serum deprivation, which inhibits cell growth, failed to induce apoptosis in *cytc* –/– cells, but resulted in active apoptosis in *cytc* +/+ cells. These *cytc* –/– cells were unable to oligomerize Apaf-1 and caspase-9 into complexes in response to stauporine. By contrast, treatment of the *cytc* –/– cells with TNFα plus CHX, which induced apoptosis by the "death ligand/receptor" pathway, revealed an even more vigorous apoptotic response than the *cytc* +/+ cells. This deficiency in mitochondrial apoptosis was not the result of the activation of the cell-survival pathways directed by NFκB, PI3K/Akt, and JNK, because they remain inactive in both *cytc* –/– and +/+ cells *(58)*. Thus, these results confirm that the release of cytochrome-*c* from the mitochondria is essential for the activation of the "mitochondrial" apoptosis pathway and that the "mitochondrial" pathway is independent of the "death ligand/receptor" pathways.

2.2.4.2. *Bax* and *Bak* Deficiency Blocks Cytochrome Release

The role of the proapoptotic Bcl2 family members BAX and BAK in activating the "mitochondrial" apoptotic pathway was confirmed by generating mice that were doubly deficient in BAX and BAK and studying the effects of apoptosis inducers on culture mouse embryo fibroblast (MEF), hepatocytes, and whole animals. In studies of *Bax* –/– and *Bak* –/– cells, transfected with tBID or treated with apoptosis initiators staurosporine, etoposide, UV light, thapsigargin, tunicamycin, and brefeldin A, it was found that apoptosis was inhibited. Comparable treatments of *Bax* –/– *Bak* +/+ or *Bax* +/+ *Bak* –/– cells induced mitochondrial apoptosis. In addition, these inducers of apoptosis did not cause the release of cytochrome-*c* from the mitochondrial in the *Bax* –/– *Bak* –/– cells, whereas injection of cytochrome-*c* into these cells initiated apoptosis. Thus, the mitochondrial apoptosis pathway downstream of cytochrome-*c* release remained intact.

In studies of animals having either an active *Bak* or *Bax* genes, injection of an agonistic antibody to Fas-stimulated apoptosis in the liver resulted in death. However, injection of the Fas antibody into the *Bax –/–* and *Bak –/–* mice failed to cause liver apoptosis. These data demonstrate that stimulation of BAX and/or BAK to oligomerize and move to the mitochondria is an essential step for initiating the release of cytochrome-*c* from the mitochondrion. Once released, the cytochrome-*c* can bind to the Apaf1 and caspase-9 complex, activating caspase-9 and initiating of apoptosis *(243)*.

2.2.4.3. Apaf1-, Cas9, & Cas3 Deficiency Blocks Apoptosis Downstream from Cytochrome-c Release

Mice deficient in Apaf-1 *(244,245)*, caspase-9 *(246,247)*, and caspase-3 *(57,248)* all show defects in the "mitochondrial" apoptosis pathway. However, treatment of the mutant cells with apoptosis initiators did not stimulate the release cytochrome-*c*. Hence, these factors act downstream of cytochrome-*c*.

Mice deficient in Apaf1, caspase-9, and caspase-3 all exhibit embryonic lethality, with only about 5–7% of the expected –/– offspring being born. Of those born, all died in the neonatal period. This embryonic and perinatal lethality was associated with a marked outgrowth of cells of the brain, involving an excessive accumulation of neurons and glia. This indicates that a major function of the "mitochondrial" apoptosis pathway during development is the removal of excess neurons, which do not make successful contacts with the appropriate target cells. Mice deficient in Apaf1 also showed a delay in the apoptotic removal of the interdigit webbing cells.

Although Apaf1-, caspase-9-, and caspase-3-deficient embryonic cells are insensitive to the induction of apoptosis by the traditional "mitochondrial" pathway inducers such as stauosporine, C6-ceramide, and UV radiation, they remained sensitive to "death ligand/receptor" pathway inducers such as TNFα + CHX. Thus, the cell-mediated apoptosis of the immune system remained intact. These results demonstrate that the mitochondrial pathway is the primary apoptosis system used for tissue remodeling during development and in response to stress, whereas the Fas pathway is primarily used to redistribute the cells of the immune system.

3. A Mitochondrial Paradigm for Degenerative Diseases, Cancer, and Aging

These observations provide strong evidence that the mitochondria play a central role in degenerative diseases, cancer, and aging. These diverse effects can be interrelated to each other through the cellular redox state as maintained by the mitochondria through oxidation and reduction of NAD^+ and $NADH + H^+$ (*see* **Fig. 3**). Dietary calories enter the mitochondria, where they provide

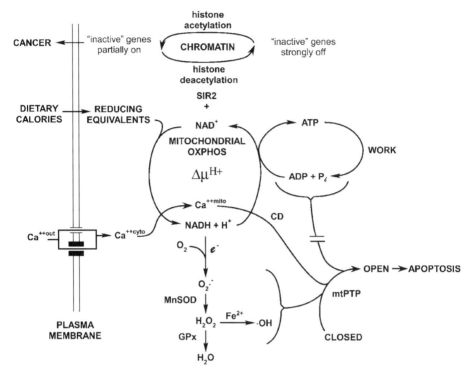

Fig. 3. Metabolic pathway showing the central role of mitochondrial NADH oxidation–reduction in the regulation of ATP production, ROS generation, apoptosis, and neoplastic transformation, leading to cancer.

reducing equivalents that reduce NAD^+ to $NADH + H^+$. $NADH + H^+$ is then reoxidized by the mitochondrial to generate $\Delta\mu^{H+}$, and $\Delta\mu^{H+}$ is used by the mitochondria to synthesize ATP from cytosolic ADP + Pi or to take up cations such as Ca^{2+}. When cellular work levels are high, ATP is actively hydrolyzed, resulting in increased cellular ADP, which is transported into the matrix by the ANT. The increased matrix ADP is rephosphorylated at the expense of $\Delta\psi$, driving the oxidation of $NADH + H^+$ to NAD^+ by the ETC. When dietary calories exceed the cellular workload, all ADP becomes phosphorylated to ATP, $\Delta\psi$ becomes hyperpolarized, and NAD^+ becomes progressively reduced to $NADH + H^+$. The excess of reducing equivalents of NADH reduce the ETC, which stimulates the transfer of electrons to O_2 to give $O_2^{\bullet-}$. The mitochondrial $O_2^{\bullet-}$, along with mitochondrial NO production, reacts with and damages mitochondrial membranes, proteins, and DNA. The increased $O_2^{\bullet-}$ is also converted to H_2O_2 by mitochondrial MnSOD, and the excess H_2O_2 diffused to the nucleus, where it mutagenizes the nDNA and activates the PARP protein to begin degrading NAD^+ *(60)*.

As somatic mtDNA mutations accumulate, they further inhibit the mitochondrial ETC and stimulate ROS production. Moreover, injury to the plasma membrane or stimulation of NMDA receptors in neurons by glutamate increases cytosolic Ca^{2+}, which is subsequently concentrated in the mitochondria. The increased Ca^{2+} binding to the cyclophilin D, elevated oxidative stress, and decreased $\Delta\mu^{H+}$ all impinge on the mtPTP, ultimately leading to permeability transition and cell loss as a result of apoptosis. This cell loss results in tissue and organ decline and system failure.

The reduction of cellular NAD^+, both by reduction to $NADH + H^+$ and degradation by PARP in response to oxidative damage of the chromatin, leads to the inhibition of the nuclear chromatin-silencing protein SIR2. SIR2 uses NAD^+ as a substrate to cleave acetyl groups from histones, thus keeping "off" genes inactive *(65)*. In the absence of active SIR2 nucleosome histones become increasingly acetylated. This results in the progressive illegitimate transcription of normally "off" genes, a characteristic feature of aging tissues. This process not only activates structural proteins but it could also reactivate inactive proto-oncogenes. This transcriptional activation, together with the associated H_2O_2 mutagenesis of the proto-oncogenes and the mitogenic stimulation of the H_2O_2, would progressively increase in probability for developing cancer as the individual ages.

This model now explains why caloric restriction not only increases longevity but also decreases cancer risks. By reducing caloric intake and balancing reducing equivalents with the work-related hydrolysis of ATP, the NAD^+ would remain oxidized. This would remove excess electrons, thus reducing production of ROS, decreasing cell loss by apoptosis, and reducing mutagenesis of the mtDNA by $O_2^{\cdot-}$ and nDNA damage by H_2O_2, thus avoiding activation of PARP and conserving NAD^+. The protection of the NAD^+ pool would also assure that SIR2 remains maximally active, thus suppressing oncogene activation and reducing neoplastic transformation.

Thus, mitochondrial disease, cancer, and aging can now be envisioned as the interaction of two mitochondrial genetic factors: (1) the inheritance of a deleterious mtDNA or nDNA mutations in a mitochondrial gene and (2) the age-related accumulation of somatic mtDNA mutations, causing mitochondrial decline, increased ROS production, and apoptosis. It is envisioned that each individual is born with an array of nDNA and mtDNA alleles that determine their initial bioenergetic capacity. If the individual inherits a strong energetic genotype, then he will have a high initial energy capacity, well above the minimum energetic thresholds required by his tissues. However, if he inherits a deleterious mutation, then his initial energetic capacity will be lower and ROS production higher. As the individual ages, somatic mtDNA mutations will accumulate in his postmitotic cells, which will further erode his tissue's energy capacities and increase ROS production. Ultimately, the combined effects of the inherited and somatic mitochondrial defects will push the tissue's energy capacity below bioenergetic thresholds, resulting in apoptosis and organ failure, and will activate nDNA proto-oncogenes, resulting in cancer.

This pathophysiological mechanism suggests that mitochondrial disease, cancer, and aging might all be treated by common strategies. These would include augmentation of energy production, removal of toxic ROS with drugs like MnTBAP, and/or inhibition of the mtPTP and postponement of cell loss resulting from apoptosis. Hopefully, such approaches might not only improve the clinical status of mitochondrial disease patients but also retard disease progression.

4. Summary

Mutations in mitochondrial genes encoded by both mitochondrial DNA (mtDNA) and nuclear DNA (nDNA have been implicated in a wide range of degenerative diseases. MtDNA base substitution and rearrangement mutations can cause myopathy, cardiomyopathy, ophthalmological defects, growth retardation, movement disorders, dementias, and diabetes. nDNA mutations can affect mtDNA replication and transcription, increase mtDNA mutations through defects in the adenine nucleotide translocator isoform 1 (ANT1), or cause Leigh's syndrome, as a result of defects in oxidative phosphorylation (OXPHOS) structural genes. Mouse models of mtDNA base substitution mutations have been created by introducing the mtDNA 16S rRNA chloramphenicol (CAP)-resistance mutation into the mouse female germline. This resulted in ophthalmological defects in chimeras and perinatal lethality resulting from myopathy and cardiomyopathy in mutant animals. Mouse models of mtDNA rearrangements have resulted in animals with myopathy, cardiomyopathy, and nephropathy. Conditional inactivation of the mouse nDNA mitochondrial transcription factor (*Tfam*) gene in the heart caused neonatal lethal cardiomyopathy, whereas its inactivation in the pancreatic β-cells caused diabetes. Mutational inactivation of the mouse *Ant1* gene resulted in myopathy, cardiomyopathy, and multiple mtDNA deletions in association with elevated reactive oxygen species (ROS) production. This suggests that multiple mtDNA deletion syndrome can be caused by increased ROS damage. The inactivation of the uncoupler protein genes (*Ucp*) 1–3 resulted in alterations in $\Delta\mu^{H+}$ and increased ROS production. Inactivation of the *Ucp2* gene, which is expressed in the pancreatic β-cells, resulted in increased islet ATP, increased serum insulin levels, and suppression of the diabetes of the *ob/ob* mouse genotype. Transgenic mice with altered β-cell ATP-sensitive K$^+$ channels (K$_{ATP}$) also developed diabetes. Mutational inactivation of the mitochondrial antioxidant genes for glutathione peroxidase (*GPx1*) and Mn superoxide dismutase (*Sod2*) caused reduced energy production and neonatal lethal dilated cardiomyopathy, respectively, the later being ameliorated by treatment with MnSOD mimics. Partial *Sod2* deficiency (+/–) resulted in mice with increased mitochondrial damage during aging, and treatment of *C. elegans* with catalytic antioxidant drugs can extend their life-span. Mice deficient in cytochrome-*c* died early in embryogenesis, but cells derived from these embryos had a complete deficiency in mitochondrial apoptosis. Mice lacking the proapoptotic *Bax* and *Bak* genes were not able to release cytochrome-*c* from the mitochondrion and were blocked in apoptosis. Mice lacking *Apaf1*, *Cas9*, and *Cas3* did release mitochondrial cytochrome-*c* and were blocked in the downstream steps of apoptosis. These animal studies confirm that alterations in mitochondrial energy generation, ROS production, and apoptosis can all contribute to the pathophysiology of mitochondrial disease.

Acknowledgments

This work was supported by National Institutes of Health grants GM46915, NS21328, NS37167, HL45572, HL64017, AG13154, and AG10130 awarded to D.C.W.

Refereneces

1. Wallace, D. C., Brown, M. D., and Lott, M. T. (1996) Mitochondrial genetics, in *Emery and Rimoin's Principles and Practice of Medical Genetics* (Rimoin, D. L., et al., eds.), Churchill Livingstone, London, pp. 277–332.
2. Wallace, D. C. (1997) Mitochondrial DNA mutations and bioenergetic defects in aging and degenerative diseases, in *The Molecular and Genetic Basis of Neurological Disease* (Rosenberg, R. N., et al., eds.), Butterworth–Heinemann, Boston, pp. 237–269.
3. Shoffner, J. M. and Wallace, D. C. (1995) Oxidative phosphorylation diseases, in *The Metabolic and Molecular Basis of Inherited Disease* (Scriver, C. R., et al., eds.), McGraw-Hill, New York, pp. 1535–1609.
4. Green, D. R. and Reed, J. C. (1998) Mitochondria and apoptosis. *Science* **281(5381),** 1309–1312.
5. Liu, X., et al. (1996) Induction of apoptotic program in cell-free extracts: requirement for dATP and cytochrome c. *Cell* **86(1),** 147–157.
6. Brustovetsky, N. and Klingenberg, M. (1996) Mitochondrial ADP/ATP carrier can be reversibly converted into a large channel by Ca^{2+}. *Biochemistry* **35(26),** 8483–8488.

7. Marzo, I., et al. (1998) *Bax* and adenine nucleotide translocator cooperate in the mitochondrial control of apoptosis. *Science* **281(5385)**, 2027–2031.
8. Wallace, D. C. (1999) Mitochondrial diseases in man and mouse. *Science* **283(5407)**, 1482–1488.
9. Wallace, D. C. (1992) Mitochondrial genetics: a paradigm for aging and degenerative diseases? *Science* **256**, 628–632.
10. Wallace, D. C. (1992) Diseases of the mitochondrial DNA. *Ann. Rev. Biochem.* **61**, 1175–1212.
11. Stepien, G., et al. (1992) Differential expression of adenine nucleotide translocator isoforms in mammalian tissues and during muscle cell differentiation. *J. Biol. Chem.* **267(21)**, 14,592–14,597.
12. Neckelmann, N., et al. (1987) cDNA sequence of a human skeletal muscle ADP/ATP translocator: lack of a leader peptide, divergence from a fibroblast translocator cDNA, and coevolution with mitochondrial DNA genes. *Proc. Natl. Acad. Sci. USA* **84(21)**, 7580–7584.
13. Houldsworth, J. and Attardi, G. (1988) Two distinct genes for ADP/ATP translocase are expressed at the mRNA level in adult human liver. *Proc. Natl. Acad. Sci. USA* **85(2)**, 377–381.
14. Giraud, S., et al. (1998) Expression of human ANT2 gene in highly proliferative cells: GRBOX, a new transcriptional element, is involved in the regulation of glycolytic ATP import into mitochondria. *J. Mol. Biol.* **281(3)**, 409–418.
15. Cozens, A. L., Runswick, M. J., and Walker, J. E. (1989) DNA sequences of two expressed nuclear genes for human mitochondrial ADP/ATP translocase. *J. Mol. Biol.* **206(2)**, 261–280.
16. Li, K., et al. (1989) A human muscle adenine nucleotide translocator gene has four exons, is located on chromosome 4, and is differentially expressed. *J. Biol. Chem.* **264(24)**, 13,998–14,004.
17. Wijmenga, C., et al. (1993) The human skeletal muscle adenine nucleotide translocator gene maps to chromosome 4q35 in the region of the facioscapulohumeral muscular dystrophy locus. *Hum. Genet.* **92(2)**, 198–203.
18. Wijmenga, C., et al. (1992) Chromosome 4q DNA rearrangements associated with facioscapulohumeral muscular dystrophy. *Nature Genet.* **2(1)**, 26–30.
19. Haraguchi, Y., et al. (1993) Genetic mapping of human heart–skeletal muscle adenine nucleotide translocator and its relationship to the facioscapulohumeral muscular dystrophy locus. *Genomics* **16(2)**, 479–485.
20. Battini, R., et al. (1987) Molecular cloning of a cDNA for a human ADP/ATP carrier which is growth-regulated. *J. Biol. Chem.* **262(9)**, 4355–4359.
21. Chen, S. T., et al. (1990) A human ADP/ATP translocase gene has seven pseudogenes and localizes to chromosome X. *Somatic Cell Mol. Genet.* **16(2)**, 143–149.
22. Ku, D. H., et al. (1990) The human fibroblast adenine nucleotide translocator gene. Molecular cloning and sequence. *J. Biol. Chem.* **265(27)**, 16,060–16,063.
23. Schiebel, K., et al. (1994) Localization of the adenine nucleotide translocase gene *ANT2* to chromosome Xq24-q25 with tight linkage to DXS425. *Genomics* **24(3)**, 605–606.

24. Schiebel, K., et al. (1993) A human pseudoautosomal gene, ADP/ATP translocase, escapes X-inactivation whereas a homologue on Xq is subject to X-inactivation. *Nature Genet.* **3(1),** 82–87.

25. Slim, R., et al. (1993) A human pseudoautosomal gene encodes the ANT3 ADP/ATP translocase and escapes X-inactivation. *Genomics* **16(1),** 26–33.

26. Levy, S. E., et al. (2000) Expression and sequence analysis of the mouse adenine nucleotide translocase 1 and 2 genes. *Gene* **254,** 57–66.

27. Graham, B., et al. (1997) A mouse model for mitochondrial myopathy and cardiomyopathy resulting from a deficiency in the heart/skeletal muscle isoform of the adenine nucleotide translocator. *Nature Genet.* **16(3),** 226–234.

28. Mills, K. A., Ellison, J. W., and Mathews, K. D. (1996) The *Ant1* gene maps near *Klk3* on proximal mouse chromosome 8. *Mammal. Genome* **7(9),** 707.

29. Ellison, J. W., et al. (1996) Rapid evolution of human pseudoautosomal genes and their mouse homologs. *Mammal. Genome* **7(1),** 25–30.

30. Cassard, A. M., et al. (1990) Human uncoupling protein gene: structure, comparison with rat gene, and assignment to the long arm of chromosome 4. *J. Cell. Biochem.* **43(3),** 255–264.

31. Kozak, L. P., et al. (1988) The mitochondrial uncoupling protein gene. Correlation of exon structure to transmembrane domains. *J. Biol. Chem.* **263(25),** 12,274–12,277.

32. Jacobsson, A., et al. (1985) Mitochondrial uncoupling protein from mouse brown fat. Molecular cloning, genetic mapping, and mRNA expression. *J. Biol. Chem.* **260(30),** 16,250–16,254.

33. Fleury, C., et al. (1997) Uncoupling protein-2: a novel gene linked to obesity and hyperinsulinemia [see comments]. *Nature Genet.* **15(3),** 269–272.

34. Gong, D. W., et al. (1997) Uncoupling protein-3 is a mediator of thermogenesis regulated by thyroid hormone, beta3-adrenergic agonists, and leptin. *J. Biol. Chem.* **272(39),** 24,129–24,132.

35. Solanes, G., et al. (1997) The human uncoupling protein-3 gene. Genomic structure, chromosomal localization, and genetic basis for short and long form transcripts. *J. Biol. Chem.* **272(41),** 25,433–25,436.

36. Vidal-Puig, A., et al. (1997) UCP3: an uncoupling protein homologue expressed preferentially and abundantly in skeletal muscle and brown adipose tissue. *Biochem. Biophys. Res. Commun.* **235(1),** 79–82.

37. Boss, O., et al. (1997) Uncoupling protein-3: a new member of the mitochondrial carrier family with tissue-specific expression. *FEBS Lett.* **408(1),** 39–42.

38. Ksenzenko, M., et al. (1983) Effect of electron transfer inhibitors on superoxide generation in the cytochrome bc1 site of the mitochondrial respiratory chain. *FEBS Lett.* **155(1),** 19–24.

39. Turrens, J. F., Alexandre, A., and Lehninger, A. L. (1985) Ubisemiquinone is the electron donor for superoxide formation by complex III of heart mitochondria. *Arch. Biochem. Biophys.* **237(2),** 408–414.

40. Boveris, A. and Turrens, J. F. (1980) Production of superoxide anion by the NADH-dehydrogenase of mamalian mitochondria, in *Chemical and Biochemical Aspects of*

Superoxide and Superoxide Dismutase. Developments in Biochemistry (Bannister, J. V. and Hill, H. A. O., eds.), Elsevier–North Holland, New York, pp. 84–91.

41. Turrens, J. F. and Boveris, A. (1980) Generation of superoxide anion by the NADH dehydrogenase of bovine heart mitochondria. *Biochem. J.* **191(2)**, 421–427.

42. Bandy, B. and Davison, A. J. (1990) Mitochondrial mutations may increase oxidative stress: implications for carcinogenesis and aging? *Free Radical Biol. Med.* **8(6)**, 523–539.

43. Goldhaber, J. I. and Weiss, J. N. (1992) Oxygen free radicals and cardiac reperfusion abnormalities. *Hypertension* **20(1)**, 118–127.

44. Yan, L. J., Levine, R. L., and Sohal, R. S. (1997) Oxidative damage during aging targets mitochondrial aconitase [published erratum appears in Proc. Natl. Acad. Sci. US 1998 Feb 17;95(4):1968]. *Proc. Natl. Acad. Sci. USA* **94(21)**, 11,168–11,172.

45. Rotig, A., et al. (1997) Aconitase and mitochondrial iron–sulphur protein deficiency in Friedreich ataxia. *Nature Genet.* **17(2)**, 215–217.

46. Boveris, A., Oshino, N., and Chance, B. (1972) The cellular production of hydrogen peroxide. *Biochem. J.* **128(3)**, 617–630.

47. Boveris, A. (1984) Determination of the production of superoxide radicals and hydrogen peroxide in mitochondria. *Methods Enzymol.* **105**, 429–435.

48. Cadenas, E. and Boveris, A. (1980) Enhancement of hydrogen peroxide formation by protophores and ionophores in antimycin-supplemented mitochondria. *Biochem. J.* **188(1)**, 31–37.

49. Chance, B., Sies, H., and Boveris, A. (1979) Hydroperoxide metabolism in mammalian organs. *Physiol. Rev.* **59(3)**, 527–605.

50. Loschen, G., et al. (1974) Superoxide radicals as precursors of mitochondrial hydrogen peroxide. *FEBS Lett.* **42(1)**, 68–72.

51. Petit, P. X., et al. (1996) Mitochondria and programmed cell death: back to the future. *FEBS Lett.* **396(1)**, 7–13.

52. Zoratti, M. and Szabo, I. (1995) The mitochondrial permeability transition. *Biochim. Biophys. Acta* **1241(2)**, 139–176.

53. Mancini, M., et al. (1998) The caspase-3 precursor has a cytosolic and mitochondrial distribution: implications for apoptotic signaling. *J. Cell Biol.* **140(6)**, 1485–1495.

54. Earnshaw, W. C. (1999) Apoptosis. A cellular poison cupboard [news]. *Nature* **397(6718)**, 387–389.

55. Susin, S. A., et al. (1999) Molecular characterization of mitochondrial apoptosis-inducing factor. *Nature* **397(6718)**, 441–446.

56. Susin, S. A., et al. (1999) Mitochondrial release of caspase-2 and -9 during the apoptotic process. *J. Exp. Med.* **189(2)**, 381–394.

57. Woo, M., et al. (1998) Essential contribution of caspase 3/CPP32 to apoptosis and its associated nuclear changes. *Genes Dev.* **12(6)**, 806–819.

58. Li, K., et al. (2000) Cytochrome c deficiency causes embryonic lethality and attenuates stress-induced apoptosis. *Cell* **101(4)**, 389–399.

59. Szabo, C. (1997) Role of poly(ADP-ribose) synthetase activation in the suppression of cellular energetics in response to nitric oxide and peroxynitrite. *Biochem. Soc. Trans.* **25(3)**, 919–924.

60. Snyder, S. H., Jaffrey, S. R., and Zakhary, R. (1998) Nitric oxide and carbon monoxide: parallel roles as neural messengers. *Brain Res. Rev.* **26(2–3)**, 167–175.
61. Eliasson, M. J., et al. (1997) Poly(ADP-ribose) polymerase gene disruption renders mice resistant to cerebral ischemia. *Nature Med.* **3(10)**, 1089–1095.
62. Takahashi, K., et al. (1999) Post-treatment with an inhibitor of poly(ADP-ribose) polymerase attenuates cerebral damage in focal ischemia. *Brain Res.* **829(1–2)**, 46–54.
63. Pieper, A. A., et al. (1999) Poly(ADP-ribose) polymerase-deficient mice are protected from streptozotocin-induced diabetes. *Proc. Natl. Acad. Sci. USA* **96(6)**, 3059–3064.
64. Schuler, M., et al. (2000) *p53* induces apoptosis by caspase activation through mitochondrial cytochrome c release. *J. Biol. Chem.* **275(10)**, 7337–7342.
65. Lin, S. J., Defossez, P. A., and Guarente, L. (2000) Requirement of NAD and SIR2 for life-span extension by calorie restriction in *Saccharomyces cerevisiae* [see comments]. *Science* **289(5487)**, 2126–2128.
66. Suh, Y. A., et al. (1999) Cell transformation by the superoxide-generating oxidase *Mox1*. *Nature* **401(6748)**, 79–82.
67. Arnold, R. S., Shi, J., Murad, E., Whalen, A. M., et al. (2001) Hydrogen peroxide mediates the cell growth and transformation caused by the mitogenic oxidase Nox1. *Proc. Natl. Acad. Sci. USA* **98**, 5550–5555.
68. Chang, D. D. and Clayton, D. A. (1987) A mammalian mitochondrial RNA processing activity contains nucleus-encoded RNA. *Science* **235(4793)**, 1178–1184.
69. Chang, D. D. and Clayton, D. A. (1987) A novel endoribonuclease cleaves at a priming site of mouse mitochondrial DNA replication. *EMBO J.* **6(2)**, 409–417.
70. Chang, D. D. and Clayton, D. A. (1989) Mouse RNAase MRP RNA is encoded by a nuclear gene and contains a decamer sequence complementary to a conserved region of mitochondrial RNA substrate. *Cell* **56(1)**, 131–139.
71. Clayton, D. A. (1994) A nuclear function for RNase MRP. *Proc. Natl. Acad. Sci. USA* **91(11)**, 4615–4617.
72. Fisher, R. P., Topper, J. N., and Clayton, D. A. (1987) Promoter selection in human mitochondria involves binding of a transcription factor to orientation-independent upstream regulatory elements. *Cell* **50(2)**, 247–258.
73. Dairaghi, D. J., Shadel, G. S., and Clayton, D. A. (1995) Addition of a 29 residue carboxyl-terminal tail converts a simple HMG box-containing protein into a transcriptional activator. *J. Mol. Biol.* **249(1)**, 11–28.
74. Ghivizzani, S. C., et al. (1994) In organello footprint analysis of human mitochondrial DNA: human mitochondrial transcription factor A interactions at the origin of replication. *Mol. Cell. Biol.* **14(12)**, 7717–7730.
75. Ghivizzani, S. C., Madsen, C. S., and Hauswirth, W. W. (1993) In organello footprinting. Analysis of protein binding at regulatory regions in bovine mitochondrial DNA. *J. Biol. Chem.* **268(12)**, 8675–8682.
76. Ikeda, S., Sumiyoshi, H., and Oda, T. (1994) DNA binding properties of recombinant human mitochondrial transcription factor 1. *Cell. Mol. Biol.* **40(4)**, 489–493.

77. Bunn, C. L., Wallace, D. C., and Eisenstadt, J. M. (1974) Cytoplasmic inheritance of chlormaphenicol resistance in mouse tissue culture cells. *Proc. Natl. Acad. Sci. USA* **71(5),** 1681–1685.

78. Wallace, D. C., Bunn, C. L., and Eisenstadt, J. M. (1975) Cytoplasmic transfer of chloramphenicol resistance in human tissue culture cells. *J. Cell Biol.* **67(1),** 174–188.

79. Wallace, D. C. (1982) Cytoplasmic inheritance of chloramphenicol resistance in mammalian cells, in *Techniques in Somatic Cell Genetics* (Shay, J. W., ed.), Plenum, New York, pp. 159–187.

80. Ziegler, M. L. and Davidson, R. L. (1981) Elimination of mitochondrial elements and improved viability in hybrid cells. *Somatic Cell Genet.* **7(1),** 73–88.

81. Trounce, I. and Wallace, D. C. (1996) Production of transmitochondrial mouse cell lines by cybrid rescue of rhodamine-6G pre-treated L-cells. *Somatic Cell Mol. Genet.* **22(1),** 81–85.

82. King, M. P. and Attardi, G. (1989) Human cells lacking mtDNA: repopulation with exogenous mitochondria by complementation. *Science* **246(4929),** 500–503.

83. Trounce, I. A., et al. (1996) Assessment of mitochondrial oxidative phosphorylation in patient muscle biopsies, lymphoblasts, and transmitochondrial cell lines. *Methods Enzymol.* **264,** 484–509.

84. Blanc, H., et al. (1981) Mitochondrial DNA of chloramphenicol-resistant mouse cells contains a single nucleotide change in the region encoding the 3′ end of the large ribosomal RNA. *Proc. Natl. Acad. Sci. USA* **78(6),** 3789–3793.

85. Blanc, H., Adams, C. W., and Wallace, D. C. (1981) Different nucleotide changes in the large rRNA gene of the mitochondrial DNA confer chloramphenicol resistance on two human cell lines. *Nucleic Acids Res.* **9(21),** 5785–5795.

86. Wallace, D. C., et al. (2001) Mitochondria and neuro-ophthalmological diseases, in *The Metabolic and Molecular Basis of Inherited Disease* (Scriver, C. R., et al., eds.), McGraw-Hill, New York, pp. 2425–2512.

87. Wallace, D. C., et al. (1988) Familial mitochondrial encephalomyopathy (MERRF): genetic, pathophysiological, and biochemical characterization of a mitochondrial DNA disease. *Cell* **55(4),** 601–610.

88. Shoffner, J. M., et al. (1990) Myoclonic epilepsy and ragged-red fiber disease (MERRF) is associated with a mitochondrial DNA tRNA[Lys] mutation. *Cell* **61(6),** 931–937.

89. Goto, Y., Nonaka, I., and Horai, S. (1990) A mutation in the tRNA[Leu(UUR)] gene associated with the MELAS subgroup of mitochondrial encephalomyopathies [see comments]. *Nature* **348(6302),** 651–653.

90. van den Ouweland, J. M., et al. (1994) Maternally inherited diabetes and deafness is a distinct subtype of diabetes and associates with a single point mutation in the mitochondrial tRNA[Leu(UUR)] gene. *Diabetes* **43(6),** 746–751.

91. van den Ouweland, J. M., et al. (1992) Mutations in mitochondrial tRNA genes: non-linkage with syndromes of Wolfram and chronic progressive external ophthalmoplegia. *Nucleic Acids Res.* **20(4),** 679–682.

92. van den Ouweland, J. M., et al. (1992) Mutation in mitochondrial tRNA[Leu(UUR)] gene in a large pedigree with maternally transmitted type II diabetes mellitus and deafness. *Nature Gen.* **1**, 368–371.

93. Shoffner, J. M., et al. (1995) Mitochondrial encephalomyopathy associated with a single nucleotide pair deletion in the mitochondrial tRNA[Leu(UUR)] gene. *Neurology* **45(2)**, 286–292.

94. Hutchin, T., et al. (1993) A molecular basis for human hypersensitivity to aminoglycoside antibiotics. *Nucleic Acids Res.* **21(18)**, 4174–4179.

95. Prezant, T. R., et al. (1993) Mitochondrial ribosomal RNA mutation associated with both antibiotic-induced and non-syndromic deafness. *Nature Genet.* **4(3)**, 289–294.

96. Ballinger, S. W., et al. (1992) Maternally transmitted diabetes and deafness associated with a 10.4 kb mitochondrial DNA deletion. *Nature Genet.* **1**, 11–15.

97. Ballinger, S. W., et al. (1994) Mitochondrial diabetes revisited [letter]. *Nature Genet.* **7(4)**, 458–459.

98. Holt, I. J., Harding, A. E., and Morgan-Hughes, J. A. (1988) Deletions of muscle mitochondrial DNA in patients with mitochondrial myopathies. *Nature* **331(6158)**, 717–719.

99. Shoubridge, E. A., Karpati, G., and Hastings, K. E. M. (1990) Deletion mutants are functionally dominant over wild-type mitochondrial genomes in skeletal muscle fiber segments in mitochondrial disease. *Cell* **62(1)**, 43–49.

100. Mitani, I., et al. (1998) Detection of mitochondrial DNA nucleotide 11778 point mutation of Leber hereditary optic neuropathy from archival stained histopathological preparations. *Acta Ophthalmol. Scand.* **76(1)**, 14–19.

101. Pearson, H. A., et al. (1979) A new syndrome of refractory sideroblastic anemia with vacuolization of marrow precursors and exocrine pancreatic function. *J. Pediatr.* **95(6)**, 976.

102. Kapsa, R., et al. (1994) A novel mtDNA deletion in an infant with Pearson syndrome. *J. Inherited Metab. Dis.* **17(5)**, 521–526.

103. Cormier, V., et al. (1991) Pearson's syndrome. Pancytopenia with exocrine pancreatic insufficiency: new mitochondrial disease in the first year of childhood. *Arch. Fr. Pediatr.* **48(3)**, 171–178.

104. McShane, M. A., et al. (1991) Pearson syndrome and mitochondrial encephalomyopathy in patient with a deletion of mtDNA. *Am. J. Hum. Genet.* **48(1)**, 39–42.

105. Poulton, J., et al. (1995) Mitochondrial DNA, diabetes and pancreatic pathology in Kearns–Sayre syndrome. *Diabetologia* **38(7)**, 868–871.

106. Rotig, A., et al. (1995) Spectrum of mitochondrial DNA rearrangements in the Pearson marrow–pancreas syndrome. *Hum. Mol. Genet.* **4(8)**, 1327–1330.

107. Heddi, A., et al. (1999) Coordinate induction of energy gene expression in tissues of mitochondrial disease patients. *J. Biol. Chem.* **274(33)**, 22,968–22,976.

108. Heddi, A., et al. (1993) Mitochondrial DNA expression in mitochondrial myopathies and coordinated expression of nuclear genes involved in ATP production. *J. Biol. Chem.* **268(16)**, 12,156–12,163.

109. Heddi, A., et al. (1994) Steady state levels of mitochondrial and nuclear oxidative phosphorylation transcripts in Kearns-Sayre syndrome. *Biochim. Biophys. Acta* **1226(2)**, 206–212.

110. Stachowiak, O., et al. (1998) Mitochondrial creatine kinase is a prime target of peroxynitrite-induced modification and inactivation. *J. Biol. Chem.* **273(27)**, 16,694–16,699.

110a. Suomalainen, A. and Kaukonen, J. (2001) Diseases caused by nuclear genes affecting mtDNA stability. *Am. J. Med. Genet.* **106**, 53–61.

110b. Orth M. and Schapira, A. H. (2001) Mitochondria and degenerative disorders. *Am. J. Med. Genet.* **106**, 27–36.

110c. Shoubridge, E. A. (2001) Cytochrome c oxidase deficiency. *Am. J. Med. Genet.* **106**, 46–52.

110d. Triepels, R. H., Van Den Heuvel, L. P., Trijbels, J. M., Smeitink, J. A. (2001) Respiratory chain complex I deficiency. *Am. J. Med. Genet.* **106**, 37–45.

111. Ridanpaa, M., et al. (2001) Mutations in the RNA component of RNase MRP cause a pleiotropic human disease, cartilage–hair hypoplasia. *Cell* p. 195–203.

112. Tritschler, H.-J., et al. (1992) Mitochondrial myopathy of childhood associated with depletion of mitochondrial DNA. *Neurology* **42(1)**, 209–217.

113. Mazziotta, M. R., et al. (1992) Fatal infantile liver failure associated with mitochondrial DNA depletion. *J. Pediatr.* **121(6)**, 896–901.

114. Moraes, C. T., et al. (1991) MtDNA depletion with variable tissue expression: a novel genetic abnormality in mitochondrial diseases. *Am. J. Hum. Genet.* **48(3)**, 492–501.

115. Dahl, H. H. (1998) Getting to the nucleus of mitochondrial disorders: identification of respiratory chain-enzyme genes causing Leigh syndrome [editorial; comment]. *Am. J. Hum. Genet.* **63(6)**, 1594–1597.

116. Zhu, Z., et al. (1998) SURF1, encoding a factor involved in the biogenesis of cytochrome c oxidase, is mutated in Leigh syndrome. *Nature Genet.* **20(4)**, 337–343.

117. Tiranti, V., et al. (1998) Mutations of SURF-1 in Leigh disease associated with cytochrome c oxidase deficiency. *Am. J. Hum. Genet.* **63(6)**, 1609–1621.

118. Valnot, I., et al. (2000) Mutations in SCO1 gene causes mitochondrial cytochrome c oxidase deficiency presenting as neonatal-onset hepatic failure and encephalopathy. *Am. J. Hum. Genet.* **67(4 Suppl. 2)**, A20 (abstract).

119. Papadopoulou, L. C., et al. (1999) Fatal infantile cardioencephalomyopathy with COX deficiency and mutations in SCO2, a COX assembly gene. *Nature Gen.* **23(3)**, 333–337.

120. Dickinson, E. K., et al. (2000) A human SCO2 mutation helps define the role of Sco1p in the cytochrome oxidase assembly pathway. *J. Biol. Chem.* **275(35)**, 26,780–26,785.

121. Jin, H., et al. (1996) A novel X-linked gene, *DDP*, shows mutations in families with deafness (*DFN-1*), dystonia, mental deficiency and blindness. *Nature Genet.* **14(2)**, 177–180.

122. Wallace, D. C. and Murdock, D. G. (1999) Mitochondria and dystonia: the movement disorder connection? *Proc. Natl. Acad. Sci. USA* **96(5)**, 1817–1819.
123. Koehler, C. M., et al. (1999) Human deafness dystonia syndrome is a mitochondrial disease. *Proc. Natl. Acad. Sci. USA* **96(5)**, 2141–2146.
124. Wilson, R. B. and Roof, D. M. (1997) Respiratory deficiency due to loss of mitochondrial DNA in yeast lacking the frataxin homologue. *Nature Genet.* **16(4)**, 352–357.
125. Koutnikova, H., et al. (1997) Studies of human, mouse and yeast homologues indicate a mitochondrial function for frataxin. *Nature Genet.* **16(4)**, 345–351.
126. Zeviani, M., et al. (1989) An autosomal dominant disorder with multiple deletions of mitochondrial DNA starting at the D-loop region. *Nature* **339(6222)**, 309–311.
127. Zeviani, M., et al. (1990) Nucleus-driven multiple large-scale deletions of the human mitochondrial genome: a new autosomal dominant disease. *Am. J. Hum. Genet.* **47**, 904–914.
128. Cormier, V., et al. (1991) Autosomal dominant deletions of the mitochondrial genome in a case of progressive encephalomyopathy. *Am. J. Hum. Genet.* **48(4)**, 643–648.
129. Moraes, C. T., et al. (1993) Atypical clinical presentations associated with the MELAS mutation at position 3243 of human mitochondrial DNA. *Neuromuscul. Dis.* **3(1)**, 43–50.
130. Checcarelli, N., et al. (1994) Multiple deletions of mitochondrial DNA in sporadic and atypical cases of encephalomyopathy. *J. Neurol. Sci.* **123(1–2)**, 74–79.
131. Suomalainen, A., et al. (1992) Multiple deletions of mitochondrial DNA in several tissues of a patient with severe retarded depression and familial progressive external ophthalmoplegia. *J. Clin. Invest.* **90**, 61–66.
132. Suomalainen, A., et al. (1995) An autosomal locus predisposing to deletions of mitochondrial DNA. *Nature Genet.* **9(2)**, 146–151.
133. Zeviani, M., et al. (1995) Searching for genes affecting the structural integrity of the mitochondrial genome. *Biochim. Biophys. Acta* **1271(1)**, 153–158.
134. Kaukonen, J., et al. (2000) Role of adenine nucleotide translocator 1 in mtDNA maintenance. *Science* **289(5480)**, 782–785.
135. Nishino, I., Spinazzola, A., and Hirano, M. (1999) Thymidine phosphorylase gene mutations in MNGIE, a human mitochondrial disorder. *Science* **283(5402)**, 689–692.
136. Boffoli, D., et al. (1994) Decline with age of the respiratory chain activity in human skeletal muscle. *Biochim. Biophys. Acta* **1226(1)**, 73–82.
137. Cooper, J. M., Mann, V. M., and Schapira, A. H. V. (1992) Analyses of mitochondrial respiratory chain function and mitochondrial DNA deletion in human skeletal muscle: effect of ageing. *J. Neurol. Sci.* **113(1)**, 91–98.
138. Trounce, I., Byrne, E., and Marzuki, S. (1989) Decline in skeletal muscle mitochondrial respiratory chain function: possible factor in ageing. *Lancet* **1**, 637–639.
139. Yen, T. C., et al. (1989) Liver mitochondrial respiratory functions decline with age. *Biochem. Biophys. Res. Commun.* **165**, 944–1003.

140. Bowling, A. C., et al. (1993) Age-dependent impairment of mitochondrial function in primate brain. *J. Neurochem.* **60(5)**, 1964–1967.

141. Cortopassi, G. A. and Arnheim, N. (1990) Detection of a specific mitochondrial DNA deletion in tissues of older humans. *Nucleic Acids Res.* **18(23)**, 6927–6933.

142. Cortopassi, G. A., et al. (1992) A pattern of accumulation of a somatic deletion of mitochondrial DNA in aging human tissues. *Proc. Natl. Acad. Sci. USA* **89(16)**, 7370–7374.

143. Cortopassi, G. A., Pasinetti, G., and Arnheim, N. (1992) Mosaicism for levels of a somatic mutation of mitochondrial DNA in different brain regions and its implications for neurological disease, in *Progress in Parkinson's Disease Research II* (Hatefi, F. and Weiner, W. J., eds.), Futura, Mt. Kisco, NY.

144. Linnane, A. W., et al. (1990) Mitochondrial gene mutation: the aging process and degenerative diseases. *Biochem. Int.* **22(6)**, 1067–1076.

145. Zhang, C., et al. (1992) Multiple mitochondrial DNA deletions in an elderly human individual. *FEBS Lett.* **297**, 4–8.

146. Corral-Debrinski, M., et al. (1992) Mitochondrial DNA deletions in human brain: regional variability and increase with advanced age. *Nature Genet.* **2(4)**, 324–329.

147. Soong, N. W., et al. (1992) Mosaicism for a specific somatic mitochondrial DNA mutation in adult human brain. *Nature Genet.* **2**, 318–323.

148. Corral-Debrinski, M., et al. (1991) Hypoxemia is associated with mitochondrial DNA damage and gene induction. *JAMA* **266(13)**, 1812–1816.

149. Corral-Debrinski, M., et al. (1992) Association of mitochondrial DNA damage with aging and coronary atherosclerotic heart disease. *Mutat. Res.* **275(3–6)**, 169–180.

150. Simonetti, S., et al. (1992) Accumulation of deletions in human mitochondrial DNA during normal aging: analysis by quantitative PCR. *Biochim. Biophys. Acta* **1180(2)**, 113–122.

151. Jazin, E. E., et al. (1996) Human brain contains high levels of heteroplasmy in the noncoding regions of mitochondrial DNA. *Proc. Natl. Acad. Sci. USA* **93(22)**, 12,382–12,387.

152. Michikawa, Y., et al. (1999) Aging-dependent large accumulation of point mutations in the human mtDNA control region for replication. *Science* **286(5440)**, 774–779.

153. Murdock, D. G., Christacos, N. C., and Wallace, D. C. (2000) The age-related accumulation of a mitochondrial DNA control region mutation in muscle, but not brain, detected by a sensitive PNA-directed PCR clamping based method. *Nucleic Acids Res.* **28(21)**, 4350–4355.

154. Trounce, I., et al. (2000) Cloning of neuronal mtDNA variants in cultured cells by synaptosome fusion with mtDNA-less cells. *Nucleic Acids Res.* **28(10)**, 2164–2170.

155. Richter, C., Park, J. W. and Ames, B. N. (1988) Normal oxidative damage to mitochondrial and nuclear DNA is extensive. *Proc. Natl. Acad. Sci. USA* **85(17)**, 6465–6467.

156. Ames, B. N., Shigenaga, M. K. and Hagen, T. M. (1993) Oxidants, antioxidants, and the degenerative diseases of aging. *Proc. Natl. Acad. Sci. USA* **90(17),** 7915–7922.

157. Hayakawa, M., et al. (1993) Age-associated damage in mitochondrial DNA in human hearts. *Mol. Cell. Biochem.* **119(1–2),** 95–103.

158. Mecocci, P., et al. (1993) Oxidative damage to mitochondrial DNA shows marked age-dependent increases in human brain. *Ann. Neurol.* **34(4),** 609–616.

159. Muller-Hocker, J. (1990) Cytochrome c oxidase deficient fibres in the limb muscle and diaphragm of man without muscular disease: an age-related alteration. *J. Neurol. Sci.* **100,** 14–21.

160. Muller-Hocker, J., et al. (1992) Progressive loss of cytochrome c oxidase in the human extraocular muscles in ageing—a cytochemical–immunohistochmeical study. *Mutat. Res.* **275,** 115–124.

161. Muller-Hocker, J., et al. (1993) Different in situ hybridization patterns of mitochondrial DNA in cytochrome c oxidase-deficient extraocular muscle fibres in the elderly. *Virchows Arch A, Pathol. Anat. Histopathol.* **422,** 7–15.

162. Khrapko, K., et al. (1999) Cell-by-cell scanning of whole mitochondrial genomes in aged human heart reveals a significant fraction of myocytes with clonally expanded deletions. *Nucleic Acids Res.* **27(11),** 2434–2441.

163. Melov, S., et al. (1997) Multi-organ characterization of mitochondrial genomic rearrangements in ad libitum and caloric restricted mice show striking somatic mitochondrial DNA rearrangements with age. *Nucleic Acids Res.* **25(5),** 974–982.

164. Lee, C. K., et al. (1999) Gene expression profile of aging and its retardation by caloric restriction. *Science* **285(5432),** 1390–1393.

165. Lee, C. K., Weindruch, R., and Prolla, T. A. (2000) Gene-expression profile of the ageing brain in mice. *Nature Genet.* **25(3),** 294–297.

166. Masoro, E. J. (1993) Dietary restriction and aging. *J. Am. Geriatr. Soc.* **41(9),** 994–999.

167. Masoro, E. J., et al. (1992) Dietary restriction alters characteristics of glucose fuel use (published erratum appears in *J. Gerontol.* 1993 Mar;48(2):B73). *J. Gerontol.* **47(6),** B202–B208.

168. McCarter, R. J. and Palmer, J. (1992) Energy metabolism and aging: a lifelong study of Fischer 344 rats. *Am. J. Physiol.* **263(3 Pt. 1),** E448–E452.

169. Sohal, R. S., et al. (1994) Oxidative damage, mitochondrial oxidant generation and antioxidant defenses during aging and in response to food restriction in the mouse. *Mech. Ageing Dev.* **74(1–2),** 121–133.

170. Harrison, D. E. and Archer, J. R. (1987) Genetic differences in effects of food restriction on aging in mice. *J. Nutr.* **117(2),** 376–382.

171. Yan, L. J. and Sohal, R. S. (1998) Mitochondrial adenine nucleotide translocase is modified oxidatively during aging. *Proc. Natl. Acad. Sci. USA* **95(22),** 12,896–12,901.

172. Sohal, R. S. and Dubey, A. (1994) Mitochondrial oxidative damage, hydrogen peroxide release, and aging. *Free Radical Biol. Med.* **16(5),** 621–626.

173. Dorner, G. and Mohnike, A. (1976) Further evidence for a predominantly maternal transmission of maturity-onset type diabetes. *Endokrinologie* **68(1)**, 121–124.
174. Dorner, G., Mohnike, A., and Steindel, E. (1975) On possible genetic and epigenetic modes of diabetes transmission. *Endokrinologie* **66(2)**, 225–227.
175. Dorner, G., Plagemann, A., and Reinagel, H. (1987) Familial diabetes aggregation in type I diabetics: gestational diabetes an apparent risk factor for increased diabetes susceptibility in the offspring. *Exp. Clin. Endocrinol.* **89(1)**, 84–90.
176. Freinkel, N., et al. (1986) Gestational diabetes mellitus: a syndrome with phenotypic and genotypic heterogeneity. *Horm. Metab. Res.* **18(7)**, 427–430.
177. Pimentel, E. (1979) Some aspects of the genetics and etiology of spontaneous diabetes mellitus. *Acta Diabetol. Latina* **16(3)**, 193–201.
178. Alcolado, J. C. and Thomas, A. W. (1995) Maternally inherited diabetes mellitus: the role of mitochondrial DNA defects. *Diabetic Med.* **12(2)**, 102–108.
179. Gerbitz, K. D., et al. (1995) Mitochondrial diabetes mellitus: a review. *Biochim. Biophys. Acta* **1271(1)**, 253–260.
180. Gerbitz, K. D., Gempel, K., and Brdiczka, D. (1996) Mitochondria and diabetes. Genetic, biochemical, and clinical implications of the cellular energy circuit. *Diabetes* **45(2)**, 113–126.
181. Alcolado, J. C., et al. (1994) Insulin resistance and impaired glucose tolerance [letter; comment]. *Lancet* **344(8932)**, 1293–1294.
182. Kadowaki, T., et al. (1995) A subtype of diabetes mellitus associated with a mutation in the mitochondrial gene. *Muscle Nerve* **3(41)**, S137–S141.
183. Suzuki, S., et al. (1994) Pancreatic beta-cell secretory defect associated with mitochondrial point mutation of the tRNA$^{Leu(UUR)}$ gene: a study in seven families with mitochondrial encephalomyopathy, lactic acidosis and stroke-like episodes (MELAS). *Diabetologia* **37(8)**, 818–825.
184. Kanamori, A., et al. (1994) Insulin resistance in mitochondrial gene mutation. *Diabetes Care* **17(7)**, 778–779.
185. Kanamori, A., et al. (1995) Response to Walker et al. (Insulin sensitivity and mitochondrial gene mutation). *Diabetes Care* **18(2)**, 274–275.
186. Odawara, M., et al. (1995) Mitochondrial gene mutation as a cause of insulin resistance. *Diabetes Care* **18(2)**, 275.
187. Koster, J. C., et al. (2000) Targeted overactivity of beta cell K(ATP) channels induces profound neonatal diabetes. *Cell* **100(6)**, 645–654.
188. German, M. S. (1993) Glucose sensing in pancreatic islet beta cells: the key role of glucokinase and the glycolytic intermediates. *Proc. Natl. Acad. Sci. USA* **90(5)**, 1781–1785.
189. Gidh-Jain, M., et al. (1993) Glucokinase mutations associated with non-insulin-dependent (type 2) diabetes mellitus have decreased enzymatic activity: implications for structure/function relationships. *Proc. Natl. Acad. Sci. USA* **90(5)**, 1932–1936.
190. Stoffel, M., et al. (1992) Human glucokinase gene: isolation, characterization, and identification of two missense mutations linked to early-onset non-insulin-dependent (type 2) diabetes mellitus [published erratum appears in Proc. Natl. Acad. Sci. USA 89(21):10,562]. *Proc. Natl. Acad. Sci. USA* **89(16)**, 7698–7702.

191. Stoffel, M., et al. (1992) Missense glucokinase mutation in maturity-onset diabetes of the young and mutation screening in late-onset diabetes. *Nature Genet.* **2(2)**, 153–156.

192. Stoffel, M., et al. (1993) Identification of glucokinase mutations in subjects with gestational diabetes mellitus. *Diabetes* **42(6)**, 937–940.

193. Gelb, B. D., et al. (1992) Targeting of hexokinase 1 to liver and hepatoma mitochondria. *Proc. Natl. Acad. Sci. USA* **89(1)**, 202–206.

194. Malaisse-Lagae, F. and Malaisse, W. J. (1988) Hexose metabolism in pancreatic islets: regulation of mitochondrial hexokinase binding. *Biochem. Med. Metab. Biol.* **39(1)**, 80–89.

195. Adams, V., et al. (1991) Porin interaction with hexokinase and glycerol kinase: metabolic microcompartmentation at the outer mitochondrial membrane. *Biochem. Med. Metab. Biol.* **45(3)**, 271–291.

196. McCabe, E. R. (1994) Microcompartmentation of energy metabolism at the outer mitochondrial membrane: role in diabetes mellitus and other diseases. *J. Bioenerg. Biomembr.* **26(3)**, 317–325.

197. Wallace, D. C. (1994) Mitochondrial DNA mutations in diseases of energy metabolism. *J. Bioenerg. Biomembr.* **26(3)**, 241–250.

198. Wollheim, C. B. (2000) Beta-cell mitochondria in the regulation of insulin secretion: a new culprit in type II diabetes. *Diabetologia* **43(3)**, 265–277.

199. Eto, K., et al. (1999) Role of NADH shuttle system in glucose-induced activation of mitochondrial metabolism and insulin secretion. *Science* **283(5404)**, 981–985.

200. Velho, G. and Froguel, P. (1998) Genetic, metabolic and clinical characteristics of maturity onset diabetes of the young. *Eur. J. Endocrinology* **138(3)**, 233–239.

201. Altshuler, D., et al. (2000) The common PPARgamma Pro12Ala polymorphism is associated with decreased risk of type 2 diabetes. *Nature Genet.* **26(1)**, 76–80.

202. Gebhart, S. S., et al. (1996) Insulin resistance associated with maternally inherited diabetes and deafness. *Metabolism* **45(4)**, 526–531.

203. Nishikawa, T., et al. (2000) Normalizing mitochondrial superoxide production blocks three pathways of hyperglycaemic damage. *Nature* **404(6779)**, 787–790.

204. Sligh, J. E., et al. (2000) Maternal germ-line transmission of mutant mtDNAs from embryonic stem cell-derived chimeric mice. *Proc. Natl. Acad. Sci. USA* **97(26)**, 14,461–14,466.

205. Inoue, K., et al. (2000) Generation of mice with mitochondrial dysfunction by introducing mouse mtDNA carrying a deletion into zygotes, *Nature Genet.* **26**, 176–181.

206. Watanabe, T., Dewey, M. J., and Mintz, B. (1978) Teratocarcinoma cells as vehicles for introducing specific mutant mitochondrial genes into mice. *Proc. Natl. Acad. Sci. USA* **75(10)**, 5113–5117.

207. Levy, S. E., et al. (1999) Transfer of chloramphenicol-resistant mitochondrial DNA into the chimeric mouse. *Transgen. Res.* **8(2)**, 137–145.

208. Marchington, D. R., Barlow, D., and Poulton, J. (1999) Transmitochondrial mice carrying resistance to chloramphenicol on mitochondrial DNA: developing the first mouse model of mitochondrial DNA disease. *Nature Med.* **5(8)**, 957–960.

209. Ferris, S. D., Sage, R. D., and Wilson, A. C. (1982) Evidence from mtDNA sequences that common laboratory strains of inbred mice are descended from a single female. *Nature* **295(5845)**, 163–165.

210. Pinkert, C. A., et al. (1997) Mitochondria transfer into mouse ova by microinjection. *Transgen. Res.* **6(6)**, 379–383.

211. Irwin, M. H., Johnson, L. W., and Pinkert, C. A. (1999) Isolation and microinjection of somatic cell-derived mitochondria and germline heteroplasmy in transmitochondrial mice. *Transgen. Res.* **8(2)**, 119–123.

212. Jenuth, J. P., et al. (1996) Random genetic drift in the female germline explains the rapid segregation of mammalian mitochondrial DNA [see comments]. *Nature Genet.* **14(2)**, 146–151.

213. Jenuth, J. P., Peterson, A. C., and Shoubridge, E. A. (1997) Tissue-specific selection for different mtDNA genotypes in heteroplasmic mice. *Nature Genet.* **16(1)**, 93–95.

214. Meirelles, F. V. and Smith, L. C. (1998) Mitochondrial genotype segregation during preimplantation development in mouse heteroplasmic embryos. *Genetics* **148(2)**, 877–883.

215. Meirelles, F. V. and Smith, L. C. (1997) Mitochondrial genotype segregation in a mouse heteroplasmic lineage produced by embryonic karyoplast transplantation. *Genetics* **145(2)**, 445–451.

216. Larsson, N. G., et al. (1998) Mitochondrial transcription factor A is necessary for mtDNA maintenance and embryogenesis in mice [see comments]. *Nature Genet.* **18(3)**, 231–236.

217. Wang, J., et al. (1999) Dilated cardiomyopathy and atrioventricular conduction blocks induced by heart-specific inactivation of mitochondrial DNA gene expression. *Nature Genet.* **21(1)**, 133–137.

218. Silva, J. P., et al. (2000) Impaired insulin secretion and beta-cell loss in tissue-specific knockout mice with mitochondrial diabetes. *Nature Genet.* **26(3)**, 336–340.

219. Murdock, D., et al. (1999) Up-regulation of nuclear and mitochondrial genes in the skeletal muscle of mice lacking the heart/muscle isoform of the adenine nucleotide translocator. *J. Biol. Chem.* **274(20)**, 14,429–14,433.

220. Esposito, L. A., et al. (1999) Mitochondrial disease in mouse results in increased oxidative stress. *Proc. Natl. Acad. Sci. USA* **96(9)**, 4820–4825.

221. Ridley, R. G., et al. (1986) Complete nucleotide and derived amino acid sequence of cDNA encoding the mitochondrial uncoupling protein of rat brown adipose tissue: lack of a mitochondrial targeting presequence. *Nucleic Acids Res.* **14(10)**, 4025–4035.

222. Reichling, S., et al. (1987) Loss of brown adipose tissue uncoupling protein mRNA on deacclimation of cold-exposed rats. *Biochem. Biophys. Res. Commun.* **142(3)**, 696–701.

223. Nicholls, D. G. and Locke, R. M. (1984) Thermogenic mechanisms in brown fat. *Physiol. Rev.* **64(1)**, 1–64.

224. Thomas, S. A. and Palmiter, R. D. (1997) Thermoregulatory and metabolic phenotypes of mice lacking noradrenaline and adrenaline. *Nature* **387(6628),** 94–97.

225. Enerback, S., et al. (1997) Mice lacking mitochondrial uncoupling protein are cold-sensitive but not obese. *Nature* **387(6628),** 90–94.

226. Arsenijevic, D., et al. (2000) Disruption of the uncoupling protein-2 gene in mice reveals a role in immunity and reactive oxygen species production. *Nature Genet.* **26(4),** 435–439.

227. Vidal-Puig, A. J., et al. (2000) Energy metabolism in uncoupling protein 3 gene knockout mice. *J. Biol. Chem.* **275(21),** 16,258–16,266.

228. Zhang, C., et al. (2001) Uncoupling protein-2 negatively regulates insulin secretion and is a major link between obesity, beta cell dysfunction, and type 2 diabetes. *Cell* **105(6),** 745–755.

229. Polonsky, K. S. and Semenkovich, C. F. (2001) The pancreatic beta cell heats up. UCP2 and insulin secretion in diabetes. *Cell* **105(6),** 705–707.

230. Esposito, L. A., et al. (2000) Mitochondrial oxidative stress in mice lacking the glutathione peroxidase-1 gene. *Free Radical Biol. Med.* **28(5),** 754–766.

231. Cortopassi, G. and Wang, E. (1995) Modelling the effects of age-related mtDNA mutation accumulation; complex I deficiency, superoxide and cell death. *Biochim. Biophys. Acta* **1271(1),** 171–176.

232. Li, Y., et al. (1995) Dilated cardiomyopathy and neonatal lethality in mutant mice lacking manganese superoxide dismutase. *Nature Genet.* **11(4),** 376–381.

233. Lebovitz, R. M., et al. (1996) Neurodegeneration, myocardial injury, and perinatal death in mitochondrial superoxide dismutase-deficient mice. *Proc. Natl. Acad. Sci. USA* **93(18),** 9782–9787.

234. Melov, S., et al. (1998) A novel neurological phenotype in mice lacking mitochondrial manganese superoxide dismutase [see comments]. *Nature Genet.* **18(2),** 159–163.

235. Reaume, A. G., et al. (1996) Motor neurons in Cu/Zn superoxide dismutase-deficient mice develop normally but exhibit enhanced cell death after axonal injury. *Nature Genet.* **13(1),** 43–47.

236. Carlsson, L. M., et al. (1995) Mice lacking extracellular superoxide dismutase are more sensitive to hyperoxia. *Proc. Natl. Acad. Sci. USA* **92(14),** 6264–6268.

237. Melov, S., et al. (1999) Mitochondrial disease in superoxide dismutase 2 mutant mice. *Proc. Natl. Acad. Sci. USA* **96(3),** 846–851.

238. Kokoszka, J. E., et al. (2001) Increased mitochondrial oxidative stress in the Sod2 (+/–) mouse results in the age-related decline of mitochondrial function culminating in increased apoptosis. *Proc. Natl. Acad. Sci. USA* **98(5),** 2278–2283.

239. Cai, J., et al. (2000) Separation of cytochrome c-dependent caspase activation from thiol-disulfide redox change in cells lacking mitochondrial DNA. *Free Radical Biol. Med.* **29(3–4),** 334–342.

240. Nakahara, H., et al. (1998) Mitochondrial dysfunction in the senescence accelerated mouse (SAM). *Free Radical Biol. Med.* **24(1),** 85–92.

241. Williams, M. D., et al. (1998) Increased oxidative damage is correlated to altered mitochondrial function in heterozygous manganese superoxide dismutase knockout mice. *J. Biol. Chem.* **273(43),** 28,510–28,515.

242. Melov, S., et al. (2000) Extension of life-span with superoxide dismutase/catalase mimetics. *Science* **289(5484),** 1567–1569.

243. Wei, M. C., Zong, W. X., Cheng, E. H., Lindsten, T., et al. (2001) Proapoptotic BAX and BAK: a requisite gateway to mitochondrial dysfunction and death. *Science* **292(5517),** 727–730.

244. Yoshida, H., et al. (1998) *Apaf1* is required for mitochondrial pathways of apoptosis and brain development. *Cell* **94(6),** 739–750.

245. Cecconi, F., et al. (1998) *Apaf1* (CED-4 homolog) regulates programmed cell death in mammalian development. *Cell* **94(6),** 727–737.

246. Kuida, K., et al. (1998) Reduced apoptosis and cytochrome c-mediated caspase activation in mice lacking caspase 9. *Cell* **94(3),** 325–337.

247. Hakem, R., et al. (1998) Differential requirement for caspase 9 in apoptotic pathways in vivo. *Cell* **94(3),** 339–352.

248. Kuida, K., et al. (1996) Decreased apoptosis in the brain and premature lethality in CPP32-deficient mice. *Nature* **384(6607),** 368–372.

2

Identification of Mutations in mtDNA from Patients Suffering Mitochondrial Diseases

Eric A. Schon, Ali Naini, and Sara Shanske

1. Introduction

The human mitochondrial genome (*see* **Fig. 1**) is a 16,569-bp (base pair) circle of double-stranded DNA *(1)*. It contains genes encoding 2 ribosomal RNAs, 22 transfer RNAs, and 13 structural genes, all of which are subunits of the respiratory chain complexes. Of the 13 structural genes, 7 encode subunits of complex I (NADH-CoQ oxidoreductase), 1 encodes the cytochrome-*b* subunit of complex III (CoQ–cytochrome-*c* oxidoreductase), 3 encode subunits of complex IV (cytochrome-*c* oxidase, or COX), and 2 encode subunits of complex V (ATP synthase). Each of these complexes also contains subunits encoded by nuclear genes, which are imported from the cytoplasm and assembled, together with the mtDNA-encoded subunits, into the respective holoenzymes, which are embedded in the mitochondrial inner membrane. Complex II (succinate dehydrogenase–CoQ oxidoreductase), of which succinate dehydrogenase (SDH) is a component, is encoded entirely by nuclear genes.

Because mitochondria (and mtDNAs) are maternally inherited *(2)*, defects in mtDNA-encoded subunits should result in pedigrees exhibiting maternal inheritance (i.e., the disease should pass only through females) and essentially all children (both boys and girls) should inherit the error. Moreover, because there are hundreds or even thousands of mitochondria in each cell, with an average of five mtDNAs per organelle *(3)*, mutations can result in two populations of mtDNAs (wild-type and mutated), a condition known as heteroplasmy.

Both mtDNA replication and mitochondrial division are stochastic processes unrelated to the cell cycle or to the timing of nuclear DNA replication. Thus, a dividing cell may potentially donate a different complement of organelles

From: *Methods in Molecular Biology, vol. 197: Mitochondrial DNA: Methods and Protocols*
Edited by: W. C. Copeland © Humana Press Inc., Totowa, NJ

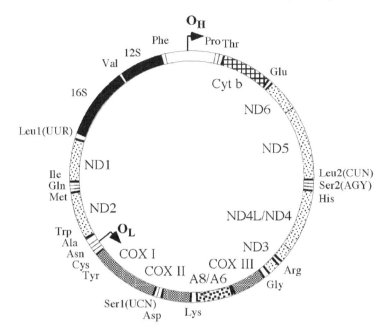

Fig. 1. The human mitochondrial genome. The structural genes for the mtDNA-encoded 12S and 16S ribosomal RNAs, the subunits of NADH–coenzyme Q oxidoreductase (ND), cytochrome-*c* oxidase (COX), cytochrome-*b* (Cyt b), and ATP synthase (A), and 22 tRNAs, are shown, as are the origins of heavy- (O_H) and light- (O_L) -strand replication.

and genomes to its progeny (i.e., mitotic segregation). This process becomes clinically important if an individual contains a heteroplasmic population of both wild-type and mutated mtDNAs, causing a mitochondrial disease. The phenotypic expression of a mutation may vary in both space (among cells or tissues) and time (during development or during the course of a life-span), based merely on the random processes of mitotic segregation. Of course, there may also be active selection processes going on as well, in which certain cells may either eliminate or concentrate a population of mutant mtDNAs. These effects will combine to generate, for example, a respiratory chain deficiency in some tissues but not others, but only if the number of mutant mtDNAs exceeds a certain threshold. This threshold varies from tissue to tissue and is related to the requirements for aerobic respiration and energy production, with brain, retina, and muscle exhibiting the highest-energy requirements.

Beginning in 1988, an ever-increasing number of maternally inherited mitochondrial diseases with distinct clinical phenotypes have been associated with mutations in mtDNA (reviewed in **ref. 4**), almost all of which result in neurological or neuromuscular disorders. These errors fall into three major classes: (1) large-scale rearrangements of mtDNA; (2) point mutations in mtDNA; (3) depletion of mtDNA.

1.1. Large-Scale Rearrangements of mtDNA

Giant deletions of mtDNA (Δ-mtDNAs) have been observed in ocular myopathy (OM) and the Kearns–Sayre syndrome (KSS), two mitochondrial disorders associated with progressive external ophthalmoplegia (PEO) *(5,6)*, as well as in a hematopoetic disorder called Pearson's marrow/pancreas syndrome *(7)*. In these disorders, up to 80% of total mtDNA can be deleted. The generation of Δ-mtDNAs in these syndromes is almost always *spontaneous*, with no affected mothers or siblings. In addition, the deletions are *unique* in each patient, with the particular type of deletion differing among patients. Taken together, these results imply that the Δ-mtDNA population in any one KSS patient is a clonal expansion of a single spontaneous deletion event occurring early in oogenesis or embryogenesis. *Duplications* of mtDNA, both sporadic *(8)* and maternally inherited *(9)*, have also been identified in these disorders. In addition to these "clonal" deletions, Mendelian-inherited disorders have been described in which affected family members harbor large quantities of *multiple* species of deletions of mtDNA in their tissues, which are apparently generated during the life-span of the patient *(10–12)*.

1.2. Point Mutations in mtDNA

Numerous mtDNA point mutations have been described, located in all regions of the genome. About two-thirds of them are in tRNA and rRNA genes, with the rest in the polypeptide-coding genes. The more notable disorders are described next.

Leber's hereditary optic neuropathy (LHON) was the first mitochondrial disease to be defined at the molecular level: Wallace and co-workers found a G→A transition at mtDNA position 11778, converting Arg-340 to His in the ND4 gene of complex I *(13)*. At least three other "primary" mutations have now been found to be associated with LHON, all in complex I genes *(14)*.

Myoclonus epilepsy with ragged red fibers (MERRF) is characterized by myoclonus, ataxia, weakness, generalized seizures, mental retardation, and hearing loss *(15)*. Muscle biopsies show ragged red fibers, or RRF, which are a morphological hallmark of massive mitochondrial proliferation. MERRF is most frequently associated with mutations in the tRNA^{Lys} gene, usually at nt-8344 *(16)* or nt-8356 *(17)*. The mutations are always heteroplasmic in patients.

Mitochondrial encephalomyopathy with lactic acidosis and strokelike episodes (MELAS) is characterized by seizures, migrainelike headaches, lactic acidosis, episodic vomiting, short stature, mental retardation, and recurrent cerebral strokelike episodes, causing hemiparesis, hemianopia, or cortical blindness *(18)*. MELAS is associated with a number of point mutations, mainly in the tRNA[Leu(UUR)] gene, with the most frequent mutation at nt-3243 *(19)*. A few MELAS mutations are in polypeptide-coding genes *(20–22)*. All of the mutations in MELAS are heteroplasmic, with the proportion of mutated mtDNAs usually exceeding 80% in muscle *(23,24)*. The MELAS-3243 mutation in tRNA[Leu(UUR)] can also cause maternally inherited PEO *(25)*.

Three types of maternally inherited disorders, all related to each other, are associated exclusively with mutations in the ATPase 6 gene. Neuropathy, ataxia, and retinitis pigmentosa (NARP) *(26)* and maternally inherited Leigh syndrome (MILS) *(27)* are associated with both a T→G transversion *(26)* and a T→C transition *(28)* at nt-8893 of ATPase 6. Patients are always heteroplasmic. A related disorder is associated with both a T→G transversion *(29)* and a T→C transition *(30)* at nt-9176, and has also been found to cause MILS *(31)*.

A T→G transition at nt-1555 in the 12*S* rRNA gene is associated with deafness as a result of sensitivity to aminoglycoside antibiotics, such as gentamycin, paromomycin, kanamycin, and streptomycin *(32)*.

1.3. Depletions of mtDNA

All known mtDNA depletion syndromes are recessive Mendelian disorders that cause the partial or even complete loss of mtDNA in specific tissues *(33,34)*. Depletion of mtDNA can also be caused by environmental toxins that interfere with mtDNA replication, such as azidothymidine (AZT), which inhibits DNA replication of the mitochondrial DNA polymerase γ *(35)*.

When clinical features, maternal inheritance, muscle histochemistry, or biochemical results suggest an mtDNA-related disorder, the mutation in the mtDNA should be defined. Guided by the clinical picture, one should screen for the most common mutations associated with that particular syndrome, and failing that, extend the search to the rarer mutations.

2. Materials

All reagents were from Sigma-Aldrich (Boston, MA), except where noted.

1. 0.5 *M* ethylenediaminetetraacetic acid (EDTA), pH 8.0.
2. 10X TBE buffer: dissolve 108 g Tris base and 55 g boric acid in 700 mL deionized water. Add 20 mL of 0.5 *M* EDTA, pH 8.0, and bring the volume up to 1 L with deionized water.
3. 10X TAE buffer: dissolve 48.4 g Tris base in 700 mL of deionized water, then add 11.42 mL glacial acetic acid and 20 mL of 0.5 *M* EDTA, and bring the volume up to 1 L with deionized water.
4. Puregene DNA Extraction Kit from Gentra (Research Triangle Park, NC).

5. Ethidium bromide (EtBr), 10 mg/mL.
6. Expand High Fidelity PCR System from Roche-Boehringer Mannheim (Indianapolis, IN).
7. QIAquick Gel Extraction Kit from QIAGEN (Valencia, CA).
8. QIAquick PCR Purification Kit from QIAGEN (Valencia, CA).
9. Random Primed DNA Labeling Kit from Roche-Boehringer Mannheim (Indianapolis, IN).
10. [α-^{32}P]-dATP (3000 Ci/mmol) from New England Nuclear (Boston, MA).
11. ProbeQuant G-50 Micro Columns from Amersham-Pharmacia Biotech (Piscataway, NJ).
12. 6X Gel-loading buffer (w/v): 0.25% bromophenol blue, 0.25% xylene cyanol, 40% sucrose.
13. Agarose LE from Roche-Boehringer Mannheim (Indianapolis, IN).
14. Denaturing solution: 1.5 M NaCl, 0.5 M NaOH.
15. Neutralizing solution: 3 M NaCl, 0.5 M Tris-HCl, pH 7.4.
16. 10X SSC buffer: 1.5 M NaCl, 0.15 M Na$_3$ citrate 2H$_2$O; adjust pH to 7.0 with 1 N HCl.
17. Zeta-Probe nylon membrane from Bio-Rad (Hercules, CA).
18. Phosphate buffers, 0.04 M, pH 7.2, containing 5%, 1%, and 0.5% (w/v) sodium dodecyl sulfate (SDS).
19. 3MM paper from Whatman (Clifton, NJ).

3. Methods

3.1. Southern Blot Hybridization Analysis to Detect Partial Deletions of mtDNA

In order to determine whether or not there are rearrangements (deletions or duplications) in mitochondrial DNA (mtDNA), total DNA is digested with restriction enzymes, separated on an agarose gel, transferred to a membrane (nitocellulose or nylon), and hybridized to a labeled mtDNA probe *(5)*. The size(s) of the bands, as detected by autoradiography, indicates whether only the normal-sized mtDNA is present or whether there are also deleted or duplicated species.

3.1.1. DNA Preparation

Prepare DNA from blood or solid tissue using Puregene DNA extraction kit and following manufacturer's recommended protocols, or according to the standard method for DNA mini-prep preparation, as described *(36)*.

3.1.2. Probe Preparation

1. Amplify a 2.5-kb fragment of mitochondrial DNA in a total volume of 50 μL using the Expand High Fidelity PCR System kit and primers.
   a. Forward primer: 5'-ccactccaccttactaccagac-3' (nt 1690–1711) *(1)*.
   b. Reverse primer: 5'-gtaatgctagggtgagtggtagg-3' (nt 4207–4185) *(1)*.

2. Load the entire polymerase chain reaction (PCR) product onto a 0.8% agarose in 1X TBE containing 0.5 µg/mL EtBr.
3. Cut the 2.5-kb fragment from the gel and purify using Gel Extraction Kit reagents and recommended protocol.

3.1.3. Labeling of the Probe

1. Using the Random Primed DNA Labeling Kit from Boehringer Mannheim, label the probe as follows: In a 500-µL microfuge tube, boil 2-µL probe (approx 25 ng) +8 µL water for 10 min. Immediately chill on ice.
2. To the above 10-µL solution, add the following:
   a. 1 µL of 0.5 mM dTTP.
   b. 1 µL of 0.5 mM dCTP.
   c. 1 µL of 0.5 mM dGTP.
   d. 5 µL of [α-^{32}P]-dATP (i.e., 50 µCi; 3000 Ci/mmol).
   e. 2 µL of 10X random hexanucleotide kit mixture.
   f. Add 1 µL Klenow enzyme (2 units), mix and incubate 30 min at 37°C.
3. Add 2 µL of 0.2 M EDTA, pH 8.0, to stop the reaction.
4. Purify using G-50 microcolumns.
5. Add 1 µL probe to 10 mL Scintisol; mix and count the radioactivity (should be approximately 1×10^6 cpm, with a specific activity of approx 10^9 cpm/µg).

3.1.4. Preparation of Restriction Enzyme Digest

1. In a 500-µL microfuge tube, prepare a digestion mix as follows:
   a. DNA   5–10 µg
   b. Restriction enzyme (10 units/µL)*   1 µL
   c. Buffer (10X)   4 µL
   d. RNAase (10 mg/ml)   1 µL
   e. H$_2$O to make a total volume of   40 µL
   f. Incubate at 37°C for 1–2 h
2. Check digestion
   a. Remove 4 µL solution from **step 1** (continue to incubate the remaining mix).
   b. Add 6 µL H$_2$O and 2 µL 6X dye (bromphenol blue/xylene cyanol).
   c. Load a 0.8% or 1% agarose gel; run at approximately 100 mA. Check digestion and amount of DNA.
   d. Add 1 µL of appropriate restriction enzyme (if not completely cut) or DNA 1–2 µg (if DNA is not enough), continue incubation as needed (overnight is OK).

*Routine analysis is usually performed using *Pvu*II and *Bam*HI, because there is only one site in mtDNA for *Pvu*II and *Bam*HI, so each enzyme will linearize the mtDNA.

3.1.5. Agarose Gel Electrophoresis

1. Prepare a 20-cm, 0.8% agarose gel containing 1 µg/mL ethidium bromide in 1X TAE buffer.
2. Load 20–40 µL digested DNA plus 4–8 µL 6X dye per well (try to apply similar amounts of DNA in each well); load markers (usually λ digested with *Hin*dIII) in one well.
3. Run gel at approximately 140 mA for 8 h, or at approximately 80 mA overnight.
4. Stop when bottom dye is about three-fourths of the way down (e.g., 16 cm for a 20-cm gel).
5. Remove gel, place gel in glass dish on top of Saran Wrap, and photograph with a ultraviolet (UV)-visible ruler; cut one corner for orientation.
6. Rinse with double distilled (dd) H_2O for 5 min.
7. Add denaturing solution, shake very gently for 30–40 min in a shaker.
8. Replace with neutralizing solution, shake very gently for 15–30 min.

3.1.6. DNA Transfer

1. Set up transfer apparatus using "transfer buffer"—10X SSC and Whatman #3 filter paper.
2. Cut the membrane (Zeta-Probe [Bio-Rad]) to size of gel—wet in ddH_2O, then wet with "transfer buffer" right before applying.
3. Invert gel onto filter paper; remove bubbles.
4. Put wet membrane on gel; remove bubbles.
5. Put wet Whatman #3 filter on top of the membrane; remove bubbles.
6. Add several dry Whatman #3 filters and layers of paper towels; "seal" with Saran Wrap so that buffer does not evaporate.
7. After transfer, cut the membrane in one corner for orientation, carefully remove and put on filter paper and let it dry in oven at 80°C for approx 45–60 min under vacuum. Alternatively, the DNA can be bound to the filter in a UV crosslinker (254-nm wavelength at 1.5 J/cm^2).

3.1.7. Hybridization

1. Roll membrane and place in a hybridization tube.
2. Prehybridize membrane in 7% SDS + 0.25 *M* Na_2HPO_4 for 15 min at 65°C in hybridization oven. Rotate gently on a rotary shaker (speed approx 5 rpm).
3. Boil the probe in water for 10 min to denature and immediately chill on ice.
4. Add the probe (approx 0.5–1X 10^6 cpm/mL [specific activity, approx 10^9 cpm/µg] hybridization solution).
5. Hybridize overnight at 65°C.

3.1.8. Washing the Membrane

1. Remove the probe (it can be kept at –20°C and reused for up to 2 wk).
2. Wash for 30 min in 0.02 *M* Na_2HPO_4 with 5% SDS, two times.
3. Wash for 30 min in 0.02 *M* Na_2HPO_4 with 1% SDS, two times.
4. Wash for 30 min in 0.02 *M* Na_2HPO_4 with 0.5% SDS.

3.1.9. Film Developing

1. Keep the membrane wet in case it is necessary to wash again or to do a second hybridization.
2. Place filter between two pieces of Whatman 3MM paper to dry.
3. Place dry filter in sealable bag or place on a 20-cm × 20-cm sheet of Whatman 3MM and cover with plastic wrap (Saran Wrap is preferred).
4. Autoradiograph with Kodak XAR-5 film.
5. Expose overnight and develop.

3.1.10. Analysis of Results

Normal pattern: One hybridizing band migrating at 16.6 kb (*see* **Notes 1–4**). Patients with deletions show one or more additional hybridizing bands.

3.2. Southern Blot Hybridization Analysis to Detect Partial Duplications of mtDNA

In any one patient harboring mitochondrial genomes with deletions and duplications, the two rearranged species are topologically related: the duplicated mtDNA is composed of two mtDNA molecules—a wild-type mtDNA and a deleted mtDNA—arranged head to tail (see example in **Fig. 2A**) *(37,38)*. The only novel sequence in the duplicated mtDNA as compared to wild-type mtDNA is at the boundary of the duplicated region, which is the same as the boundary present in the corresponding deleted mtDNA (dashed line in **Fig. 2A**). Because the two molecules are related, a "standard" Southern blot analysis will not be able to distinguish between the two species.

Figure 2 shows an example of how these ambiguities can arise and how they can be circumvented. As shown in **Fig. 2A**, a patient with a 7813-bp deletion *(37)* harbored, in addition to the 16.6-kb wild-type mtDNA, a partially deleted genome 8.8 kb long (i.e., 7.8 kb was deleted) and a partially duplicated genome 25.3 kb long (i.e., 8.8 kb was inserted into a wild-type genome). Note that *Pvu*II cuts wild-type and deleted mtDNA once, but cuts duplicated mtDNA twice (**Fig. 2A**). Digestion with *Pvu*II is therefore expected to result in a 16.6-kb band corresponding to wild-type mtDNA, in a smaller 8.8-kb band corresponding to deleted mtDNA, and to two bands derived from duplicated mtDNA, one of 16.6 kb (corresponding to the wild-type region of the molecule) and the other of 8.8 kb (corresponding to the duplicated region of the molecule). When hybridized with a probe located outside the deleted region (i.e., probe 1), both the 16.6-kb and 8.8-kb fragments will hybridize, but in a heteroplasmic sample, it will not be clear from which of the three types of molecules these bands are derived. When hybridized with a probe located inside the deleted region (i.e., probe 2), there is no signal from the 8.8-kb band (derived from either the deleted or duplicated mtDNAs). Only a 16.6-kb fragment will

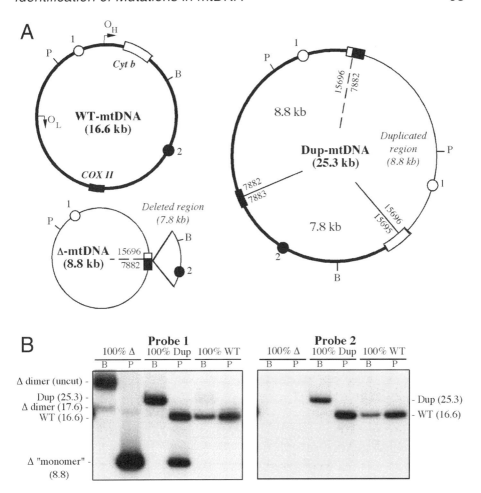

Fig. 2. A typical large-scale mtDNA rearrangement, showing the relationship among wild-type, deleted, and duplicated species, using a patient with a 7.8-kb deletion *(37)* as an example. (**A**) The deletion (protruding "pie-shaped" segment) removes 7813 bp between the COX II [solid box] and cytochrome-*b* [open box] genes). The breakpoint is indicated by a dashed line; the nucleotides associated with the rearrangement straddle the "pie slices." Numbers denote nucleotide positions. The locations of the *Pvu*II (P) and *Bam*HI (B) restriction sites, and probes 1 (open circle) and 2 (shaded circle) used in the Southern blot analyses (see text) are also shown. (**B**) Autoradiograms of Southern blot analyses of cybrid cell lines containing 100% wild-type (WT), 100% duplicated (Dup), and 100% deleted (Δ) mtDNAs, following digestion with *Pvu*II (P) or *Bam*HI (B) and hybridization with probes 1 or 2. The identity of each hybridizing fragment and its size (in kb) is indicated at the sides. The 8.8-kb band represents either linearized full-length Δ-mtDNA (in deleted lines) or the duplicated region in dup-mtDNA (in duplicated lines).

hybridize, but it will not be clear whether this fragment derives from a wild-type molecule, from a duplicated molecule, or from both.

Unlike *Pvu*II, *Bam*HI has no cutting site on the deleted mtDNA, leaving this species in nonlinearized circular forms. In addition, *Bam*HI cuts both wild-type mtDNA and duplicated mtDNA only *once*, yielding a 16.6- and a 25.3-kb fragment, respectively (**Fig. 2B**). Thus, when *Bam*HI-digestion products are hybridized with probe 1, bands corresponding to all three species will be visualized: a 16.6-kb band corresponding to linearized wild-type mtDNA, a 25.3-kb band corresponding to linearized duplicated mtDNA, and one or more bands, migrating elsewhere on the gel, corresponding to the topological conformations of the uncut deleted mtDNA (e.g., supercoils and nicked circles). When hybridized with probe 2, however, *none* of the topological forms of the uncut deleted molecules will be visualized; only the 16.6- and 25.3-kb bands, corresponding to the wild-type and duplicated mtDNAs, will appear (*37*).

Thus, by using a combination of the two restriction enzymes and the two regional probes, one can distinguish (and quantitate) unambiguously all three mtDNA species, even if they are present in a heteroplasmic mixture.

3.2.1. Method

The method is identical to the procedure described in **Subheading 3.1.**, except that the DNA is cut with multiple diagnostic restriction enzymes, and is hybridized with probes representing segments of mtDNA located "inside" and "outside" the deleted region. (*See* **Notes 5–7**.)

3.3. Southern Blot Hybridization Analysis to Detect Very Low Levels of mtDNA Deletions

In some cases, one wishes to use Southern blot hybridization to detect a deleted species that is present in extremely low levels (i.e., less than 5%). If a standard *Pvu*II digestion is used, the intensity of the 16.6-kb wt-mtDNA fragment will often be so great that the signal will obscure any deleted fragment migrating on the gel below, but close to, the wild-type band. We have found that digestion with a different enzyme that cuts two or three times in the mtDNA (e.g. *Eco*RI, *Hind*III, and *Pst*I) can help solve this problem. Following digestion with such an enzyme, the deleted fragment will often be *larger* than the wild-type fragment (owing to loss of one of the restriction sites in the deleted region). Upon electrophoresis and hybridization with an appropriate region-specific probe, the minority of deleted molecules will migrate *above* the wild-type band and can be identified and quantified relatively easily, with little background present. The method is identical to the procedure described in **Subheading 3.1.**

3.4. Southern Blot Hybridization Analysis to Detect and Quantitate Depletions of mtDNA

In order to quantify the mtDNA content in tissues, one can use a second probe on the Southern blot filter—a fragment containing a nuclear-encoded multicopy DNA gene—as an internal control to correct for differences in the amount of DNA loaded in each lane *(39)*. Total DNA is prepared as above and is digested with *Pvu*II, electrophoresed through a 0.8% agarose gel, and transferred to nitrocellulose. The filters are then hybridized with *two* probes, either sequentially or simultaneously. The first probe detects mtDNA, as above. The second probe detects the nuclear-encoded 18S ribosomal DNA genes, which are present at about 400 copies per cell. We use clone pB *(40)*, containing nuclear-encoded 18S rDNA sequences on a 5.8-kb *Eco*RI fragment. Digestion of the ribosomal DNA repeat regions containing the 18S rDNA genes with *Pvu*II will release hybridizable fragments approximately 12 kb in size. Thus, in order to quantitate the mtDNA and rDNA signals, we scan the nitrocellulose filters and count the mtDNA (16.6-kb) and rDNA (12.0-kb) signals (as counts per minute). The ratio mtDNA:rDNA (dimensionless) is calculated by dividing the mtDNA signal by the rDNA signal, after correcting for background signal on the filter. The method is identical to the procedure described above (*see* **Note 8**).

3.5 Detection of mtDNA Point Mutations Using PCR/RFLP Analysis

Known point mutations in mtDNA are screened for by amplifying an appropriate fragment of mtDNA by PCR, digesting with a diagnostic restriction enzyme, and analyzing the fragment sizes on a gel. This involves the following steps: (1) oligonucleotide primers are designed to allow amplification of mtDNA encompassing the mutated site; (2) appropriate mtDNA fragments are amplified by PCR amplification in the presence of ^{32}P-dATP; (3) PCR products are digested with a restriction enzyme that will cleave normal versus mutated sequences differentially (i.e., restriction fragment length polymorphism, or RFLP); (4) digestion products are electrophoresed through nondenaturing polyacrylamide gels; and (5) fragment sizes are detected by autoradiography.

3.5.1. Preparation of DNA

Prepare DNA from blood or solid tissue using the Puregene DNA extraction kit and following the manufacturer's recommended protocols.

3.5.2. Polymerase Chain Reaction Amplification

1. Prepare PCR cocktail as follows:
   a. ddH$_2$O   76 µL
   b. l0X *Taq* buffer   10 µL
   c. dNTP mix (1.25 m*M* each dNTP)   10 µL
   d. Forward primer (100 pmol/µL)   1 µL
   e. Backward primer (100 pmol/µL)   1 µL
   f. Template DNA   1 µL
2. Denature template mixture in a thermal cycler at 94°°C for 10 min. Place on ice.
3. Add 0.5 µL (2.5 units) *Taq* polymerase and quickly add 2 drops of mineral oil.
4. Set appropriate PCR program and run.
5. Remove PCR product, check 10 µL (+2 µL 6X loading dye) on a 0.8% agarose gel in 1X TAE buffer (40 m*M* Tris-acetate, 1 m*M* EDTA). Run the gel at l00 V.

3.5.3. Digestion and Electrophoresis of PCR Products

1. "Last cycle hot" labeling of PCR products: to 90 µL of PCR reaction, add 0.5–1 µL [α-^{32}P]-dATP and do *one* extension using the same PCR program as in **Subheading 3.5.2.**
2. Digest 10 µL of labeled PCR product with a diagnostic restriction enzyme at the appropriate temperature for approximately 2 h or overnight, as follows:
   a. PCR product   10 µL
   b. Buffer (10X)   4 µL
   c. Restriction enzyme   1 µL
   d. H$_2$O   25 µL
3. To 40 µL restriction digest, add 8 µL dye, load approximately 20 µL on a 12% acrylamide gel (37.5:1 ratio of acrylamide:bis) in 1X TBE (90 m*M* Tris-borate, 1 m*M* EDTA). Run at 300 V, 15 W for about 1 h.
4. Remove the gel and place between two plastic sheet protectors.
5. Autoradiograph with Kodak XAR-5 film. Expose overnight and develop.

3.5.4. Analysis of Results

Detect and quantitate relevant fragments in a phosphorimager.

3.6. Detection of Specific Point Mutations

More than 115 mtDNA point mutations have been reported to date *(41,42)*, but fewer than two dozen are encountered with any frequency. The primer sequences, amplification conditions, and RFLP analyses required to detected these more common mutations are shown in **Table 1**. (*See* **Notes 9–14.**)

3.7 Single-Cell and Single-Fiber PCR

Individual muscle fibers can be dissected out of thick muscle sections and can be used to determine by PCR the levels of a given mtDNA mutation or the

Table 1
Conditions for PCR/RFLP Analyses of Selected Pathogenic mtDNA Point Mutations

Mutation		Forward primer		Backward primer		PCR conditions				Size	RFLP analysis		
nt	Mut	Range	Sequence, 5'->3'	Range	Sequence, 5'->3'	Cyc	Denat.	Anneal	Extend		Enz	Normal	Mutant
1555	A→G	1009-1032	CACAAAATAGACTACGAAAGTGGC	1575-1556	ACTTACCATGTTACGACTTG**G**G	25	1 min @94	1 min @55	1 min @72	566	Hae III	455, **111**	455, **91**, **20**
1644	G→T	1412-1431	GTGGAAGGTGGGATTTAGCAG	1664-1645	CGGTCAAGTTAAGTTGA**G**AT	30	1 min @94	1 min @60	1 min @72	252	Eco RV	**252**	232, 20
3243	A→G	3116-3134	CCTCCCTGTACCAAAGGAC	3353-3333	GCGATTAGAATGGGTACAATG	25	1 min @94	1 min @55	45 sec @72	238	Hae III	169, 37, 32	97, 72, 37, 32
3250	C→T	3225-3249	GGTTTGTTAAGATGGCAGAG**G**CCGG	3324-3404	CAACGTTGGGGCCTTTGCGTA	25	1 min @94	1 min @55	1 min @72	199	Nae I	199	177, 22
3251	A→G	3230-3253	GTTAAGATGGCAGAGCCCGGTA**CT**	3353-3333	GCGATTAGAATGGGTACAATG	30	1 min @94	1 min @55	1 min @72	123	Rsa I	88, 20, 15	108, 15
3252	A→G	3230-3256	GTTAAGATGGCAGAGCCCGGT**G**ATCGC	3353-3333	GCGATTAGAATGGGTACAATG	30	1 min @94	1 min @55	1 min @72	123	Dpn II	103, 20	123
3254	C→G	3230-3258	GTTAAGATGGCAGAGCCCGGTAATCGC**CT**	3353-3333	GCGATTAGAATGGGTACAATG	30	1 min @94	1 min @55	1 min @72	123	Hae III	86, 37	61, 37, 25
3256	C→T	3230-3255	GTTAAGATGGCAGAGCCCGGTAAG**CG**	3353-3333	GCGATTAGAATGGGTACAATG	25	1 min @94	1 min @55	1 min @72	123	Hin PI	99, 13, 11	110, 13
3260	A→G	3145-3170	AGGCCTACTTCACAAAGCCGCCTTCCC	3310-3261	GTATGTTGTTAAGAAGAGGAATTGAACCTCTGACTGTAAA **GGAATAAGTT**	25	1 min @94	1 min @55	1 min @72	166	Xmn I	166	116, 50
3271	T→C	3148-3169	CCTACTTCACAAAGCCGCCTTCC	3295-3272	GAGGAATTGAACCTCTGACT**C**TAA	25	1 min @94	1 min @55	45 sec @72	147	Dde I	103, 44	79, 44, **24**
3271	Del T	3148-3169	CCTACTTCACAAAGCCGCCTTCC	3295-3272	GAGGAATTGAACCTCTGACT**C**TAA	25	1 min @94	1 min @55	45 sec @72	147	Dde I	103, 44	79, 44, **24**
3291	T→C	3264-3290	TTAAAACTTTACAGTCAGAGGTTCG**A**TTC	4349-4329	GTTCGATTCTCATAGTCCTAG	25	1 min @94	1 min @55	2 min @72	1068	Dpn II	658, **394**, 34	658, **370**, 34, **24**
3302	A→G	3231-3248	TAAGATGGCAGAGCCCG	3422-3404	ACGTTGGGGCCTTTGCGTA	25	1 min @94	1 min @55	1 min @72	191	Msp I	124, **35**, 32	159, 32
3303	C→T	3277-3300	GTCAGAGGTTCAATTCCTTCTT**GT**	3353-3333	GCGATTAGAATGGGTACAATG	25	1 min @94	1 min @55	1 min @72	76	Hpa I	50, 26	76
3460	G→A	3116-3134	CCTCCCTGTACCAAAGGAC	4349-4329	CTCCCTGTACCAAAGGAC	25	1 min @94	1 min @55	2 min @72	1233	Acy I	**889**, **344**	1233
4269	A→G	4185-4207	CCTACCACTCACCCTAGCATTAC	4292-4270	CAAAGTAACTCTTTTATCAG**GT**A	25	1 min @94	1 min @55	1 min @72	107	Rsa I	107	84, 23
4300	A→G	4281-4298	AGAGTTACTTTGATAG**GG**	4544-4523	GTGCGAGCTTAGCGCTGTGATG	25	1 min @94	1 min @55	1 min @72	263	Hph I	263	235, 28
4320	C→T	4185-4207	CCTACCACTCACCCTAGCATTAC	4544-4523	GTGCGAGCTTAGCGCTGTGATG	25	1 min @94	1 min @55	1 min @72	359	Mnl I	282, 77	216, 77, 66
4409	T→C	389-4410	CACCCCATCCTAAAGTAAGTGTC	4544-4523	GTGCGAGCTTAGCGCTGTGATG	30	1 min @94	1 min @55	1 min @72	155	Mae III	137, 18	155
5537	InsT	5514-5536	AAATTTAGGTTAAATACAGACG**A**	5798-5777	GCAAATTCGAAGAAGCAGCTTC	25	1 min @94	1 min @60	1 min @72	284	Mbo II	250, 34	284
5703	G→A	5472-5491	CTACTCCTACCTATCTCCCC	5798-5777	GCAAATTCGAAGAAGCAGCTTC	25	1 min @94	1 min @55	1 min @72	326	Dde I	99,80,60,41,34,**12**	111,80,60,41,34
5814	T→C	5685-5709	CAGTTAGTTAACAGTAAGCACCC	5982-5960	CAGCTCATGCGCGAATAATAGG	25	1.5 min @94	1 min @55	1 min @72	297	Hph I	121, 98, **78**	199, 98
8344	A→G	8278-8296	CTACCCCCTCTAGAGCCCAC	8385-8345	GTAGTATTTAGTTGGGGCATTTCACTGTAAAG**C**G**C**GTGTTGG	25	1.5 min @94	1.5 min @55	1 min @72	108	Bgl I	108	73, 35
8356	T→C	8239-8263	CTTTGAAATAGGGGCCCGTATTTACC	8380-8357	ATTTAGTTGGGGCATTTCACTTA	25	1 min @94	1 min @55	1 min @72	141	Dra I	119, 22	141
8363	G→A	8241-8361	AGAACCAACACCTCTTTAC**GG**	8582-8561	GGCTGAGAGGGCCCCTGTTAGG	25	1 min @94	1 min @55	1 min @72	240	Hph I	113, **53**, 44, **30**	113, **83**, 44
8851	T→C	8829-8852	CCACCCAACAATGACTAATCAAACTAACC	9278-9256	GGCTGAGAGGGCCCCTGTTAGG	25	1 min @94	1 min @60	2 min @72	450	Rsa I	280, **170**	280, **148**, **22**
8993	T→G	8657-8685	CCACCCAACAATGACTAATCAAACTAACC	9278-9256	GGCTGAGAGGGCCCCTGTTAGG	25	1 min @94	1 min @60	2 min @72	621	Msp I	621	336, 285
8993	T→C	8657-8685	CCACCCAACAATGACTAATCAAACTAACC	9278-9256	GGCTGAGAGGGCCCCTGTTAGG	25	1 min @94	1 min @60	2 min @72	621	Msp I	621	337, 284
9176	T→C	8896-8914	GCCCTAGCCCACTTCTTAC	9278-9256	GTGTTGTCGTGCAGGTAGAGGCTT**C**T	25	1 min @94	1 min @60	1 min @72	308	Scr FI	211, 97	184, 97, **27**
9957	T→C	9933-9956	CATTTGTAGATGTGTTTGAG**TA**	10269-10238	GGGCAATTCTAGATCAAATAATAAGAAGG	25	1 min @94	1 min @55	1 min @72	336	Rsa I	259, **77**	259, **54**, **23**
9997	T→C	9910-9931	CTTTGGCTTCGAAGCCGCGCC	10269-10238	GGGCAATTCTAGATCAAATAATAAGAAGG	25	1 min @94	1 min @55	1 min @72	367	Bfa I	162, **128**, 65, 12	162, **95**, 65, **33**, 12
10044	A→G	9910-9931	CTTTGGCTTCGAAGCCGCGCC	10269-10238	GGGCAATTCTAGATCAAATAATAAGAAGG	25	1 min @94	1 min @55	1 min @72	367	Acl I	367	225, 142
11084	A→G	1059-11083	CTCCCTACAAATCTCCTTAATT**CT**A	11272-11247	TAGGGTGTTGTGAGTGTAAATTAGTG	25	1 min @94	1 min @60	1 min @72	213	Bfa I	161, 52	139, 52, **22**
11778	G→A	12413-12393	CCAACCCCCTGAAGCTTCACCGGCGCAG	12413-12393	GGGTTAACGAGGGTGGTAAGG	25	1 min @94	1 min @60	1 min @72	745	Sfa NI	**634**, **111**	745
14459	G→A	14430-14458	ATGCCTCAGGATACTCCTCAATAGCCG**TC**	14874-14855	AGGATCAGGCAGGCCAAG	25	1 min @94	1 min @55	1 min @72	444	Mae III	444	419, 25
14484	T→C	14463-14484	TAGTATATCCAAAGACAACA**GA**	14810-14790	TTTGGGGAAGTTATATGGG	25	1 min @94	1 min @55	1 min @72	75	Dpn III	54, 21	75
14709	T→C	14688-14707	CTACAACCGACCAATGA**CA**	14967-14924	TTTACGTCTCGAGTGATGTGGCG	25	1 min @94	1 min @50	1 min @72	279	Nla III	124, 70, 62, 23	147, 70, 62
15590	C→T	15803-15825	GTAGCATCCGTAGTATACTTCAC	16014-15991	GTTTAAATTAGAATCTTAGCTCC**CG**	25	1 min @94	1 min @55	1 min @72	211	Msp I	123, **66**, **22**	123, **88**

Underlined letters denote mismatched nucleotides in primer. Underlined numbers indicate the PCR fragments associated with the RFLP.

relative amounts of mtDNA and nuclear DNA. The single-fiber PCR technique is a very useful tool to document the pathogenicity of a novel mutation when it shows that the mutation is more abundant in affected RRFs or COX-negative fibers than in nonaffected fibers (non-RRF, COX-positive fibers). The method for dissecting out the fibers has been described elsewhere *(43,44)*; the basic PCR/RFLP analysis is identical to that described here.

4. Notes

1. When using a "full-length" mtDNA probe, be sure that the intensities of the wild-type and deleted bands are corrected for probe length when quantitating the fragments for the purpose of measuring percent heteroplasmy *(5)*. Alternatively, one can use a shorter probe that covers both the wt-mtDNA and an *undeleted* region of the Δ-mtDNA (e.g., a probe located in the "minor arc" in the rRNA region), for which no length correction is required. In this case, of course, the intensity of the signal with a shorter probe will be lower and exposure times will be longer, but in our experience, this is not a major problem.

2. When using *Pvu*II for the restriction digestion, the *normal* pattern sometimes shows two bands—one migrating at approx 13.5 kb and the other at approx 3 kb. This pattern is the result of the presence of a second, polymorphic, *Pvu*II site resulting from an A→G mutation at nt-16241 in some populations *(45)*. Note that a rare *Pvu*II (or *Bam*HI) polymorphism can be distinguished from an authentic deletion by measuring accurately the sizes of the hybridizing bands: With the neutral polymophisms, the sum of the two (or more) "aberrantly migrating" bands will equal 16.6 kb, whereas a deletion produces bands whose sizes sum to greater than 16.6 kb.

3. Most mitochondrial deletions are located in the 10.8-kb "major arc" located between O_H and O_L and usually map to the region between nucleotides 8000 and 15,000. Because most deletions are greater than 3 or 4 kb in size, they remove a segment of mtDNA harboring restrictions sites for three "single cutters": *Sph*I (at nt-8997), *Sna*BI (at nt-10734), and *Pme*I (at bp-10414). Thus, a confirmatory Southern blot using one of these enzymes will linearize wt-mtDNA but should not cut the Δ-mtDNA.

4. The loss of unique restriction sites located in the deleted region allows one to enrich for intact deleted circles: digestion of total DNA with, say, *Pme*I, will linearize wt-mtDNA but not the Δ-mtDNA (of course, the nuclear DNA will be cleaved into thousands of fragments). Upon digestion with Bal31, a double-stranded DNA exonuclease, the wt-DNA (and nuclear DNA), but not the circular Δ-mtDNA, will be digested *(46)*.

5. It is advisable to hybridize the membrane first with probe 2 (the "inside" probe) and then with probe 1 (the "outside" probe). It is also advisable to strip the first probe off the membrane prior to hybridizing with the second probe.

6. In the example shown in **Fig. 2**, the Southern blots were performed on DNA isolated from cybrid cells containing homoplasmic levels of wild-type, deleted,

and duplicated mtDNAs, in order to demonstrate more clearly how the use of two restriction enzymes and two regional probes can be used to distinguish among the three species. In a "real-life" situation, however, patient cells will almost certainly be heteroplasmic and will contain more than one species of mtDNA. It can be appreciated that in such a situation, the resulting Southern blot pattern will be more complicated, as it will be, in essence, a "superposition" of the two panels shown in **Fig. 2B**. Nevertheless, all three species can still be distinguished, but great care must be taken to keep track of the intensities among the relevant hybridizing fragments after hybridization with both the "inside" and "outside" probes. A change in relative intensity of the "deleted" band (8.8-kb band in **Fig. 2B**) following hybridization with the two probes is indicative of the presence of both deleted and duplicated species.

7. The example shown in **Fig. 2** also highlights a second phenomenon often encountered in this analysis, namely the presence of multimeric forms of the various mtDNA species. The first lane in the left panel in **Fig. 2B** shows that the majority of deleted mtDNA was present as a 17.6-kb dimer (uncut by *Bam*HI, but migrating on the gel as randomly linearized and nicked circles that were likely formed during the isolation of the DNA), rather than as an 8.8-kb monomer. However, in other samples from this patient (not shown), uncut supercoiled monomers were also found *(37,38)*.

8. In most standard Southern blot hybridizations, the probe is in excess and the quantitation of hybridizing bands is straightforward. In the case of blots for quantitating mtDNA, however, the probe *often may not be in excess*, especially when multiple samples are present on one filter. It is therefore critical to be sure that the hybridization conditions, especially the ratio of probe to "hybridizable DNA" on the filter, allow for all hybridizable bands to be visualized in a quantitative manner. Moreover, *quantitative results obtained from one blot cannot be compared to those obtained from a different blot* unless great care is taken to be sure that the two, in fact, are indeed comparable. Minimally, the probe lengths and specific activities in the two blots must be known, so that the signals can be normalized. Furthermore, it is extremely useful to include, in each blot, "common samples" (e.g. one normal and one sample from a patient with known mtDNA depletion) that can be used as a standard—both internal and external—for comparison. Finally, the amount of mtDNA present in a particular normal tissue can vary considerably among subjects. Thus, it is important to have more than one normal sample on a blot, in order to be more confident that the percent depletion measured in a patient is biologically meaningful.

9. When looking for point mutations in tRNA genes, blood is usually an adequate source of tissue. However, some mutations—and in particular, the amount of the A3243G mutation in the tRNA$^{Leu(UUR)}$ gene most commonly associated with MELAS—is much lower in blood than in other tissues. For this reason, muscle DNA provides more secure molecular diagnoses for all mutations (and is *required* for the detection of single deletions in patients with isolated PEO). This is particularly true when analyzing for some point mutations in structural genes

[e.g., sporadic mutations in the cytochrome-*b* gene *(47)*] and when studying oligosymptomatic or asymptomatic maternal relatives of patients with the MELAS point mutation. Hair and urinary epithelial cells are recommended if a less invasive procedure is required to obtain a DNA sample.

10. When designing a protocol for PCR/RFLP analysis, one should first look for a naturally occurring polymorphism at the site of the mutation. Often, such a convenient site is not available, in which case a "mismatched" primer can be synthesized to create an appropriate restriction site. In either case, it is better to perform the analysis so that the mutant allele "gains" the polymorphic site rather than "loses" it, for the simple reason that a gain-of-site mutation is unambiguous—it generates a new fragment on the gel—whereas loss of a site cannot be distinguished from an uncut PCR fragment. In addition, it is useful to design the PCR/RFLP protocol so that at least two cutting sites are present on the amplified DNA: one site that is present on either the wild-type or mutated allele, but not both (i.e., it is diagnostic for the mutation), whereas the other site is present on both alleles. Cutting at the latter site thus serves as an internal control for completeness of digestion and can assist in the quantitation of the percent heteroplasmy.

11. During the amplification of a heteroplasmic population of mtDNAs, three species of products are formed after each cycle of denaturation/renaturation: wild-type homoduplexes, mutated homoduplexes, and heteroduplexes consisting of one wild-type strand annealed to one mutated strand. Cleavage of this mixture in the RFLP step will result in an underestimation of the proportion of mutation, because only the perfect homoduplexes will be cut. This problem can be circumvented by the addition of a labeled nucleotide in the last extension cycle of the PCR reaction ["last-cycle-hot" PCR *(48,49)*], as the only species that are visualized (and cut) are those that have incorporated the label in a pefectly complementary "daughter" strand. In other words, the heretofore uncuttable heteroduplexes can now be cut and visualized on the gel, because the heteroduplexes were denatured and converted to pairs of perfect duplexes in the final extension step.

12. If label was added at the beginning of the PCR reaction (i.e., "last-cycle-hot" PCR was not performed), it is prudent to estimate the "true" percent mutation by taking the square root of the raw proportion of the mutation. For example, if 64% of the alleles is calculated to be mutated, the true proportion of mutated molecules is actually 80% (i.e., $[0.64]^{1/2} = 0.8$).

13. If the proportion of heteroplasmy is calculated with unlabeled fragments (e.g., by densitometry of ethidium bromide-stained fragments on a gel), it is absolutely critical to amplify a set of serially diluted known samples as a standard. The standard can be DNA from homoplasmic cells or can be from templates subcloned into plasmids.

14. The use of RFLP analysis to detect point mutations may sometimes result in a false-positive result. This is especially true in the case of loss-of-site mutations, not only because of the problem of partial or failed digestion but also because a

neutral mutation elsewhere within the restriction enzyme's recognition site will render the site uncleavable, *even if the target mutation is absent.* This problem was first uncovered in the case of the use of *Sfa*NI to detect mutations at nt-11778 in LHON patients *(50–52).* Note that false positives may also exist in gain-of-site mutations when a new restriction site is generated by a polymorphism, especially when located in a region near the diagnostic RFLP site *(53).*

Acknowledgments

This work was supported by grants from the U.S. National Institutes of Health (NS28828, NS11766, NS39854, and HD32062) and the Muscular Dystrophy Association.

References

1. Anderson, S., Bankier, A. T., Barrell, B. G., de Bruijn, M. H. L., Coulson, A. R., Drouin, J., et al. (1981) Sequence and organization of the human mitochondrial genome. *Nature* **290,** 457–465.
2. Giles, R. E., Blanc, H., Cann, H. M., et al. (1980) Maternal inheritance of human mitochondrial DNA. *Proc. Natl. Acad. Sci. USA* **77,** 6715–6719.
3. Satoh, M. and Kuroiwa, T. (1991) Organization of multiple nucleoids and DNA molecules in mitochondria of a human cell. *Exp. Cell Res.* **196,** 137–140.
4. Schon, E. A. (2000) Mitochondrial genetics and disease. *Trends Biochem. Sci.* **25,** 555–560.
5. Zeviani, M., Moraes, C. T., DiMauro, S., Nakase, H., Bonilla, E., Schon, E. A., et al. (1988) Deletions of mitochondrial DNA in Kearns–Sayre syndrome. *Neurology* **38,** 1339–1346.
6. Moraes, C. T., DiMauro, S., Zeviani, M., Lombes, A., Shanske, S., Miranda, A. F., et al. (1989) Mitochondrial DNA deletions in progressive external ophthalmoplegia and Kearns–Sayre syndrome. *N. Engl. J. Med.* **320,** 1293–1299.
7. Rötig, A., Cormier, V., Koll, F., Mize, C. E., Saudubray, J.-M., Veerman, A., et al. (1991) Site-specific deletions of the mitochondrial genome in the Pearson marrow–pancreas syndrome. *Genomics* **10,** 502–504.
8. Poulton, J., Deadman, M. E., and Gardiner, R. M. (1989) Tandem direct duplications of mitochondrial DNA in mitochondrial myopathy: analysis of nucleotide sequence and tissue distribution. *Nucl. Acids Res.* **17,** 10,223–10,229.
9. Rötig, A., Bessis, J.-L., Romero, N., Cormier, V., Saudubray, J.-M., Narcy, P., et al. (1992) Maternally inherited duplication of the mitochondrial genome in a syndrome of proximal tubulopathy, diabetes mellitus, and cerebellar ataxia. *Am. J. Hum. Genet.* **50,** 364–370.
10. Zeviani, M., Servidei, S., Gellera, C., Bertini, E., DiMauro, S., and DiDonato, S. (1989) An autosomal dominant disorder with multiple deletions of mitochondrial DNA starting at the D-loop region. *Nature* **339,** 309–311.
11. Nishino, I., Spinazzola, A., and Hirano, M. (1999) Thymidine phosphorylase gene mutations in MNGIE, a human mitochondrial disorder. *Science* **283,** 689–692.

12. Kaukonen, J., Juselius, J. K., Tiranti, V., Kyttala, A., Zeviani, M., Comi, G. P., et al. (2000) Role of adenine nucleotide translocator 1 in mtDNA maintenance. *Science* **289**, 782–785.

13. Wallace, D. C., Singh, G., Lott, M. T., Hodge, J. A., Schurr, T. G., Lezza, A. M. S., et al. (1988) Mitochondrial DNA mutation associated with Leber's hereditary optic neuropathy. *Science* **242**, 1427–1430.

14. Wallace, D. C. (1992) Diseases of the mitochondrial DNA. *Annu. Rev. Biochem.* **61**, 1175–1212.

15. Fukuhara, N., Tokigushi, S., Shirakawa, K., and Tsubaki, T. (1980) Myoclonus epilepsy associated with ragged-red fibers (mitochondrial abnormalities): Disease entity or syndrome? Light and electron microscopic studies of two cases and review of the literature. *J. Neurol. Sci.* **47**, 117–133.

16. Shoffner, J. M., Lott, M. T., Lezza, A. M. S., Seibel, P., Ballinger, S. W., and Wallace, D. C. (1990) Myoclonic epilepsy and ragged-red fiber disease (MERRF) is associated with a mitochondrial DNA tRNA[Lys] mutation. *Cell* **61**, 931–937.

17. Silvestri, G., Moraes, C. T., Shanske, S., Oh, S. J., and DiMauro, S. (1992) A new mtDNA mutation in the tRNA[Lys] gene associated with myoclonic epilepsy and ragged-red fibers (MERRF). *Am. J. Hum. Genet.* **51**, 1213–1217.

18. Hirano, M., Ricci, E., Koenigsberger, M. R., Defendini, R., Pavlakis, S. G., DeVivo, D. C., et al. (1992) MELAS: an original case and clinical criteria for diagnosis. *Neuromusc. Disord.* **2**, 125–135.

19. Goto, Y.-i., Nonaka, I., and Horai, S. (1991) A new mtDNA mutation associated with mitochondrial myopathy, encephalopathy, lactic acidosis and stroke-like episodes (MELAS). *Biochim. Biophys. Acta* **1097**, 238–240.

20. Manfredi, G., Schon, E. A., Moraes, C. T., Bonilla, E., Berry, G. T., Sladky, J. T., et al. (1995) A new mutation associated with MELAS is located in a mitochondrial DNA polypeptide-coding gene. *Neuromusc. Disord.* **5**, 391–398.

21. Santorelli, F. M., Tanji, K., Kulikova, R., Shanske, S., Vilarinho, L., Hays, A. P., et al. (1997) Identification of a novel mutation in the mtDNA ND5 gene associated with MELAS. *Biochem. Biophys. Res. Commun.* **238**, 326–328.

22. Corona, P., Antozzi, C., Carrara, F., D'Incerti, L., Lamantea, E., Tiranti, V., et al. (2001) A novel mtDNA mutation in the ND5 subunit of complex I in two MELAS patients. *Ann. Neurol.* **49**, 106–110.

23. Ciafaloni, E., Ricci, E., Servidei, S., Shanske, S., Silvestri, G., Manfredi, G., et al. (1991) Widespread tissue distribution of a tRNA[Leu(UUR)] mutation in the mitochondrial DNA of a patient with MELAS syndrome. *Neurology* **41**, 1663–1665.

24. Goto, Y.-i., Horai, S., Matsuoka, T., Koga, Y., Nihei, K., Kobayashi, M., et al. (1992) Mitochondrial myopathy, encephalopathy, lactic acidosis, and stroke-like episodes (MELAS): a correlative study of the clinical features and mitochondrial DNA mutation. *Neurology* **42**, 545–550.

25. Moraes, C. T., Ciacci, F., Silvestri, G., Shanske, S., Sciacco, M., Hirano, M., et al. (1993) Atypical clinical presentations associated with the MELAS mutation at position 3243 of human mitochondrial DNA. *Neuromusc. Disord.* **3**, 43–49.

26. Holt, I. J., Harding, A. E., Petty, R. K. H., and Morgan-Hughes, J. A. (1990) A new mitochondrial disease associated with mitochondrial DNA heteroplasmy. *Am. J. Hum. Genet.* **46,** 428–433.

27. Tatuch, Y., Christodoulou, J., Feigenbaum, A., Clarke, J. T. R., Wherret, J., Smith, C., et al. (1992) Heteroplasmic mtDNA mutation (T→G) at 8993 can cause Leigh's disease when the percentage of abnormal mtDNA is high. *Am. J. Hum. Genet.* **50,** 852–858.

28. de Vries, D. D., van Engelen, B. G. M., Gabreëls, F. J. M., Ruitenbeek, W., and van Oost, B. A. (1993) A second missense mutation in the mitochondrial ATPase6 gene in Leigh's syndrome. *Ann. Neurol.* **34,** 410–412.

29. Thyagarajan, D., Shanske, S., Vasquez-Memije, M., De Vivo, D., and DiMauro, S. (1995) A novel mitochondrial ATPase 6 point mutation in familial bilateral striatal necrosis. *Ann. Neurol.* **38,** 468–472.

30. Carrozzo, R., Murray, J., Santorelli, F. M., and Capaldi, R. A. (2000) The T9176G mutation of human mtDNA gives a fully assembled but inactive ATP synthase when modeled in *Escherichia coli. FEBS Lett.* **486,** 297–299.

31. Dionisi-Vici, C., Seneca, S., Zeviani, M., Fariello, G., Rimoldi, M., Bertini, E., et al. (1998) Fulminant Leigh syndrome and sudden unexpected death in a family with the T9176C mutation of the mitochondrial ATPase 6 gene. *J. Inherit. Metab. Dis.* **21,** 2–8.

32. Fischel-Ghodsian, N. (1999) Genetic factors in aminoglycoside toxicity. *Ann. NY Acad. Sci.* **884,** 99–109.

33. Moraes, C. T., Shanske, S., Tritschler, H.-J., Aprille, J. R., Andreetta, F., Bonilla, E., et al. (1991) Mitochondrial DNA depletion with variable tissue expression: a novel genetic abnormality in mitochondrial diseases. *Am. J. Hum. Genet.* **48,** 492–501.

34. Tritschler, H.-J., Andreetta, F., Moraes, C. T., Bonilla, E., Arnaudo, E., Danon, M. J., et al. (1992) Mitochondrial myopathy of childhood associated with depletion of mitochondrial DNA. *Neurology* **42,** 209–217.

35. Arnaudo, E., Dalakas, M., Shanske, S., Moraes, C. T., DiMauro, S., and Schon, E. A. (1991) Depletion of mitochondrial DNA in AIDS patients with zidovudine-induced myopathy. *Lancet* **337,** 508–510.

36. Sambrook, J. and Russell, D. W. (2001) *Molecular Cloning, A Laboratory Manual, 3rd ed.*, Cold Spring Harbor Laboratory Press, Cold Spring Harbor, NY.

37. Tang, Y., Manfredi, G., Hirano, M., and Schon, E. A. (2000) Maintenance of human rearranged mtDNAs in long-term-cultured transmitochondrial cell lines. *Mol. Biol. Cell* **11,** 2349–2358.

38. Tang, Y., Schon, E. A., Wilichowski, E., Vazquez-Memije, M. E., Davidson, E., and King, M. P. (2000) Rearrangements of human mitochondrial DNA (mtDNA): new insights into the regulation of mtDNA copy number and gene expression. *Mol. Biol. Cell* **11,** 1471–1485.

39. Moraes, C. T., Shanske, S., Tritschler, H. J., Aprille, J. R., Andreetta, F., Bonilla, E., et al. (1991) MtDNA depletion with variable tissue expression: a novel genetic abnormality in mitochondrial diseases. *Am. J. Hum. Genet.* **48,** 492–501.

40. Wilson, G., Hollar, B., Waterson, J., and Schmickel, R. (1978) Molecular analysis of cloned human 18S ribosomal DNA segments. *Proc. Natl. Acad. Sci. USA* **75**, 5367–5371.

41. Schon, E. A., Hirano, M., and DiMauro, S. (2001) Molecular genetic basis of the mitochondrial encephalomyopathies, in *Mitochondrial Disorders in Neurology II* (Schapira, A. H. V. and DiMauro, S., eds.), Butterworth–Heinemann, Boston, pp. 69–113.

42. Servidei, S. (2001) Mitochondrial encephalopathies: gene mutation. *Neuromusc. Disord.* **11**, 121–130.

43. Johnson, M. A., Bindoff, L. A., and Turnbull, D. M. (1993) Cytochrome *c* oxidase activity in single muscle fibers: assay techniques and diagnostic applications. *Ann. Neurol.* **33**, 28–35.

44. Moraes, C. T. and Schon, E. A. (1996) Detection and analysis of mitochondrial DNA and RNA in muscle by *in situ* hybridization and single-fiber PCR. *Methods Enzymol.* **264**, 522–540.

45. Torroni, A., Schurr, T. G., Cabell, M. F., Brown, M. D., Neel, J. V., Larsen, M., et al. (1993) Asian affinities and continental radiation of the four founding native American mtDNAs. *Am. J. Hum. Genet.* **53**, 563–590.

46. Baumer, A., Zhang, C., Linnane, A. W., and Nagley, P. (1994) Age-related human mtDNA deletions: a heterogeneous set of deletions arising at a single pair of directly repeated sequences. *Am. J. Hum. Genet.* **54**, 618–630.

47. Andreu, A. L., Hanna, M. G., Reichmann, H., Bruno, C., Penn, A. S., Tanji, K., et al. (1999) Exercise intolerance due to mutations in the cytochrome b gene of mitochondrial DNA. *N. Engl. J. Med.* **341**, 1037–1044.

48. Moraes, C. T., Ricci, E., Bonilla, E., DiMauro, S., and Schon, E. A. (1992) The mitochondrial tRNA$^{Leu(UUR)}$ mutation in MELAS: genetic, biochemical, and morphological correlations in skeletal muscle. *Am. J. Hum. Genet.* **50**, 934–949.

49. Petruzzella, V., Moraes, C. T., Sano, M. C., Bonilla, E., DiMauro, S., and Schon, E. A. (1994) Extremely high levels of mutant mtDNAs co-localize with cytochrome c oxidase-negative ragged-red fibers in patients harboring a point mutation at nt-3243. *Hum. Mol. Genet.* **3**, 449–454.

50. Johns, D. R. and Neufeld, M. J. (1993) Pitfalls in the molecular genetic diagnosis of Leber hereditary optic neuropathy (LHON). *Am. J. Hum. Genet.* **53**, 916–920.

51. Mashima, Y., Hiida, Y., Saga, M., Oguchi, Y., Kudoh, J., and Shimizu, N. (1995) Risk of false-positive molecular genetic diagnosis of Leber's hereditary optic neuropathy. *Am. J. Ophthalmol.* **119**, 245–246.

52. Yen, M. Y., Wang, A. G., Chang, W. L., Hsu, W. M., Liu, J. H., and Wei, Y. H. (2000) False positive molecular diagnosis of Leber's hereditary optic neuropathy. *Zhonghua Yi Xue Za Zhi (Taipei)* **63**, 864–868.

53. Kirby, D. M., Milovac, T., and Thorburn, D. R. (1998) A false-positive diagnosis for the common MELAS (A3243G) mutation caused by a novel variant (A3426G) in the ND1 gene of mitochondria DNA. *Mol. Diagn.* **3**, 211–215.

3

Screening for Aging-Dependent Point Mutations in mtDNA

Yuichi Michikawa and Giuseppe Attardi

1. Introduction

Recently, clear evidence has been obtained in our laboratory of a large accumulation of specific point mutations at the same critical control sites for replication in mitochondrial DNA (mtDNA) of human fibroblasts *(1)* and skeletal muscles *(2)*. This demonstration was based on the use of a very sensitive method for denaturing gradient gel electrophoresis (DGGE) *(3,4)* that can identify 2–4% single mismatches in artificially produced heteroduplexes and that is suitable for simultaneous screening of a large number of samples, in combination with cloning and sequencing. The overall approach is illustrated schematically in **Fig. 1**. The approach requires the use of mtDNA that has been highly purified from total cell DNA, in order to minimize the possible interference by nuclear mtDNA pseudogenes, which have complicated the identification of mtDNA mutations in previous investigations *(5,6)*. As shown in **Fig. 2** for the main mtDNA control region, the particular mtDNA region chosen for analysis in our work is subdivided into several segments, each exhibiting a single melting domain, thus becoming analyzable by DGGE (*see* **Fig. 3**). The various segments are then screened by a highly sensitive new version of the latter mentioned technique *(4)*, using a large number of DNA samples from differently aged individuals. The segments that show evidence of heterogeneity by DGGE analysis, indicative of the presence of mutations, will be selected for further study. This will involve cloning of the polymerase chain reaction (PCR) product of each segment in *Escherichia coli*, the isolation of 48 plasmids, and PCR amplification of the cloned segment from their DNAs. The PCR products will then be subjected to a second DGGE step, after

From: *Methods in Molecular Biology, vol. 197: Mitochondrial DNA: Methods and Protocols*
Edited by: W. C. Copeland © Humana Press Inc., Totowa, NJ

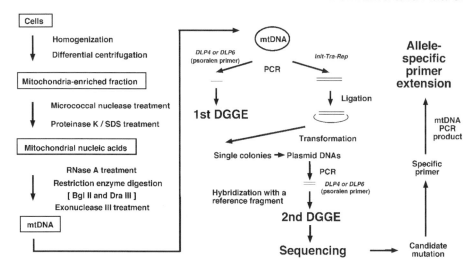

Fig. 1. Scheme of the overall approach. This comprises the following three procedures: (1) isolation of highly purified mtDNA; (2) preliminary screening of the mtDNA samples for the presence of mutations by first-round DGGE, then cloning of the whole *Init-Tra-Rep* fragment (Fig. 2), followed by second-round DGGE and sequencing; (3) allele-specific termination of primer extension to carry out a large-scale screening of mtDNA samples for the identified mutations.

hybridization with one or more of them, randomly chosen as a reference clone. All presumptive mutant clones and several of the presumptive wild-type clones identified by the second DGGE analysis will then be sequenced. This screening should identify most of the mutations in the chosen mtDNA segments. If any of these mutations occurs in more than one individual, therefore qualifying itself as a site-specific mutation, a rapid screening of a large number of samples from different individuals can be carried out by allele-specific termination of primer extension *(7)*, after choosing an appropriate primer (**Fig. 2**). In this chapter, we describe in detail the protocols used in our laboratory for the application of the above-outlined approach to the detection of point mutations in a portion of the mtDNA control region of human fibroblasts and muscles. The same protocols can be applied to the analysis of other mtDNA regions in the same cells and in other cell types. Furthermore, it can be anticipated that the same approach is applicable to the analysis of multiple mutations of a given nuclear gene in the same tissue, which may be accumulated during the aging process or in other situations.

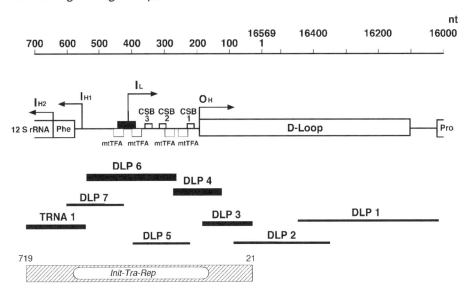

Fig. 2. Main control region of human mtDNA. This region contains the initiation sites for rDNA (I_{H1}), whole heavy-strand (P_{H2}) and light-strand transcription (I_L), and the primary origin (O_{H1}) and secondary origin (O_{H2}) of heavy-strand DNA synthesis. The map positions of the CSB1, CSB2, and CSB3 (conserved sequence blocks 1, 2, and 3, respectively), of mtTFA binding sites, of the DLP segments and TRNA1 segment chosen for DGGE analysis, and of the large fragment, encompassing all the initiation sites for transcription and the primary and secondary origins of heavy-strand DNA synthesis (*Init-Tra-Rep*), used for cloning are also shown.

2. Materials

2.1. Preparation of Highly Purified mtDNA

Unless otherwise specified, all solutions should be autoclaved. Gloves should be worn at all times.

1. TD buffer: 25 mM Tris-HCl, pH 7.4 (at 25°C), 135 mM NaCl, 5 mM KCl, 0.5 mM Na_2HPO_4.
2. Trypsin–EDTA–TD buffer: 5 g/L trypsin (from porcine pancreas; ICN Biomedicals, Irvine, CA), 3.2 g/L penicillin, 0.2 g/L streptomycin, 0.003% (w/v) phenol red in TD buffer containing 10 mM EDTA; sterilize by filtration.
3. NKM buffer: 1 mM Tris-HCl, pH 7.4, 0.13 M NaCl, 5 mM KCl, 7.5 mM $MgCl_2$.
4. Homogenization buffer: 10 mM Tris-HCl, pH 6.7, 10 mM KCl, 0.15 mM $MgCl_2$.

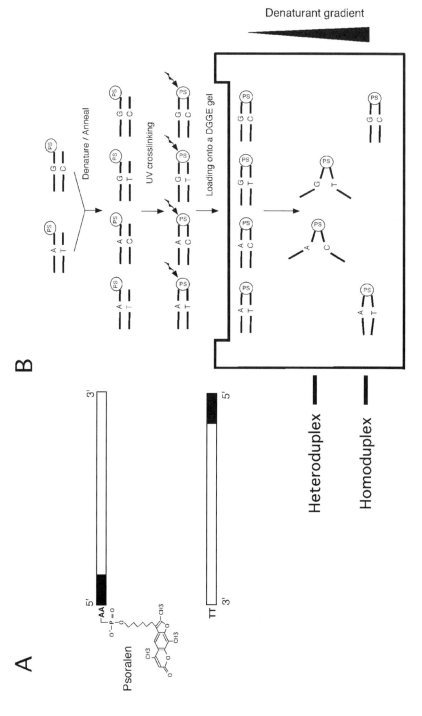

Fig. 3. Structure of the PCR product with attached psoralen used for DGGE (A), and principle of mutation detection by DGGE (B). For details, see ref. 4.

5. Thomas homogenizer, size 0 (A.H. Thomas, Swedesboro, NJ), with motor-driven pestle.
6. 2 *M* Sucrose: sterilize by filtration, store frozen at –20°C.
7. Mitochondria suspension buffer: 10 m*M* Tris-HCl, pH 6.7, 0.25 m*M* sucrose, 0.15 m*M* MgCl$_2$: sterilize by filtration, store frozen at –20°C.
8. Micrococcal nuclease (Worthington Biochemical Corporation, Lakewood, NJ). Dissolve in autoclaved distilled water at concentration of 100 units/μL. Store frozen at –20°C.
9. Micrococcal nuclease buffer: 0.25 *M* sucrose, 0.01 *M* Tris-HCl, pH 8.0, 0.001 *M* CaCl$_2$: sterilize by filtration, store frozen at –20°C.
10. 0.25 *M* EGTA.
11. Micrococcal nuclease reaction dilution buffer: 0.25 *M* sucrose, 0.01 *M* Tris-HCl, pH 6.7, 0.01 *M* EDTA: sterilize by filtration, store frozen at –20°C.
12. Mitochondrial nucleic acids extraction buffer: 0.01 *M* Tris-HCl, pH 7.4, 0.15 *M* NaCl, 0.001 *M* EDTA.
13. 10 mg/mL Proteinase K (Roche Molecular Biochemicals, Mannheim, Germany): prepare a fresh solution each time.
14. 10% Sodium dodecyl sulfate (SDS).
15. Phenol–chloroform–isoamylalcohol mixture (25 : 24 : 1), saturated with 0.1 *M* Tris, pH 8.0.
16. Ethanol.
17. 70% Ethanol.
18. TE: 0.01 *M* Tris-HCl, pH 7.4, 0.001 *M* EDTA.
19. 10 mg/mL Pancreatic RNase A in 10 m*M* Tris-HCl, pH 7.5, 15 m*M* NaCl: store frozen at –20°C.
20. Restriction enzyme *Bgl*II (10,000 units/mL; New England Biolab, Beverly, MA).
21. Restriction enzyme *Dra*III (20,000 units/mL; New England Biolab, Beverly, MA).
22. 10X NE buffer 3: 50 m*M* Tris-HCl, pH 7.9, 10 m*M* MgCl$_2$, 100 m*M* NaCl, 1 m*M* dithiothreitol (DTT) (New England Biolab, Beverly, MA): sterilize by filtration.
23. 100X Bovine serum albumin (BSA, 10 mg/mL; New England Biolab, Beverly, MA).
24. Exonuclease III (100,000 units/mL; New England Biolab, Beverly, MA).
25. Glycogen for molecular biology (20 mg/mL; Roche Diagnostics GmbH, Mannheim, Germany).
26. TRI Reagent (Molecular Research Center, Inc., Cincinnati, OH).
27. Brinkmann homogenizer (model PT10/35, Brinkman Instrument Co., Westbury, NY).

2.2. DGGE–Cloning–Sequencing

1. MacMelt1.0 (Bio-Rad Laboratories, Hercules, CA); see also **ref. 8**.
2. DLP6.For primer [positions 261–280 in Cambridge sequence *(9)*]: 5'-CCAC TTTCCACACAGACATC.

3. DLP6.Rev (+ psoralen C6 phosphoramidite + AA) primer (positions 541–524 in Cambridge sequence): 5′-psoralen-AA-GTATGGGGTTAGCAGCGG.
4. DLP4.For primer (positions 111–132 in Cambridge sequence): 5′-ACCCTATG TCGCAGTATCTGTC.
5. DLP4.Rev (+ psoralen C6 phosphoramidite + AA) primer (positions 271–253 in Cambridge sequence): 5′-psoralen-AA-GTGGAAAGTGGCTGTGCAG.
6. 10X Expand HF buffer, without $MgCl_2$ (Roche Molecular Biochemicals, Mannheim, Germany).
7. High Fidelity PCR Enzyme Mixture (3.5 units/μL, Roche Molecular Biochemicals, Mannheim, Germany).
8. 25 mM $MgCl_2$ (Roche Molecular Biochemicals, Mannheim, Germany).
9. Deoxynucleotide mixture (10 mM of each) (Sigma Biosciences, St. Louis, MO).
10. Water for PCR (Sigma Biosciences, St. Louis, MO).
11. Mineral oil (Sigma Biosciences, St. Louis, MO).
12. DNA Thermal Cycler Model 480 (Perkin-Elmer, Foster City, CA).
13. Chloroform.
14. 10X TBE: 0.9 M Tris, 0.9 M boric acid, 0.02 M EDTA, pH 8.0.
15. PCR sample loading buffer: 2X TBE, 15% Ficoll PM400 (Amersham Pharmacia Biotech, Piscataway, NJ), 0.0025% xylene cyanol FF.
16. 10 mg/mL Ethidium bromide: sterilize by filtration.
17. Kodak Electrophoresis Documentation and Analysis 120 System (Eastman Kodak Company, Rochester, NY).
18. Blak-Ray ultraviolet (UV) lamp (Ultra-Violet Products Inc., San Gabriel, CA).
19. Denaturing gradient gel electrophoresis system (C.B.S. Scientific Company, Del Mar, CA).
20. 20X TAE running buffer: 0.8 M Tris base, 0.4 M sodium acetate, 20 mM EDTA, pH 7.4.
21. DGGE sample loading buffer: 6X TAE running buffer, 15% Ficoll PM400 (Amersham Pharmacia Biotech, Piscataway, NJ), 0.25% bromophenol blue, 0.25% xylene cyanol FF.
22. 40% (w/v) Acrylamde–bisacrylamide (29:1). Deionize the solution by Amberlite MB-150 mixed-bed exchanger (15 g for 500 mL solution; Sigma Biosciences, St. Louis, MO) for 15 min, then filter it through Whatman #1 filter paper. Store at 4°C.
23. 10% Ammonium persulfate.
24. TEMED (Sigma Biosciences, St. Louis, MO).
25. Deionized formamide (Sigma Biosciences, St. Louis, MO), packaged under nitrogen in amber glass bottle. Store at 4°C.
26. Ultraviolet transilluminator Model TM-36E (UVP, Upland, CA).
27. MP4+ instant camera system (Polaroid, Cambridge, MA).
28. Polapan 55 PN black and white instant sheet films (Polaroid, Cambridge, MA).
29. DLP3.For primer (positions 21–41 in Cambridge sequence): 5′-ATTAACCACT CACGGGAGCTC.

30. TRNA1.Rev primer (positions 719–702 in Cambridge sequence): 5′-CTCACT GGAACGGGGATG.
31. 0.5% Methylene blue in distilled water.
32. QIAEX II gel extraction kit (Qiagen, Valencia, CA).
33. pGEM-T Easy Vector System (Promega, Madison, WI).
34. QIAprep 8 Miniprep kit (Qiagen, Valencia, CA).
35. *Taq* DNA polymerase (5 units/µL, Promega, Madison, WI).
36. 10X Thermophilic DNA polymerase buffer (Promega, Madison, WI).
37. DNA sequencing kit ABI Prism™ (Dye Terminator Cycle Sequencing Core Kit; Perkin-Elmer Applied Biosystems, Warrington, UK).
38. 3 *M* Sodium acetate, pH 5.2.
39. ABI PRISM Sequencer Model 373 (Perkin-Elmer Applied Biosystems, Warrington, UK).

2.3. Allele-Specific Termination of Primer Extension

1. T4 polynucleotide kinase (10,000 units/mL; New England Biolabs, Beverly, MA).
2. 10X NE Buffer for T4 polynucleotide kinase (New England Biolabs, Beverly, MA).
3. T414G primer (positions 434–417 in Cambridge sequence): 5′-GGTGACTGT TAAAAGTGC.
4. γ-^{32}P ATP (10 mCi/ml, 3000 Ci/mmol; NEN Life Science Products, Boston, MA).
5. 0.5 *M* EDTA, pH 8.0.
6. DE81 filters (Whatman, England).
7. 0.5 *M* Na_2HPO_4, pH 7.0.
8. Sequencing gel loading buffer: 95% formamide, 20 m*M* EDTA, pH 8.0, 0.05% bromophenol blue, 0.05% xylene cyanol FF.
9. Kodak Scientific Imaging Film X-OMAT™ AR (Eastman Kodak Company, Rochester, NY).
10. Glycogen for molecular biology. (*see* **Subheading 2.1., item 25**).
11. TE: 10 m*M* Tris-HCl, pH 7.4, 1 m*M* EDTA.
12. 10% SDS.
13. DLP3.For primer (*see* **Subheading 2.2., item 29**).
14. TRNA1.Rev primer (*see* **Subheading 2.2., item 30**).
15. Sequenase Version 2.0 DNA polymerase (13 units/µL; USB, Cleveland, OH).
16. 5X Sequenase reaction buffer: 0.2 *M* Tris-HCl, pH 7.5, 0.1 *M* $MgCl_2$, 0.25 *M* NaCl.
17. 0.1 *M* DTT.
18. Enzyme dilution buffer: 0.01 *M* Tris-HCl, pH 7.5, 0.005 *M* DTT.
19. Dideoxy CTP (100 m*M*; Sigma Biosciences, St. Louis, MO).
20. Deoxy ATP, deoxy GTP, deoxy TTP (100 m*M* each; Sigma Biosciences, St. Louis, MO).
21. Model 583 Gel Dryer Filter Paper (Bio-Rad Laboratories, Hercules, CA).
22. Molecular Dynamics Storage Phosphor Screen (Amersham Pharmacia Biotech, Piscataway, NJ).

23. Molecular Dynamics Phosphorimager 445 SI (Amersham Pharmacia Biotech., Piscataway, NJ).

3. Methods
3.1. Preparation of Mitochondrial DNA

This section describes a protocol for isolation of highly purified mtDNA, free of nuclear-mitochondrial pseudogenes, from tissue culture cells and skeletal muscle. This mtDNA will be used as a template to prepare the PCR product for the first DGGE analysis, cloning, and the allele-specific termination of primer extension.

3.1.1. Skin Fibroblast Cultures

The protocol described in the following subsections is for the amount of cells grown to near-confluency in a single 10-cm Petri dish. For larger amounts of cells, the volumes and the concentrations of enzymes and other reagents should be increased appropriately.

3.1.1.1 FRACTIONATION OF MITOCHONDRIA (*SEE* **NOTE 1**)

1. Grow cells on a Petri culture dish (100 mm × 20 mm) up to 80–90% confluency.
2. Wash the plate with 5 mL of TD buffer.
3. Detach the cells by incubating in 5 mL of trypsin–EDTA–TD buffer at room temperature for 5 min.
4. Collect the cell suspension and transfer to a 15-mL centrifuge tube containing 5 mL of TD buffer with 5% calf serum placed on ice. Count number of cells.
5. Centrifuge the cell suspension at $500g_{av}$ at 4°C for 5 min.
6. Suspend the pelleted cells in 5 mL NKM buffer, centrifuge at $500g_{av}$ at 4°C for 5 min, then remove the supernatant. Repeat this step twice.
7. Suspend the cells in 0.5 mL NKM buffer, transfer the cells to a 0.5-mL microfuge tube, place the tube in an appropriate larger tube or adapter, centrifuge at $500g_{av}$ at 4°C for 5 min, then remove the supernatant. Estimate the approximate volume of packed cells.
8. Place the cells at –20°C for 15 min to destabilize the cell membrane.
9. Place the tube in ice and suspend the cells in 10X the volume of homogenization buffer (4°C). Wait 3 min to let cells swell.
10. Transfer the cells into a Thomas homogenizer, size 0, on ice and homogenize them with 30 strong strokes, using a pestle rotating at 1500 rpm. Check the breakage of the cells in the phase-contrast microscope. The proportion of broken cells should be around 70%.
11. Pour the homogenate into a 1.5-mL microfuge tube containing one-sixth volume of 2 *M* sucrose, and mix well by pipetting.

12. Centrifuge at $1200g_{av}$ at 4°C for 3 min to pellet nuclei, large debris, and unbroken cells. Transfer the supernatant to a new 1.5-mL microfuge tube.
13. Repeat **step 12**.
14. Centrifuge the supernatant at $13,000g_{av}$ at 4°C for 1 min. Discard the supernatant.
15. Add 1 mL of mitochondria suspension buffer, resuspend the pellet, centrifuge at $13,000g_{av}$ at 4°C for 1 min, then remove the supernatant. Repeat this step twice.
16. Suspend the pelleted mitochondria in 0.2 mL of the micrococcal nuclease buffer containing 100 units of micrococcal nuclease and incubate on ice for 30 min.
17. Add 2 μL of 0.25 *M* EGTA to inhibit the nuclease activity. Mix well by pipetting and incubate on ice for 5 min.
18. Dilute the reaction mixture with 0.8 mL of micrococcal nuclease dilution buffer.
19. Centrifuge at $13,000g_{av}$ at 4°C for 10 min. Discard the supernatant.

3.1.1.2. EXTRACTION OF MITOCHONDRIAL NUCLEIC ACIDS FROM FRACTIONATED MITOCHONDRIA

1. Resuspend the pelleted mitochondria in 0.2 mL of mitochondrial nucleic acids extraction buffer. Add 4 μL of a 10 mg/mL proteinase K solution and incubate at room temperature for 5 min.
2. Lyse the mitochondrial membranes by addition of 20 μL of 10% SDS and incubation at room temperature for 30 min.
3. Extract the nucleic acids with 200 μL of phenol–chloroform–isoamyl alcohol mixture (25 : 24 : 1).
4. Transfer the aqueous phase to a new tube. Add 400 μL of ethanol and place the tube at –20°C for 20 min. Centrifuge at $13,000g_{av}$ at 4°C for 30 min. Discard the supernatant. Add 500 μL of 70% ethanol and centrifuge at $13,000g_{av}$ at 4°C for 5 min. Discard the supernatant. Dry the precipitate by removing the cap of the tube, leaving it open at room temperature for 20 min.
5. Dissolve the precipitated nucleic acids in TE (100 μL per 2×10^5 cells). Store at 4°C.

3.1.1.3 DIGESTION OF CONTAMINATING NUCLEAR DNA AND RNA

1. Treat the mitochondrial nucleic acids with 10 μg of RNase A, 10 units of *Bgl*II and 20 units of *Dra*III in 1× NE buffer 3 with 1X BSA at 37°C for 2 h.
2. Add 100 units of Exonuclease III and incubate at 37°C for 2 h.
3. Terminate the reaction by incubating at 65°C for 10 min. Phenol–chloroform–isoamyl alcohol extract, ethanol precipitate, and dissolve in TE (100 μL per 2×10^5 cells).

3.1.2. Skeletal Muscle

The protocol described in the following subsections is for 100–200 mg of wet weight of muscle. For larger amounts, the volumes and the concentrations of enzymes and other reagents should be increased appropriately.

3.1.2.1. Extraction of Total DNA

1. Homogenize the frozen biopsied or autopsied muscle samples in the TRI reagent (1 mL per 50–100 mg [wet weight] of tissue) with a Brinkmann homogenizer, using a power unit set at 3 to 4 for 1 min.
2. Extract total DNA from the homogenate according to the manufacturer's recommendations.

3.1.2.2. Digestion of Contaminating Nuclear DNA and RNA

Follow the procedure described in **Subheading 3.1.1.3.**, using 0.25–0.5 mg DNA.

3.2. DGGE–Cloning–Sequencing

This section describes a protocol for the primary screening of the samples for the purpose of detecting the presence of mutations in the main control region of mtDNA from different tissues of aged human subjects and for the subsequent identification and quantification of the sequence variations observed. Isolation of single molecules of the DNA fragment analyzed, which is essential for the identification of individual mutations by sequencing, is performed by bacterial cloning of the PCR products.

3.2.1. PCR Amplification

1. Use the DLP6.For and DLP6.Rev (+PS) primer pair for amplification of the DLP6 segment of the mtDNA main control region (**Fig. 2**) and the DLP4.For and DLP4.Rev (+PS) primer pair for amplification of the DLP4 segment (*see* **Note 2**).
2. Prepare 50 µL of a PCR reaction mixture containing expand HF buffer, 0.2 mM of each dNTP, 0.7 mM MgCl$_2$ for DLP6 or 0.65 mM MgCl$_2$ for DLP4 (*see* **Note 3**), 0.2 µM of each primer, 1.75 units of High Fidelity PCR Enzyme Mixture, and 1 µL of highly purified mtDNA (*see* **Note 4**). Overlay 50 µL of mineral oil (*see* **Note 5**).
3. Temperature cycling in the DNA Thermal Cycler Model 480: 94°C for 45 sec, 57°C for 45 sec, 72°C for 90 sec; number of cycles 35.

3.2.2. Heteroduplex Formation of PCR Products and UV Crosslinking

1. Remove the mineral oil overlying each PCR product by adding to each tube 50 µL of chloroform, mixing, centrifuging at 13,000g_{av} for 1 min, and, finally, discarding the bottom chloroform/mineral oil layer.
2. Determine the concentration of PCR products as follows (*see* **Note 6**). Mix 5 µL of each sample with 5 µL of PCR sample loading buffer, then load onto a 1.5% TBE–agarose gel containing 0.1 mg/mL ethidium bromide. Use TBE containing 0.1 mg/mL ethidium bromide as a running buffer. Run electrophoresis at 75 mA

for 30 min, then use the Kodak Electrophoresis Documentation and Analysis 120 System to analyze each PCR product.

3. In order to form internal heteroduplexes, heat 100 ng of each PCR product at 100°C for 5 min and then let it cool gradually (–1°C/min) to 25°C. Use 20 μL of mineral oil to avoid evaporation of H_2O from the samples.

4. Irradiate the samples by UV. Place each sample on the cap of a microfuge tube 1 cm below a UV source (365 nm), consisting of two 15-W lamps, and irradiate at 4°C for 15 min.

3.2.3. First DGGE

1. Make a 20-cm-long 8% polyacrylamide gel in TAE containing a 0–90% broad-range denaturant gradient parallel to the direction of electrophoresis (100% denaturant = 7 M urea + 40% [v/v] formamide).

2. For preparation of the 0% denaturant gel (four gels), mix 3 mL of 20X TAE running buffer, 12 mL of 40% (w/v) acrylamide–bisacrylamide (29:1), and distilled H_2O (dH_2O) up to 60 mL. Aliquot 12-mL portions in four 15-mL centrifuge tubes on ice; after addition to each tube of 30 μL of 10% ammonium persulfate and 15 μL of TEMED, and thorough mixing, pour 11.5 mL at a time into the gradient maker.

3. For preparation of the 90% denaturant gel (four gels), mix 21.6 mL of deionized formamide, 22.7 g of urea, 3 mL of 20X TAE buffer, 12 mL of 40% acrylamide–bisacrylamide (29:1), and dH_2O up to 60 mL, and filter through Whatman #1 filter paper. Prepare a fresh solution each time. Aliquot 12-mL portions in four 15-mL centrifuge tubes on ice; after addition to each tube of 30 μL of 10% ammonium persulfate and 15 μL of TEMED, and thorough mixing, pour 11.5 mL at a time into the gradient maker, and make the gradient.

4. Heat the buffer tank to 60°C.

5. Mix each of the UV-irradiated samples with 10 μL of DGGE sample loading buffer, then load them onto the denaturant gradient gel.

6. Run the gel at constant voltage (150 V) for 4.5 h (*see* **Note 7**).

7. Stain the gel with 0.1 mg/mL ethidium bromide in dH_2O for 15 min and then destain it with dH_2O for 15 min.

8. Illuminate the gel on the ultraviolet transilluminator model TM-36E and take pictures with Polapan 55 PN black and white instant sheet films, using the MP4+ instant camera system (*see* **Note 8**).

3.2.4. Cloning of PCR Products

1. All of the DNA samples that show evidence of mutations in DLP6 and/or DLP4 in the first DGGE analysis are used to amplify an *Init-Tra-Rep* fragment, which contains all the initiation sites for transcription and replication of mtDNA and includes DLP6 and DLP4, therefore allowing the determination of a possible collinearity of mutations found in the two fragments. For PCR, use a DLP3.For

and TRNA1.Rev primer pair. Prepare 50 µL of a PCR reaction mixture as follows: Expand HF buffer, 0.2 m*M* dNTPs, 1.05 m*M* MgCl$_2$, 0.2 µ*M* of each primer, 1.75 units of High Fidelity PCR Enzyme, and 1 µL of highly purified mtDNA. Temperature cycling: the same as in **Subheading 3.2.1., step 3**.

2. Purify the PCR product (*see* **Note 9**). Mix 50 µL of the PCR product with 10 µL of PCR sample loading buffer (**Subheading 2.2., item 15**), then load 50 µL onto a 1.5% TBE–agarose gel. Run electrophoresis at 75 mA for 45 min, stain the gel with 0.1% methylene blue for 15 min and destain with dH$_2$O for 15 min, excise a slice of gel with the DNA band, then extract DNA from the gel using the QIAEX II Gel Extraction Kit.

3. Determine the concentration of the purified PCR product as described in **Subheading 3.2.2., step 2**. Use one-tenth volume of the product.

4. Ligate 20 ng of the PCR product with 20 ng of an appropriate plasmid vector, pGEM-T, then transform JM109 High Efficiency Competent *E. coli* cells.

5. Isolate plasmid DNA from overnight 3-mL cultures of 48 individual colonies, using the QIAprep 8 miniprep kit (*see* **Note 10**).

3.2.5. Second DGGE

1. Amplify by PCR a portion of the main control region from the insert of each cloned, *Init-Tra-Rep* fragment-containing plasmid DNA, using the DLP6.For and DLP6.Rev (+PS) primer pair for the DLP6 segment and the DLP4.For and DLP4.Rev (+PS) primer pair for the DLP4 segment. Prepare 50 µL PCR reaction mixtures as follows: thermophilic DNA polymerase buffer, 0.2 m*M* of each dNTP, 0.65 m*M* MgCl$_2$, 0.2 µ*M* of each primer, 2.5 units of *Taq* DNA polymerase, and 1 µL of plasmid DNA. Overlay 50 µL of mineral oil on each reaction mixture. For one of the plasmid DNAs (subsequently referred to as plasmid #1 [*see* **Note 11**]), prepare five 50 µL reaction mixtures. Temperature cycling will be the same as described in **Subheading 3.2.1., step 3**.

2. Remove mineral oil as described in **Subheading 3.2.2., step 1**, then determine the concentration of PCR products as described in **Subheading 3.2.2., step 2**.

3. Mix 50 ng of PCR product from plasmid #1 DNA with 50 ng of each of the other samples, and overlay 50 µL of mineral oil on each.

4. Heat-denature the samples at 100°C for 5 min, then cool them down gradually to 25°C at the rate of –1°C/min.

6. UV-irradiate the samples as described in **Subheading 3.2.2., step 4**.

7. Analyze the samples on a 0–90% denaturing gradient gel, as described in **Subheading 3.2.3**.

3.2.6. Sequencing

1. Amplify by PCR the insert of each plasmid DNA containing the *Init-Tra-Rep* fragment by PCR, using the DLP3.For and TRNA1.Rev primer pair. Prepare 50 µL reaction mixtures as follows: thermophilic DNA polymerase buffer, 0.2 m*M* of each dNTP, 1.25 m*M* MgCl$_2$, 0.2 µ*M* of each primer, 2.5 units of

Taq DNA polymerase, and 1 µL of plasmid DNA. Overlay 50 µL of mineral oil on each reaction mixture. Temperature cycling will be the same as described in **Subheading 3.2.1., step 3**.

2. Purify the PCR product as described in **Subheading 3.2.4., step 2**.
3. Determine the concentration of the purified PCR product as described in **Subheading 3.2.2., step 2**.
4. Carry out cycle sequencing of each sample, using the ABI PRISM™ Dye Terminator Cycle Sequencing Core Kit and the DLP4.For primer, in order to sequence together the overlapping DLP6 and DLP4 segments (**Fig. 2**). Prepare 20 µL of reaction mixture as follows: sequencing buffer, 1 µL of dNTP mixture, 0.5 µL of each of the dye terminators, 1 µL of AmpliTaq DNA polymerase FS, 200 ng of template DNA, 3.2 pmol of the DLP4.For primer. Overlay 20 µL of mineral oil. Temperature cycling: 96°C for 30 s, 50°C for 15 s, 60°C for 4 min; cycle number 25.
5. Remove the mineral oil as described in **Subheading 3.2.2., step 1**.
6. Remove the free dye terminators as follows. Add to each sample 2 µL of 3 *M* sodium acetate, pH 5.2, and 40 µL of ethanol. Place the samples at –20°C for 30 min, then centrifuge them at $13,000g_{av}$ at 4°C for 30 min. Discard the supernatants and rinse the precipitates with 70% ethanol. Dry the precipitates by removing the caps of the tubes and let them stand at room temperature for 20 min.
7. Analyze the samples by the ABI PRISM Sequencer Model 373, version 3.0.

3.3. Allele-Specific Termination of Primer Extension (Fig. 4)

This section describes a method for detecting and quantifying known aging-dependent specific point mutations in a large number of samples. We refer, as an example, to the conditions used for the T414G transversion, which was the most frequent mutation in skin fibroblasts from old human individuals. For the conditions to be used for the muscle-specific A189G and T408A mutations, *see* **ref. 2**.

3.3.1. Preparation of 5′-end 32P-Labeled Primer

1. Prepare 25 µL of ^{32}P-labeling reaction as follows: T4 polynucleotide kinase buffer, 10 pmol of [γ-^{32}P] ATP (3000 Ci/mmol), 10 pmol of T414G-specific primer (*see* **Note 12**), 15 units of T4 polynucleotide kinase.
2. Incubate the reaction mixture at 37°C for 45 min.
3. Terminate the reaction by adding 10 µL of 0.5 *M* EDTA, pH 8.0.
4. Dilute 1 µL of the reaction mixture with 49 µL of 0.05 *M* EDTA, pH 8.0. Spot 2 µL of this dilution onto each of two DE81 filters. After drying them at room temperature for 1 h, wash out the free [γ-^{32}P] ATP from one of them with 0.5 *M* Na_2HPO_4, pH 7.0, according to **ref. 10**. Measure the radioactivity of the filters and calculate the proportion of [γ-^{32}P] ATP incorporated and the specific activity of the ^{32}P-labeled primer by using the following formulas:

$$\text{Proportion of [γ-}^{32}\text{P] ATP incorporated} = \frac{\text{cpm in washed filter}}{\text{cpm in unwashed filter}} \quad (1)$$

$$\text{Specific activity} \left(\frac{\text{pmol}}{\text{cpm}} \right)$$

$$= \frac{\text{Radioactivity in washed filter} \times \dfrac{\begin{array}{c}\text{Reaction}\\ \text{mixture volume}\end{array}}{\text{Sample volume}} \times \dfrac{\text{Dilution volume}}{\text{Spotted volume}}}{\text{Amount of primer}} \quad (2)$$

$$= \frac{\text{cpm} \times \dfrac{35\ \mu L}{1\ \mu L} \times \dfrac{50\ \mu L}{2\ \mu L}}{10\ \text{pmol}}$$

$$= \text{cpm} \times \frac{87.5}{\text{pmol}}$$

5. Mix the rest of the labeling reaction mixture with 8 µL of sequencing gel loading buffer. Heat the sample at 100°C for 5 min.
6. Load the sample onto a 50-cm-long TBE–20% polyacrylamide/6 M urea gel. Run the gel at a constant 45-mA current (voltage is around 3000 V). Under these conditions, the temperature of the gel should be approx 55°C.
7. Stop electrophoresis when the xylene cyanol has migrated to a position approx 21 cm from the bottom of the gel.
8. Expose the gel to a Kodak Scientific Imaging Film X-OMAT AR at room temperature for 30 s, then develop it.
9. Cut out the gel slice with radioactivity, place it in a 1.5-mL microfuge tube, and crash the gel with a plastic rod.
10. Add 1 mL of TE containing 0.1% SDS. Rotate the tube in a rotator at room temperature for 1 h and then centrifuge it at 13,000g_{av} for 1 min. Transfer the supernatant to a 15-mL Corex tube. Repeat this extraction from the gel slice twice and combine the extracts.
11. Add 300 µL of 3 M sodium acetate, pH 5.2, 30 µL of 20 mg/mL glycogen, and 6 mL ethanol. Place the tube at –20°C for 30 min, centrifuge at 13,000g_{av} at 4°C for 30 min, and remove the supernatant. Rinse the precipitate with 70% ethanol, then dry the pellet for 20 min at room temperature.
12. Dissolve the precipitate in 100 µL of 10 mM of Tris-HCl, pH 8.0. Spot 2 µL on a DE81 filter, then dry it at room temperature for 1 h. Measure the radioactivity of the filter and calculate the concentration of the purified primer in the final solution, using the formula: concentration of primer (pmol/µL) = total cpm in purified primer ÷ 100X specific activity

3.3.2. Preparation of Template

1. Amplify by PCR the *Init-Tra-Rep* fragment from purified mtDNAs, wild-type plasmid DNA, and T414G mutation-carrying plasmid DNA (as controls) as described in **Subheading 3.2.4., step 1**.

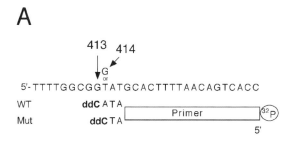

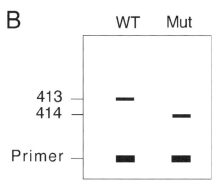

Fig. 4. Allele-specific termination of primer extension for the detection and quantification of the T414G transversion (*1*). Structures of differently extended primers (**A**) and their separation on a sequencing gel (**B**).

2. Purify the PCR products as described in **Subheading 3.2.4., step 2** to remove free deoxynucleotides.
3. Determine the concentration of purified PCR products as described in **Subheading 3.2.2., step 2**.

3.3.3. Primer Extension Reaction

1. Mix the 5′-^{32}P-labeled primer and the template at 1:1 molar ratio (0.1 pmol each) in 14 µL solution containing 0.67X Sequenase reaction buffer, 6.7 m*M* DTT, 100 µ*M dATP,* 100 µ*M* dGTP, 100 µ*M* dTTP and 100 µ*M* ddCTP. Prepare the following reactions as controls of the reaction:
   a. Primer with no template (without Sequenase).
   b. Primer with no template (with Sequenase).
   c. Primer with wild-type template (with Sequenase).
   d. Primer with T414G mutation-carrying template (with Sequenase).
   e. Primer with 1:1 mixture of wild-type and T414G mutant-carrying templates (with Sequenase).

2. Overlay 15 µL of mineral oil, heat the samples at 100°C for 5 min, slowly cool them down to 65°C (at a rate of –1°C/min), keep them at 65°C for 10 min, then chill them to 0°C. Place the samples in ice.
3. Add 1 µL of 1:8 diluted Sequenase version 2.0 (USB) to the annealed primer–template mixtures, except control (a), then incubate them at 45°C for 5 min.
4. Terminate the reactions by heating the samples at 100°C for 5 min. Add 5 µL of the sequencing gel loading buffer.

3.3.4. Separation of the Extended Products on a Sequencing Gel

1. Load 10 µL of the samples onto a sequencing gel as described in **Subheading 3.3.1., step 6**.
2. Terminate electrophoresis as described in **Subheading 3.3.1., step 7**.
3. Dry the gel on Model 583 Gel Dryer Filter Paper.
4. Expose the gel to a Molecular Dynamics Storage Phosphor Screen for 12 h.

3.3.5. Quantification of Band Intensities

1. Develop the image by the Molecular Dynamics Phosphorimager 445 SI.
2. Quantify the intensity of the bands by using the ImageQuaNT program.

4. Notes

1. Conditions are based on **ref. *11***. When starting with frozen tissue samples, avoid fractionation of mitochondria, and use, instead, standard proteinase K–sodium dodecyl sulfate treatment, followed by phenol–chloroform extraction of total cell nucleic acids. Then, continue with procedure described in **Subheading 3.1.1.3.**
2. Designing of PCR primers for the mtDNA segments to be amplified is carried out as follows. First, from the mtDNA region of interest select segments (shorter than 500 base pairs) containing a single uniform melting domain after attachment of a 60-nucleotide GC clamp at the 5′- or 3′-end, using the computer program MacMelt1.0 or Melt94 *(8)*. Then, define each segment by the PCR primers. Synthesize one oligodeoxynucleotide of each primer pair as carrying two extra A residues and a terminal psoralen C6 phosphoramidite molecule at its 5′-end *(4)*.
3. The $MgCl_2$ concentration in the reaction mixture must be optimized for each fragment to achieve high-fidelity PCR. We use the minimum concentration of $MgCl_2$ that allows an adequate yield (approximately 1 µg of PCR product/µL of highly purified mtDNA in a 50 µL reaction mixture [*see* **Subheading 3.2.1., step 2**]).
4. For the first DGGE analysis, cloning, and primer extension reaction, use a high-fidelity PCR system to reduce the rate of introduction of artificial mutations during PCR amplification. A low-fidelity PCR system can be used for the second DGGE, whose purpose is not to detect small levels of heteroplasmy.
5. Utilization of a mineral-oil-free PCR machine will save time and labor.
6. Alternative methods can be applied here.

7. Determine the optimal running time at a constant voltage (150 V) at 60°C by time-interval loading analysis. End electrophoresis after the denatured uncrosslinked PCR product has passed through the denatured crosslinked homoduplex.
8. For taking pictures of multiple bands on a DGGE gel, we prefer using the Polapan 55 PN black and white instant sheet films to using a digital camera.
9. This step may not be necessary. We do it, because the same purified product can be used later as a template for the primer extension reaction. Alternative methods to remove free deoxynucleotides from the PCR sample can be used.
10. Alternative methods can be used.
11. If the plasmid #1 DNA turned out not to carry a dominant mtDNA genotype, choose another reference among the plasmid DNAs that showed a very frequent pattern of heteroduplex band(s) in the second DGGE. Prepare five 50-μL reaction mixtures using this plasmid DNA, then carry out the second DGGE again.
12. The rationale underlying the design of this primer is described in **ref. *1***.

Acknowledgment

This work was supported by National Institute on Aging (National Institutes of Health) grant AG12117 (to G.A.).

References

1. Michikawa, Y., Mazzucchelli, F., Bresolin, N., et al. (1999) Aging-dependent large accumulation of point mutations in the human mtDNA control region for replication. *Science* **286,** 774–779.
2. Wang, Y., Michikawa, Y., Mallidis, C., et al. (2001) Muscle-specific mutations accumulate with aging in critical human mtDNA control sites for replication. *Proc. Natl. Acad. Sci. USA* **98,** 4022–4027.
3. Myers, R. M., Maniatis, T., Lerman, L. S., et al. (1987) Detection and localization of single base changes by denaturing gradient gel electrophoresis. *Methods Enzymol.* **155,** 501–527.
4. Michikawa, Y., Hofhaus, G., and Lerman, L. S. (1997) Comprehensive, rapid and sensitive detection of sequence variants of human mitochondrial tRNA genes. *Nucleic Acids Res.* **25,** 2455–2463,
5. Hirano, M., Shtilbans, A., Mayeux, R., et al. (1997) Apparent mtDNA heteroplasmy in Alzheimer's disease patients and in normals due to PCR amplification of nucleus-embedded mtDNA pseudogenes. *Proc. Natl. Acad. Sci. USA* **94,** 14,894–14,899.
6. Wallace, D. C., Stugard, C., Murdock, D., et al. (1997) Ancient mtDNA sequences in the human nuclear genome: a potential source of errors in identifying pathogenic mutations. *Proc. Natl. Acad. Sci. USA* **94,** 14,900–14,905.
7. Hofhaus, G. and Attardi, G. (1995) Efficient selection and characterization of mutations of a human cell line which are defective in mitochondrial DNA-encoded subunits of respiratory NADH dehydrogenase. *Mol. Cell. Biol.* **15,** 964–974.

8. Lerman, L. S., Silverstein, K., Fripp, B., et al. Melt94. http://wcb.mit.edu/osp/www/melt.html.
9. Anderson, S., Bankier, A. T., Barrell, B. G., et al. (1981) Sequence and organization of the human mitochondrial genome. *Nature* **290,** 457–465.
10. Sambrook, J., Fritsch, E. F., and Maniatis, T. (eds.) (1989) Measurement of radioactivity in nucleic acids (E.18 and E.19), in *Molecular Cloning: A Laboratory Manual, 2nd ed.*, Cold Spring Harbor Laboratory Press, Cold Spring Harbor, NY.
11. Fernandez-Silva, P., Micol, V., and Attardi, G. (1996) Mitochondrial DNA transcription initiation and termination using mitochondrial lysates from cultured human cells. *Methods Enzymol.* **264,** 129–138.

4

Scanning Low-Frequency Point Mutants in the Mitochondrial Genome Using Constant Denaturant Capillary Electrophoresis

Weiming Zheng, Luisa A. Marcelino, and William G. Thilly

1. Introduction

The ability to measure rare mutational events in mitochondrial genomes from human blood and tissues without resorting to phenotypic selection is invaluable. It is essential for the study of the cause(s) of mitochondrial mutation and for the identification of mutations associated with diseases. We have developed a highly sensitive technique termed *constant denaturant capillary electrophoresis* (CDCE) *(1)*, permitting measurement of mitochondrial point mutations at fractions as low as 10^6 in human cells and tissues *(2)*. CDCE is based on the earlier development of denaturing gradient gel electrophoresis (DGGE) by Fischer and Lerman *(3)* and constant denaturing gradient gel electrophoresis (CDGE) by Hovig *(4)*. When a DNA consists of two contiguous "isomelting domains" with different melting temperatures, under partially denaturing condition it alternates in rapid equilibrium between the double helix state and the partially melted state. The later form has an Y-shaped configuration and, thus, lower mobility through the gel matrix than the double helix. The distribution of the "fast" and "slow" states governs the velocity under electrophoretic conditions. Base changes in the low-melting domain result in changes in the melting behavior and, hence, the mobility of the sequence under the denaturing condition of electrophoresis, allowing variants to be separated from each other. Unlike DGGE and CDGE, CDCE employs replaceable linear (uncrosslinked) polyacrylamide and capillary electrophoresis *(1)*. A zone of elevated temperature along the capillary maintains the constant denaturing

From: *Methods in Molecular Biology, vol. 197: Mitochondrial DNA: Methods and Protocols*
Edited by: W. C. Copeland © Humana Press Inc., Totowa, NJ

condition necessary for DNA separation. The result is fast separation, improved resolution, and higher sensitivity *(1)*.

The combination of CDCE and high-fidelity polymerase chain reaction (hifi PCR) provides a powerful analytical approach capable of detecting almost all of the point mutants in a given sequence (approx 100–150 bp [base pairs]) at fractions as low as 10^{-6} *(2)*. With this methodology, we were able to determine the mitochondrial mutational spectra in human cells and tissues *(2,5,6)*, as well as spectra created by the human polymerase gamma from ex vitro DNA amplification (W. Zheng et al., unpublished data).

The specific protocol for the CDCE/hifi PCR approach depends on the desired degree of sensitivity and the error rate of the hifi PCR on the target sequence (*see* **Note 1**). To detect mutant fractions on the order of $5×10^{-4}$ to 10^{-6}, the protocol consists of the following steps (*see* **Fig. 1**): (1) preparation of the target sequence from biological samples; (2) introduction of internal standard; (3) pre-PCR mutant enrichment; (4) high-fidelity PCR; (5) post-PCR mutant enrichment; (6) display of mutational spectra and identification of individual mutants. To detect mutant fractions higher than $5×10^{-4}$, hifi PCR (**step 4**) can be directly applied after the introduction of the internal standard (**step 2**).

Suitable target sequences should have contiguous, high- (50–150 bp) and low- (100–150 bp) melting domains (*see* **Fig. 2**). The low-melting domain should have an essentially constant melting temperature *(7)*. There is no shortage of DNA sequences of the desired isomelting profile in the mitochondrial genome. Here, we use a 205-bp fragment of the human mitochondrial genome (10,011–10,215 bp) (referred to as CW7/J3) for demonstration purposes. Shown in **Fig. 2** is the CW7/J3 melting map as calculated by the MacMelt/WinMelt software (Bio-Rad, Hercules, CA) using the algorithm of Lerman and Silverstein *(8)* based on the studies by Poland *(9)*. If the sequence of interest does not have a natural high-melting domain, an artificially created clamp can be attached through ligation *(10)*. The specific protocol for the ligation step can be found in **ref. *10***.

2. Materials

2.1. Preparation of the Target Sequence

1. Human tissues or blood. Tissues are stored in 20% dimethyl sulfoxide (DMSO) (–70°C or –20°C). Blood is stored in 1 mL of ACD (anticoagulant) solution B per 6 mL of blood (–70°C or –20°C). ACD solution B: 0.48% citric acid, 1.32% sodium citrate, 1.47% glucose. In general, $4 × 10^5$ cells are needed to detect mutants at fractions of 10^{-6} with 95% statistical confidence if there are 1000 copies of mitochondrial DNA per cell *(2)*.

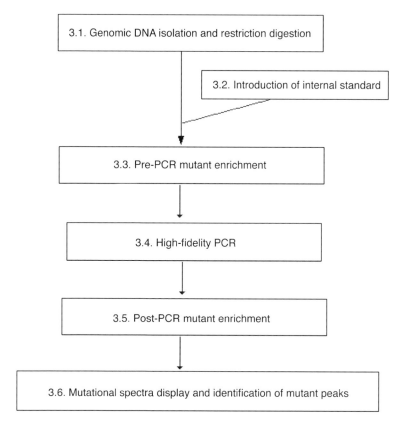

Fig. 1. Flow diagram of the CDCD/hifi PCR method for scanning rare point mutants.

2. Phosphate-buffered saline (PBS) (only for blood): dissolve 8 g of NaCl, 0.2 g of KCl, 1.44 g of Na_2HPO_4, and 0.24 g of KH_2PO_4 in 800 mL of double distilled (dd) H_2O. Adjust the pH to 7.4 with HCl. Add deionized H_2O to 1 L.
3. 1X TE buffer: 50 mM Tris-HCl, pH 8.0, 10 mM EDTA; 0.1X TE buffer.
4. Proteinase K: 20 mg/mL (Boehringher Mannheim, Indianapolis, IN).
5. Sodium dodecyl sulfate (SDS): 10%.
6. RnaseA: 10 mg/mL (Boehringher Mannaheim).
7. 5 M NaCl.
8. Ethanol: 100% and 70%. Both are chilled to −20°C before using.
9. Restriction endonucleases and 10X digestion buffers. For our CW7/J3 sequence, DNA was digested with *Rsa*I and *Dde*I (New England Biolabs, Beverly, MA), which recognize base pairs 10,009–10,012 and 10,227–10231, respectively.
10. Mortar, pestle, and surgical scalpels (only for tissues).
11. Liquid nitrogen (only for tissues).

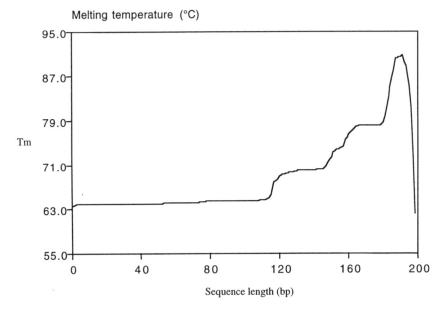

Fig. 2. Typical melting profile of a DNA sequence suitable for CDCE separation. A fragment (CW7/J3) of the human mitochondrial genome (10,011–10,210 bp). Note a constant melting temperature from 0 to 115 bp, which is essential for separating mutants within a given sequence with the CDCE technology.

2.2. Constant Denaturing Capillary Electrophoresis

1. A CDCE instrument with laser-induced fluorescence detection (*1*). In general, it consists of a CE apparatus, a detector, and a constant temperature circulator, all mounted on an optical board. A full description of the instrumental setup and operating steps was detailed in **ref. 7** and is beyond the scope of this chapter. Interested readers can contact the corresponding author for building instructions.
2. Fused-silica capillaries of 75 μm inner diameter (ID), 363 μm outer diameter (OD), and 542 μm ID, 665 μm OD (Polymicro Technologies, Phoenix, AZ).
3. Dialysis filter of VS type and 0.025-μm size (Millipore, Marlborough, MA).
4. 0.1X TBEB buffer: 0.1X TBE, 0.03 mg/mL bovine serum albumin [BSA]; 0.4X TBEB buffer and 0.8X TBEB buffer.
5. Teflon tubing of 1/16 in. OD and 0.010 in. ID (Bodman, Aston, PA) and 1.5 mm OD and 0.5 mm ID (Rainin, Woburn, MA).

2.3. High-fidelity PCR

1. Primer J (20 μ*M*): 5′-fluorescein-labeled high-melting-domain primer (20-mer). For the CW7/J3 fragment, this primer is 5′-GCG GGC GCA GGG AAA GAG GT- 3′, complementary to base pairs 10,196–10,215.

2. Primer C (20 μ*M*): low melting domain primer (20-mer). For the CW7/J3 fragment, this is 5′-ACC GTT AAC TTC CAA TTA AC- 3′, complementary to 10,011–10,031 bp.

3. Primer A (20 μ*M*): 40-mer to 50-mer low-melting-domain primer that forms a mismatch with the wild-type sequence. The mismatch is such that with primer J, it creates an artificial mutant (#1) with a lower melting temperature (T_m) than the wild type. We used 5′-ACC GTT AAC TTC CAA TTA ACT AAGT TTT GA̲T̲ AAC ATT CAA A- 3′, where the T introduces a GC ↔ AT mutation.

4. Primer B (20 μ*M*): Like primer A, this 40-mer to 50-mer low-melting-region primer forms a mismatch with the wild-type sequence and creates an artificial mutant (#2) with a higher T_m than the wild-type DNA. An AT ↔ GC or a TA ↔ GC mutation would increase the thermal stability of the DNA. The high-T_m artificial mutant is not necessary unless in the pre-PCR (polymerase chain reaction) enrichment-step (**Subheading 3.3.**) mutants are collected in homoduplex forms. See **Note 2** for the homoduplex versus the heteroduplex approach.

5. Native *Pyrococcus furiosus (pfu)* DNA polymerase (2.5 U/μL) and 10X native *pfu* buffer (Stratagene, La Jolla, CA).

6. dNTP mixture: 25 m*M* each of dATP, dTTP, dGTP, and dCTP.

7. 2X PCR master mix: 0.4 μ*M* each primer, 0.2 m*M* each dNTP, 2X PCR reaction buffer (native *pfu*), 100 μg/mL BSA. For wild-type master mix, use primer J and C. For artificial mutant master mixes #1 and #2, use primer pair J/A and J/B, respectively. Store at –20°C.

8. Air thermocycler (Idaho Technology, Idaho Falls, ID).

9. 10-μL glass capillary tubes (Idaho Technology, Idaho Falls, ID).

10. Glass cutter.

11. Gas or alcohol burner.

3. Methods

To prevent contamination with enriched mutants, all of the procedures before high-fidelity PCR (**Subheading 3.4.**) must be carried out in a separate clean room equipped with high-throughput HEPA air filters.

3.1. Preparation of the Target Sequence

3.3.1. Genomic DNA Isolation

1. For tissues: Cut the tissue sample into small pieces with a scalpel. Deep-freeze them in liquid nitrogen and grind them into fine powders with a mortar and pestle. Suspend the powdered tissue sample in 1X TE buffer at 50 mg tissue/mL *(10)*.

2. For blood: Mix 1 volume of PBS with thawed blood. Centrifuge the sample at 3500*g* for 15 min. Discard the supernatant and resuspend the pellet in 1X TE buffer at 3 mL blood (10^7 cells)/mL.

3. Add proteinase K and SDS to the sample to a final concentration of 1 mg/mL and 0.5%, respectively. Mix thoroughly and incubate in a 50°C water-bath shaker (100–200 rpm) for 3 h.

4. Add RnaseA to a final concentration of 20 μg/mL. Mix thoroughly and incubate in a 50°C water-bath shaker (100–200 rpm) for 1 h.
5. Centrifuge at 35,000*g* at 4°C for 15 min.
6. Using a Pasteur pipet, transfer the central portion of the supernatant to a new tube. Avoid the bottom pellet and top layers of white matter and fat. Repeat **step 5** and **step 6** two or three times to collect as much supernatant as possible.
7. Add 5 *M* NaCl to the pooled supernatant to a final concentration of 250 m*M* and then add 2 volumes of 100% ethanol. Mix by inverting the tube several times. A DNA spool should form.
8. After the spool compacts and floats to the surface, transfer the DNA spool to a 1.5-mL microcentrifuge tube and wash the DNA twice with 1 mL of cold 70% ethanol.
9. Air-dry DNA for 30 min. Add 0.1X TE buffer to a concentration of 2–4 mg DNA/mL. Dissolve the DNA by vortexing it vigorously.
10. Dilute 5 μL of the DNA sample in 0.5 mL of ddH$_2$O and measure A$_{260\,nm}$ and A$_{260\,nm}$ on the ultraviolet (UV) spectrophotometer to estimate the yield and purity of the sample. The typical yield is over 90% and the A$_{260\,nm}$/A$_{260\,nm}$ ratio is 1.4–1.6.

3.1.2. Restriction Digestion

Digest DNA overnight at 37°C with suitable restriction endonucleases at a DNA concentration of 2–3 mg/mL and enzyme/DNA ratio of 2–4 U/μL.

3.2. Introduction of Internal Standard

In this step, artificial mutants (**Subheading 2.3.**) are created (*see* **Note 2**). The copy number of the target sequence in the restriction-digested sample is then measured by mixing a small aliquot of the sample with a known amount of artificial mutants. The mixture is subjected to PCR followed by CDCE. The ratio of the peak area of the wild type to that of the artificial mutant gives the copy number of the target sequence. Samples are then doped with a known fraction of the artificial mutant to serve as internal standard, which allows us to determine the absolute mutant fractions of all the mutants in the sample.

3.2.1. Preparation of Internal Standard

1. In a 0.65-mL-tube mix: 1 μL of restriction-digested DNA, 5 μL of 2X PCR mutant master mix (#1 or #2), 3.4 μL of ddH$_2$O and 0.6 μL of native *pfu* polymerase.
2. Perform PCR (**Subheading 3.4.**). Use enough cycles to exhaust all of the primers. The final concentration of the artificial mutant should be 10^{11} copies/μL, which is equal to the initial primer concentration (**Subheading 2.3., item 7**).
3. Make serial dilutions of the artificial mutants: 10^{10}, 10^9, …,10^4, 10^3 copies/μL.

3.2.2. Measuring the Sample Copy Number

1. In a 0.65-mL-tube mix: 1 μL of diluted (X-fold) restriction digested sample DNA, 1 μL of diluted artificial mutant (Y copies/μL), 2X PCR wild-type master mix, 2.4 μL of ddH2O, and 0.6 μL of native *pfu* polymerase. Perform PCR (**Subheading 3.4.**).
2. Run the PCR product on CDCE to separate the wild type from the artificial mutant. The precise copy number of the target sequence in the sample is calculated using $XY (A_w + A_h)/(A_m + A_h)$, where A_w and A_m are the peak areas of the wild type and artificial mutant and A_h is the average peak area of the two heteroduplexes (*see* **Note 3**).

3.2.3. Introduction of Internal Standard(s)

Add artificial mutant(s) (*see* **Note 2**) to the restriction-digested sample at the desired mutant fraction (e.g., 10^{-4} or 10^{-5}).

3.3. Pre-PCR Mutant Enrichment

During this step, the restriction-digested sample with added internal standard is subjected to CDCE on a wide-bore capillary (542 μm ID) where mutants are separated from the wild type and collected (*11*) (*see* **Note 4**). The appropriate separation temperatures are determined in advance through test runs using the fluorescein-labeled wild type and two internal standards. When homoduplexes are to be separated, mutant homoduplexes with higher and lower melting temperatures than the wild type are collected before and after the wild type, respectively. In this way, mutants are enriched against the excess wild type by about 200-fold for mutants with higher T_m and 20-fold for mutants with lower T_m (*see* **Note 2**). When mutant/wild type heteroduplexes are to be separated from wild-type homo-duplexes, all the mutant homoduplexes are first converted into heteroduplexes by boiling and reannealing the restriction-digested sample. Heteroduplexes are then collected as a single fraction enriched against the excess wild type by about 20-fold (*see* **Note 2**). The collected heteroduplex or low-T_m homoduplex fractions is then subjected to room-temperature capillary electrophoresis (RTCE) to separate mutants from other wild-type sequences (e.g., single strands or partial double strands) that might comigrate with the mutants in the CDCE separation. The RTCE step brings about another 10- to 20-fold enrichment for the low-T_m mutant homoduplexes or mutant/wild type heteroduplexes (*10*).

3.3.1. Homoduplex Approach

1. After both internal standards (#1 and #2) are added to the restriction-digested sample at known fractions, desalt the sample against 0.1X TBE buffer using a

0.025-μm dialysis filter (**Subheading 2.2., step 3**) for about 1 h on a stirring plate. Gently stir the buffer with a stirring bar during dialysis (*see* **Note 5**).

2. Reduce the volume of the dialyzed sample to about 6 μL by speed-vacuum centrifugation.

3. Cut a piece of coated 542-μm ID capillary about 20 cm long with a detection window (*see* **ref. 7**) 7 cm away from the anode end and mount it onto the CDCE setup. The detection window (approx 0.5 cm long) is made by scraping off the capillary coating with a razor blade. A portion of the capillary passes through a water jacket connected with a constant-temperature circulator. The optimal temperature and jacket length for collecting the high-T_m as well as the low-T_m mutants were predetermined such that mutants coalesce into one fraction that is well separated from the wild type.

4. Replace about 35 μL of linear polyacrylamide matrix inside the capillary *(7)*.

5. Pipet 3 μL of the dialyzed sample into one end of a short piece of Teflon tube (about 1 cm long). Insert the cathode end of the capillary into the same end and push it gently until the sample just reaches the other end of the tube. Make sure that there is no bubble at the end of the tube and between the sample and the capillary end. Immerse the cathode end of capillary with the tube in the cathode end of the buffer reservoir. Apply a current of 80 μA for 2 min to load the sample (*see* **Note 6**).

6. Remove the Teflon tube from the capillary. Put the cathode end of the capillary back into the buffer reservoir. Run CDCE at a constant current of 80 μA (*see* **Note 6**). The temperature and jacket length is set to be optimal for collecting the high-T_m mutants.

7. Stop electrophoresis right before the high-T_m mutants reach the anode end based on a predetermined timing. Remove the anode end from the buffer reservoir and rinse it twice in ddH$_2$O. Place the anode end into 8 μL of collection buffer (0.8X TBEB) in a 0.65-mL tube with a platinum wire inside. Start electrophoresis at 80 μA to collect all of the high-T_m mutants. Then, in a separate tube with collection buffer, collect the wild-type DNA that elutes after the high-T_m mutants (*see* **Note 7**).

8. Estimate the level of mutant enrichment. An aliquot of the mutant and wild-type collection is taken to determine the wild-type copy number using artificial mutants (*see* **Subheading 3.2.2.**). The degree of enrichment is estimated as the ratio of the wild-type copy number in the wild-type fraction to that in the mutant fraction. For high-T_m mutants, the enrichment from this step is at least 200-fold because the region in front of the wild type is relatively free of carryover DNA compared to that behind the wild type, which has a significant wild-type "tail" and, consequently, less enrichment (20-fold) for the low-T_m mutants).

9. Desalt the collected high-T_m mutants against 0.1X TBE, followed by speed-vacuum centrifugation (*see* **step 2** and **step 3**).

10. Repeat **steps 4–9** for collecting all of the low-T_m mutants.

11. Turn off the temperature circulator to let the water jacket cool down to room temperature.

12. Load the desalted low-T_m mutant collection (**step 5** and **step 6**) and perform RTCE. Collect all of the double-stranded sequences in 8 μL of 0.4X TBEB for 5 min. Estimate enrichment (**step 9**). This should give another 10- to 20-fold enrichment, resulting in a total of 200-fold enrichment for the low-T_m mutants after CDCE and RTCE separation.

3.3.2. Heteroduplex Approach

1. After adding the internal standard (#1) to the restriction-digested sample, desalt the sample (**Subheading 3.3.1., step 1**).
2. For boiling and annealing, heat the sample to 100°C for 10–20 s, then adjust to 200 mM NaCl, 10 mM Tris-HCl, pH 8.0, 2 mM EDTA, and incubate at 60°C for 0.5 h.
3. Go through **steps 1–13** in **Subheading 3.3.1.** and collect after the wild type.

3.4. High-Fidelity PCR

After enrichment by wide-bore capillary, the samples are subjected to PCR. Sufficient cycles should be used to exhaust all of the primers so that all the mutants are converted into heteroduplexes with the excess wild type through mass action. All mutant/wild type heteroduplexes have lower melting temperatures than the wild-type homoduplex and, hence, can be collected as a single fraction.

1. Mix the enriched mutant collection (**Subheading 3.2.**) with 5 μL of 2X PCR and add ddH$_2$O and 0.6 μL of native *pfu* polymerase.
2. Transfer the reaction mixture to a 10-μL glass capillary through capillary action and seal both ends in the flame.
3. Perform PCR in an air thermocycler. For the CW7/J3 fragment, the PCR consists of an initial denaturing step at 94°C for 30 s, followed by 35–40 cycles of denaturation at 94°C, annealing at 50–57°C and extension at 72°C, each for about 10-s intervals. After the desired number of cycles, samples are incubated at 72°C for 2 min and then at 45°C for 15 min (*see* **Note 8**).

3.5. Post-PCR Mutant Enrichment

During this step, amplified mutants are further enriched by about 200-fold through two rounds of CDCE/hifi PCR. In each round, the sample is loaded onto a 75-μm-ID capillary, where heteroduplexes are separated from the wild-type homoduplex, collected, and amplified by PCR. Prepare a 20-cm-long capillary (75 μm ID) and set up the temperature and water jacket as described in **Subheading 3.3.1., step 3**.

1. Replace the linear polyacrylamide matrix (about 2 μL) inside the capillary.
2. Dilute the PCR product 10-fold with ddH$_2$O to a total volume of 4–5 μL in a 0.65-mL tube. Load the sample from the cathode end of the capillary from a

0.65-mL tube at 2 µA for 30 s. A platinum wire is used for electroinjection. About 10^8–10^9 copies should be loaded. Remove the sample tube and put the cathode end back into the buffer reservoir. Start electrophoresis at about 9 µA (*see* **Note 6**).

3. Collect all of the heteroduplexes from the anode end into 6 µL of 0.1X TBEB for 2 – 3 min. Also collect the wild-type homoduplex fraction. Estimate enrichment (**Subheading 3.3.1.**, **step 8** and **step 9**). The typical enrichment is about 15-fold.
4. Take 4.4 µL of the heteroduplex collection and perform high-fidelity PCR (**Subheading 3.4.**).
5. Repeat the CDCE/hifi PCR (**Subheading 3.5.**, **steps 2–5**) to achieve another 15-fold enrichment, bringing a total of 200-fold enrichment.

3.6. Display of Mutational Spectra

Once the mutants are enriched, the final step is to separate, quantify, isolate, and sequence the individual mutants. First, mutant/wild type heteroduplexes are converted into homoduplexes with a few cycles of PCR. Homoduplexes are then separated and isolated under the appropriate temperature via CDCE using a long water jacket (19 cm), high-resolution gel, and low electric field strength (100 V/cm) (*see* **Notes 9** and **10**). Figure 3 shows the mutational spectra of colon and muscle tumor. Both the high-T_m and low-T_m internal standards (#2 and #1) were added to the sample at fractions of 10^{-4} (**Subheading 3.2.**). The homoduplex approach (**Subheading 3.3.1.**) was used for the initial mutant enrichment *(2)*. The precise mutant fraction of each mutant is measured by comparing its peak area with that of the internal standard. The identity of each isolated mutant is determined by sequencing after physically collecting each mutant homoduplex CDCE peak.

1. Take 0.5 µL of the PCR product (from **Subheading 3.5.**, **step 5**) and perform PCR (**Subheading 3.4.**) for three to five cycles. Check the product copy number to ensure that the amount of unused primers exceeds that of the product. The PCR cycling can be stopped when the ratio of the peak area of the product to that of the primer reaches 1/10.
2. Set up CDCE with a 33-cm-long capillary (75 µm ID) and 19-cm water jacket (**Subheading 3.3.1.**, **step 5**).
3. Replace the linear polyacrylamide gel inside the capillary (*see* **Note 9**).
4. Load the 10-fold diluted PCR product (from **step 1**) and perform electrophoresis at 100 V/cm.
5. Optimize the running temperature to separate all the mutants from one another.
6. Determine the initial mutant fraction of each mutant by comparing its peak area to that of the internal standards.
7. Collect individual mutant into 6 µL of 0.1X TBEB at the anode end of the capillary.

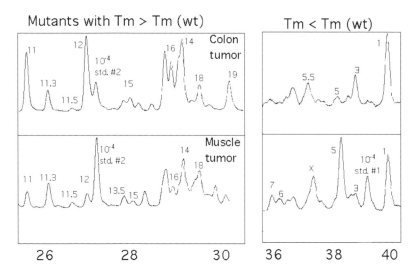

Fig. 3. Mutational spectra of colon and muscle tumor samples *(5)*. The *x* axis reflects the time since the beginning of the run at which the peak reaches the detector and the *y* axis marks the relative intensity of fluorescence. (**Left**) Mutants with melting temperatures higher than the wild type. (**Right**) Mutants with melting temperatures lower than the wild type. Two internal standards (#1 and #2) were added (**Subheading 3.2.**), both at fractions of 10^{-4}.

8. Amplify the collected mutant and check the PCR product via CDCE.
9. Repeat **step 6** and **step 7** until the collected mutant is sufficiently pure for sequencing.
10. Sequence the collected mutants.

4. Notes

1. When the efficiency of amplification is high (>85%), the homoduplex mutant fraction (MF) can be approximated by MF = *Ldr*, where *L* is the length of the target sequence in base pairs, *d* is the number of doublings, and *r* is the error rate per base pair per doubling. A more exact expression taking into account the PCR efficiency is MF = 2*nerL*/(1 + *e*), where *e* is PCR efficiency and *n* is the number of cycles *(12)*.

2. Whether or not the high-T_m internal standard (#2) is necessary depends on the how mutants are collected during the pre-PCR enrichment step (**Subheading 3.3.**). In the homoduplex approach, high-T_m mutants and low-T_m mutants are collected separately in the form of homoduplexes and both internal standards are needed. In our experiment, we used both high-T_m (#2) and low-T_m (#1) internal standards for the CW7/J3 fragment (**Fig. 3**). A variation of the procedure is the conversion of all mutant homoduplexes into heteroduplexes with the wild-type sequence by boiling and reannealing the sample. By mass action, all the mutants,

both the high T_m and low T_m, form heteroduplexes with the excess amount of wild type. Because mutant/wild type heteroduplexes have a lower melting temperature than the wild-type homoduplex, they can be easily separated from the wild type and collected as a single fraction under the appropriate temperature. In this case, only the low-T_m standard (#1) is needed.

3. It is important to make sure that the internal standard has the same amplification efficiency as the wild type. One way to test this is to mix the internal standard and wild type at different proportions and compare the ratios at different PCR cycles. A constant ratio of internal-standard PCR product to wild-type PCR product should be observed throughout the entire conditions to conclude that both templates have the same PCR efficiency.

4. In general, 4×10^5 cells are needed to detect mutants at fraction of 10^{-6} (**Subheading 2.1.**). This is equivalent to about 2.5 μg of genomic DNA for pre-PCR enrichment. At most, 80 ng of DNA can be loaded onto a 75-μm ID capillary. The maximum amount of DNA that can be loaded onto a 250-μm-ID capillary is 2 μg, whereas 2.5 μg can be loaded onto a 320-μm-ID capillary and 10 μg for a 542-μm-ID capillary *(11)*.

5. Because anions compete with DNA for the injection current, a higher salt concentration results in a lower amount of DNA being electroinjected.

6. The electric field strength of CDCE varies with the inner diameter of the capillary and the optimal parameters were empirically determined *(1,11)*. The field strength for the 75-μm-ID capillary is 250 V/cm, for the 250 μm ID capillary is 100 V/cm, for the 320 μm is 90 V/cm and for the 542μm ID capillary is 60 V/cm. For wide-bore capillary peak resolution is limited by significant joule heating, so lower field strength is required.

7. Electroelution of CDCE- and RTCE-separated mutants can be done either blindly or with fluorescein-labeled markers added into the sample. The markers serve as a reference for timing the peak collection. One marker is the wild-type target sequence with an altered high-melting domain. The second marker can be the artificial mutant sequence with an altered high-melting domain or a completely different sequence *(4)*. The principle is that the melting behaviors of the first and second markers were similar to that of the wild-type and artificial mutants, respectively, but they could not be amplified with the primers used in subsequent PCR (**Subheading 2.3.**).

8. We have observed PCR byproducts created by the *pfu* enzyme. Some of these byproducts are incomplete extension products that have a lower T_m than the wild type and, hence, interfere with the enrichment of mutants. We found that using higher *pfu* concentration and 15-min post-PCR incubation at 45°C reduces the amount of byproducts generated *(2)*.

9. The protocol for making the standard 5% linear polyancrylamide matrix is described in **ref. 7**. To improve the peak resolution for final display of the spectra and isolation of individual mutants, a variation of the protocol can be used to make the high-resolution gel with 6% of the acrylamide monomer, twofold to threefold less amount of ammonium persulfate and TEMED and 30 mM

sodium borate, pH 8.36. The use of higher monomer concentration and lower amount of initiators increases the molecular weight of the polymer matrix. Higher salt concentration increases the T_m of the DNA and, hence, the kinetic rate of the melting/reannealing processes.

Acknowledgments

We wish to thank Dr. K. Khrapko and Dr. H. Coller for sharing **Fig. 3** with us and Dr. X. Li and Ms. A. Kim for providing us with the reference materials. This work was supported by grants from the National Institute for Environmental Health Sciences (5 PO1 ES07168).

References

1. Khrapko, K., Hanekamp, J. S., Thilly, W. G., Belenkii, A., Foret, F., and Karger, B. L. (1994) Constant denaturant capillary electrophoresis (CDCE): a high resolution approach to mutational analysis. *Nucleic Acids Res.* **22,** 364–369.
2. Khrapko, K., Coller, H., Andre, P., Li, X.-C., Foret, F., Belenky, A., et al. (1997) Mutational spectrometry without phenotypic selection: human mitochondrial DNA. *Nucleic Acids Res.* **25,** 685–693.
3. Fischer, S. G. and Lerman, L. S. (1983) DNA fragments differing by single base-pair substitutions are separated in denaturing gradient gels: corresponding with melting theory. *Proc. Natl. Acad. Sci. USA* **80,** 1579–1583.
4. Hovig, E., Smith-Sørensen, B., Brøgger, A., and Børensen, A. L. (1991) Constant denaturant gel electrophoresis, a modification of denaturing gradient gel electrophoresis, in mutation detection. *Mutat. Res.* **262(1),** 63–71.
5. Khrapko, K., Coller, H. A., Andre, P. C., Li, X.-C., Hanekamp, J. S., and Thilly, W. G. (1997) Mitochondrial mutational spectra in human cells and tissues. *Proc. Natl. Acad. Sci. USA* **94,** 13,798–13,803.
6. Coller, H., Khrapko, K., Torres, A., Frampton, M. W., Utell, M. J., and Thilly, W. G. (1998) Mutational spectra of a 100-base pair mitochondrial DNA target sequence in bronchial epithelial cells: a comparison of smoking and nonsmoking twins. *Cancer Res.* **58,** 1268–1277.
7. Khrapko, K., Coller, H., Li, X.-C., Andre, P. C., and Thilly, W. G. (1999) High resolution analysis of point mutations by constant denaturant capillary electrophoresis (CDCE), in *Capillary Electrophoresis of Nucleic Acids* (Mitchelson, K. R. and Cheng, J., eds.), Humana, Totowa, NJ, Vol. 1, pp. 57–72.
8. Lerman, L. S. and Silverstein, K. (1987) Computational simulation of DNA melting and its application to denaturing gradient gel electrophoresis. *Methods Enzymol.* **155,** 482–501.
9. Poland, D. (1974) Recursion relation generation of probability profiles for specific-sequence macromolecules with long-range correlations. *Biopolymers* **13,** 1859–1871.
10. Kim, A. S., Li, X.-C., and Thilly, W. G. (1999) Application of constant denaturant capillary electrophoresis and complementary procedures: measurement of point

   mutational spectra, in *Capillary Electrophoresis of Nucleic Acids* (Mitchelson, K. R. and Cheng, J., eds.), Humana, Totowa, NJ, Vol. 1, pp. 57–72.

11. Li, X.-C. and Thilly, W. G. (1996) Use of wide-bore capillaries in constant denaturant capillary electrophoresis. *Electrophoresis* **17,** 1884–1889.

12. Sun, F. (1995) The polymerase chain reaction and branching processes. *J. Comput. Biol.* **2(1),** 63–68.

5

Identification of mtDNA Mutations in Human Cancer

Shuji Nomoto, Montserrant Sanchez-Cespedes, and David Sidransky

1. Introduction

The human mitochondrial genome is composed of a 16-kb circular, double-stranded DNA that encodes 13 polypeptides of the mitochondrial respiratory chain, 22 transfer RNAs, and 2 ribosomal RNAs required for protein synthesis *(1)*. It is generally accepted that mitochondrial DNA (mtDNA) mutations are generated during oxidative phosphorylation through pathways involving reactive oxygen species (ROS) *(2)*. In a recent breakthrough, several mtDNA mutations were found specifically in human colorectal cancer *(3)*. We and others also detected frequent mutations in bladder, head and neck, lung tumors, pancreas, and hepatocellular carcinoma *(4–6)*. These mutations were scattered throughout the coding and noncoding regions of the mtDNA in the various tumors studied. Many, but not all, coding mutations were confined to respiratory complex I *(3,7)*. Of particular interest, the noncoding displacement-loop (D-loop) region was found to be a mutational hot spot in our studies *(4)*.

There are limited proven approaches to identify mtDNA mutations in cancer cells. Although a long-range polymerase chain reaction (PCR) approach may one day allow amplification of the entire mtDNA in one fragment, it is still necessary to amplify the mtDNA fragments with overlapping primers *(3)*. Moreover, although some reports have detected mtDNA mutations in cancer tissue by analyzing short fragments (200–300 bp) using PCR-SSCP (single-stranded conformation polymorphism) *(8)*, the amplified fragments might contain nuclear-encoded pseudogenes *(9)*. It is therefore critical to carefully consider the primers for amplifying the appropriate mtDNA fragments. In this section, we describe the methods for PCR amplification of the mtDNA fragments as well as the primer sets *(3,4)*.

From: *Methods in Molecular Biology, vol. 197: Mitochondrial DNA: Methods and Protocols*
Edited by: W. C. Copeland © Humana Press Inc., Totowa, NJ

After the amplification, we used a direct sequencing (cycle sequencing) method to identify the mutations in order to avoid to reading potential PCR errors as mtDNA mutations by sequencing single molecules after cloning. Detected mutations must be confirmed by repeated amplification and sequence analysis. In **Subheading 2.**, we describe the methods of cycle sequencing and the referral mitomap web address. It is still necessary to confirm mitochondrial mutations by comparing the sequence to some normal tissue in the same patient **(Fig. 1)**. All recent mitochondrial sequencing studies have identified many new polymorphisms so that comparison to existing databases is not sufficient to call a mutation.

As mitochondrial DNA is present at extremely high levels (10^3–10^4 copies per cell) compared to nuclear DNA and most mutations are homoplasmic *(3,4)*, we were able to readily detect mtDNA mutations in paired bodily fluids *(4)*. In some cases (e.g., bronchioalveolar lavage from lung cancer patients), the low level of neoplastic cells among normal cells prohibited detection of mutant mtDNA using sequence analysis alone. For these cases, a more sensitive oligonucleotide mismatch ligation assay was performed using smaller segments *(10)*. It is critical to compare mutant signals in a positive control (mutated tissue DNA) and negative control (a cell line devoid of mtDNA). The methods for this assay are described in **Subheading 3.**

Recently, we found frequent deletions/insertions in a polymorphic cytosine mononucleotide repeat sequence between nucleotides 303 and 315 *(11)*, an area considered to be a replication start site of the closed circular mitochondrial genome *(12)*. A simple PCR fragment-length detection assay, including the D310 C-tract region, was developed for rapid detection of variants in this region. This method can also be used to detect mutant mtDNA in paired bodily fluids with the same precautions as those listed earlier for the mismatch ligation assay.

We thus describe four major assay protocols for identification of mtDNA mutations in cancer cells focused on how to amplify the mtDNA fragments, cycle sequencing, use of a oligonucleotide mismatch ligation assay, and a simple PCR fragment-length detection assay to detect common C-tract deletions/insertions. The functional relevance of identified mtDNA mutation in cancers remains to be determined.

2. Materials

2.1. Stock Solution

1. 1 *M* Tris-HCl, pH 7.5.
2. 0.5 *M* Ethylenediaminetetraacetic acid (EDTA), pH 8.0.
3. 0.1 *M* Dithiothreitol (DTT) in water. Store aliquots at –20°C.

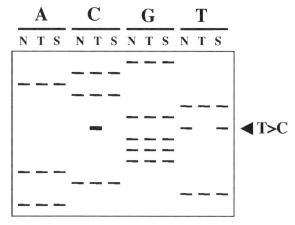

N: Normal, T: Tumor, S: Serum

Fig. 1. Cycle sequencing. The arrowhead indicates a mutation (T to C change) in mtDNA from a tumor sample.

4. 0.1 M ATP in water, adjusted to pH 7.0 with NaOH. Store aliquots at –20 °C.
5. 5X TBE buffer for electrophoresis: 535 mM Tris-borate, 10 mM EDTA.
6. 1X TAE buffer for agarose electrophoresis: 40 mM Tris-acetate, pH 8.0, 1 mM EDTA. Prepared as 50X concentrated solution: 2 M Tris base, 1 M acetic acid, and 50 mM EDTA.
7. TE (10:0.1) buffer: 10 mM Tris-HCl, pH 7.6, 0.1 mM EDTA.

2.2. Amplification of Mitochondrial DNA Fragments

1. 10X PCR buffer: 166 mM $(NH_4)_2SO_4$, 670 mM Tris-HCl, pH 8.8, 67 mM $MgCl_2$, 100 mM β-mercaptoethanol.
2. 25 mM of each dNTP.
3. PLATINUM[R] Taq DNA Polymerase (Life Technologies): This reagent provide an automatic "hot start" for *Taq* DNA polymerase.
4. Primers (*see* **Tables 1** and **2**).
5. Dimethyl sulfoxide (DMSO).
6. Gel extraction kit.

2.3. Cycle Sequencing (see Note 1)

1. 10X Cycling mix: 1 tube containing 100 sequencing units of AmpliTaq[R] DNA Polymerase, CS, in 500 mM Tris-HCl, pH 8.9, 100 mM KCl, 25 mM $MgCl_2$, 0.25% (v/v) Tween[R]-20. Store at –20°C.
2. A termination mix: 22.5 μM c^7dGTP, 10 μM each dATP, dCTP, and dTTP, and 600 μM 2,3-dideoxy-ATP in 10 mM Tris-HCl, 0.1 mM EDTA, pH 7.5. Store at –20 °C.

Table 1
Primers for Identification of mtDNA Mutations

Forward				Reverse	
Name	Position	Name	Position	Name	Position
F1	10027-10047	F31	7570-7590	R1	14700-14679
F2	3566-3586	F32	7831-7850	R2	5542-5521
F3	3818-3836	F33	8085-8106	R3	6500-6478
F4	4069-4089	F34	8349-8370	R4	14452-14432
F5	4430-4450	F35	8605-8625	R5	14207-14186
F6	4697-4716	F36	8860-8881	R6	13958-13939
F7	4946-4966	F37	9120-9139	R7	13687-13667
F8	5174-5195	F38	9374-9394	R8	13436-13417
F9	5521-5542	F39	11011-11031	R9	13192-13172
F10	5771-5791	F40	14937-14958	R10	12886-12866
F11	6038-6057	F41	15196-15216	R11	12634-12613
F12	10266-10285	F42	15467-15486	R12	12383-12361
F13	10517-10537	F43	15717-15735	R13	3331-3311
F14	10755-10775	F44	15967-15987	R14	10033-10012
F15	11255-11275	F45	6749-6769	R15	577-557
F16	11512-11533	F46	6959-6979	R16	9807-9787
F17	11761-11782	F47	14680-14701	R19	16388-16368
F18	12005-12025	F48	6710-6731	R20	16224-16204
F19	577-597	F49	6280-6301	R21	863-842
F20	822-843	F50	12238-12258	R22	14958-14937
F21	1077-1098	F52	9892-9913	R23	2057-2037
F22	13120-1340	F53	5462-5482	R24	7584-7565
F23	1578-1598	F54	10951-10971	R25	8110-8090
F24	1831-1852	F55	16347-16368	R26	8922-8902
F25	2037-2057	F56	16761-16782	R28	15221-15200
F26	2314-2334	F57	170-191	R29	15752-15731
F27	2568-2587	F58	1764-1785	CYTR	7347-7326
F28	2825-2844	CYTF	6478-6499	NDR	4450-4429
F29	3075-3095	NDF	3310-3330		
F30	7323-7342	F49N	9671-9692		

Nucleotide numbers are written according to the mtDNA database (http//www.gen.smory.edu/mitomap.html).

Table 2
Fragments and Primers for Amplification of mtDNA
and Cycle Sequencing

Name	Size (bp)	Amplification Forward	Amplification Reverse	Sequence primer
Fragment 1	1140	NDF	NDR	F2, F3, F4
Fragment 2	2070	F5	R3	F6, F7, F8, F9, F10, F11, F49, R2
Fragment 3	869	CYTF	CYTR	F48, F45, F46
Fragment 4	2607	F1	R11	F12, F13, F14, F39, F15, F16, F17, F18, F50, R12
Fragment 5	2695	F18	R1	R11, R10, R9, R8, R7, R6, R5, R4
Fragment 6	2754	F19	R13	F20, R21, F21, F22, F23, F24, F25, F26, R23, F27, F28, F29
Fragment 7	2466	F47	R15	F40, R22, F41, R28, F42, F43, R29, F44, R20, R19
Fragment 8	3264	F45	R14	F30, F31, F32, F33, F37, F38, F49N, R24, R26, R16

The position of the primers is listed in **Table 1**.

3. G Termination mix: 22.5 μM c^{7}dGTP, 10 μM each dATP, dCTP, and dTTP, and 80 μM 2,3-dideoxy-GTP in 10 mM Tris-HCl, 0.1 mM EDTA, pH 7.5. Store at −20 °C.
4. C Termination mix: 22.5 μM c^{7}dGTP, 10 μM each dATP, dCTP, and dTTP, and 300 μM 2,3-dideoxy-CTP in 10 mM Tris-HCl, 0.1 mM EDTA, pH 7.5. Store at −20 °C.
5. T Termination mix: 22.5 μM c^{7}dGTP, 10 μM each dATP, dCTP, and dTTP, and 900 μM 2,3-dideoxy-TTP in 10 mM Tris-HCl, 0.1 mM EDTA, pH 7.5. Store at −20 °C.
6. Stop solution (loading buffer): 0.8 mL of 95% formamide, 20 mM EDTA, 0.05% bromophenol blue, 0.02% xylene cyanole.
7. Labeled ATP: [γ-^{33}P]-dATP, 10000 μCi/mL
8. T4 polynucleotide kinase (10 U/μL) (New England Biolab).
9. HR-1000™ 6% denatured polyacrylamide gel (GENOMYX).

2.4. Oligonucleotide Mismatch Ligation Assay

1. 5X Ligation buffer: 250 mM Tris-HCl, pH 7.5, 50 mM MgCl, 750 mM NaCl, 5 mM spermidine, 5 mM ATP, 25 mM DTT. Store aliquots at −20 °C.
2. Mutant oligo: The mutant oligo should contain the mutant nucleotide change at the 3′-end. The length of the oligo should be 8–12 bases.

3. Common oligo: The common oligo should be polyacrylamide gel electrophoresis (PAGE) purified by the manufacture. The length of the oligo should be 8–12 bases.
4. Single-strand DNA binding protein (T4 gene 32 protein: Boehringer Mannheim).
5. Calf intestinal alkaline phosphatase (CIP, New England Biolabs).
6. Loading buffer (*see* **Subheading 2.3., item 6**).

2.5. Two Simple PCR Fragment-Length Detection Assays for C-Tract Deletion/Insertion Variants

2.5.1. Radioactivity Incorporation Method

This method is favorable for detecting small amounts of mutant DNA in samples such as serum, plasma, and saliva because of its high sensitivity.

1. 10X PCR buffer (*see* **Subheading 2.2., item 1**)
2. 25 mM dNTP.
3. Forward primer: C-tract F: 5′-TCTGACCAGCCACTTTCCA-3′.
4. Reverse primer: C-tract R: 5′-GGGTTTGGCAGAGATGTGT-3′.
5. *Taq* DNA polymerase (Life Technologies).
6. Labeled dCTP: [α-^{32}P]-dCTP, 1000 µCi/mL.
7. Loading buffer (*see* **Subheading 2.3., item 6**).

2.5.2. Primer Labeling Method

This method is effective in deriving a clear, single band when working with the higher amounts of DNA found in tumor samples.

1. 10X PCR buffer.
2. 25 mM dNTPs.
3. Forward primer: 5′-ACAATTGAATGTCTGCACAGCCACTT-3′.
4. Reverse primer: 5′-GGCAGAGATGTGTTTAAGTGCCTG-3′.
5. *Taq* DNA polymerase (Life Technologies).
6. [γ-^{32}P]-dATP, 10000 µCi/mL.
7. Loading buffer (*see* **Subheading 2.3., item 6**).

3. Methods

3.1. Amplification of Mitochondrial DNA Fragments

Mitochondrial DNAs should be amplified with overlapping primers (**Table 2**), and in order to avoid amplifications of nuclear-encoded pseudogenes (GenBank accession number, AK000414), DNA extracted from a mtDNA-negative cell line (ρ^0) should be included as a negative control.

1. Prepare the master mix: 1X PCR buffer, 1 mM of each dNTP, 6% DMSO, 0.2 µM of forward and reverse primers (**Table 2**), 1 unit of Platinum Taq DNA polymerase, ddH$_2$O (e.g., to final volume of 20 µL).

2. Add 200 ng of genomic DNA of interest. Overlay oil and set up the step-down PCR: initial denaturing at 94 °C for 5 min, followed by 94°C for 30 s, 64°C for 1 min, 70°C for 3 min, for three cycles; 94°C for 30 s, 61°C for 1 min, 70°C for 3 min, for three cycles; 94°C for 30 s, 58°C for 1 min, 70°C for 3.5 min, 25 cycles, and a final extension at 70°C for 5 min.

3. Gel-purify the PCR products using the Qiagen gel extraction kit (Qiagen Inc., Valencia, CA) following the manufacturer's protocol. Final volume of the PCR products should be 25–40 µL.

3.2. Cycle Sequencing (Fig. 1)

1. End-label the primer. For setting up 10 samples: 3 µL of H_2O, 1 µL of T4 polynucleotide kinase buffer, 2 µL of primer (30 ng/µL), 3 µL of $[\gamma\text{-}^{33}P]ATP$, 1 µL of T4 polynucleotide kinase: total 10 µL.

2. Incubate for 30 min at 37 °C.

3. Add 110 µL of H_2O and 20 µL of 10X cycling mix to the labeled primer: total 140 µL (master mix).

4. Add 14 µL of master mix to 1.3 µL of each amplified DNA (**Subheading 3.1.**) samples.

5. Transfer 3.4 µL DNA/master mix to the termination (ACGT) well (96-well plate).

6. Overlay oil and set up the following PCR: 95°C for 30 s, 52°C for 1 min, 70°C for 1 min, for 30 cycles. (It is not necessary to perform an initial denaturing and final extension step.)

7. Add 2.5 µL of the stop solution to each well. If the samples cannot be analyzed immediately, they can be stored for up to 1 wk at –20 °C.

8. Denature the samples by incubating at 90–94 °C for 3 min, and place on ice immediately. Do not leave samples at room temperature.

9. Carefully dispense 3 µL of the denatured DNA from each of the four termination reactions of one template into separate adjacent wells on the gel.

3.3. Oligonucleotide Mismatch Ligation Assay (Fig. 2)

The principles of the ligation are based on the fact that two adjacent short oligos are preferentially ligated by T4 DNA ligase in the presence of a perfectly matched template *(10)*. This method is very sensitive and can be used to detect mutant mtDNA in bodily fluids.

1. Design and order the appropriate oligos. The 5′ oligo (mutant oligo) should contain the mutant nucleotide change at the 3′-end. Have the 3′ common oligo PAGE purified by the vendor.

2. PCR-amplify the DNA fragment containing the mutant mtDNA and controls. Ethanol-precipitate and concentrate the product. Otherwise, directly use 1–2 µL of the PCR reaction product for each ligation assay.

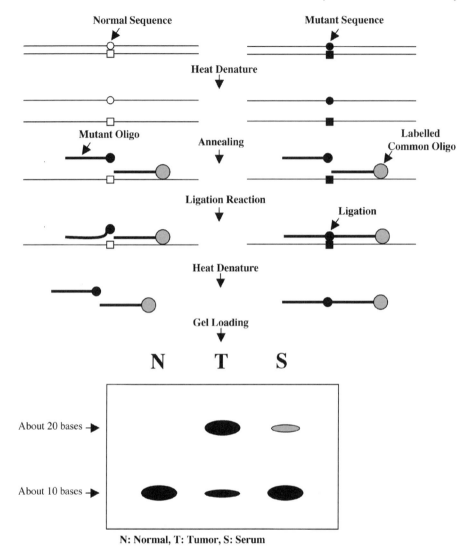

Fig. 2. Schematic diagram of the oligonucleotide mismatch ligation assay. See text for details. Typical reactions for fragments having a normal sequence and a mutant nucleotide change. DNA obtained from a serum sample contains diluted mutant DNA.

3. End-label the common 3' oligo: 1X T4 polynucleotide kinase buffer, 1 µL T4 polynucleotide kinase, 500 ng common oligo, ddH$_2$O (e.g. to final volume 20 µL), incubate at 37 °C for 30 min.
4. Standerd ethanol precipitation with 3 µL glycogen. Resuspend the oligomer in LoTE to a final concentration of 4 ng/µL. The specific activity of the probes is usually approx 108 cpm/µg DNA.

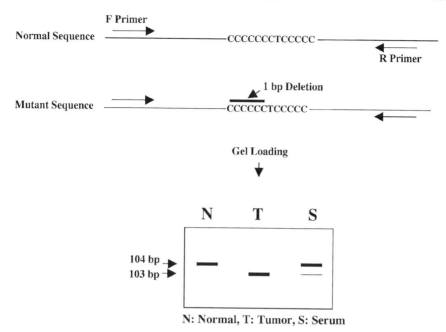

Fig. 3. A simple PCR fragment-length detection assay including the C-tract region. Tumor DNA has a 1-bp deletion in the C-tract and the serum DNA contains diluted mutant DNA.

5. Mix the following (to make a master mix with most components except DNA): 50 ng amplified DNA template, 4 ng mutant 5′ oligo, 4 μL of 5X ligation buffer, 3 μg single-strand DNA binding protein (USB), and add H_2O a final volume of 18 μL.
6. Add 1 μL of kinase-labeled 3′ common oligo to get a final volume of 19 μL.
7. Add 1 drop mineral oil and heat denature the mixture to 95 °C for 5 min, then allowed to cool to room temperature for 15 min.
8. Add 1 μL of T4 ligase (1U/μL, GIBCO BRL) and incubate at 37°C for 1 h.
9. Terminate the reaction by heat inactivation at 68°C for 10 min.
10. Add unit of alkaline phosphatase and incubate at 37 °C for 15–30 min to reduce the intensity of the substrate signal.
11. Add 10–15 μL of sequencing loading buffer and after denaturing at 95°C for 5 min, load 5 μL on to 12% polyacrylamide gel.

3.4. A Simple PCR Fragment-Length Detection Assay for C-Tract Deletion/Insertion Variants (Fig. 3)

Recently, we found frequent deletions/insertions in a polymorphic cytosine mononucleotide repeat sequence (CCCCCCCTCCCCC) between nucleotide 303 and 315. The biologic reason for these alterations is unclear, but this region is the start site of replication for the closed circular mitochondrial genome. Although we can detect these frameshift mutations by direct sequencing methods, it is easier to use a simple PCR assay, based on the amplification of a 104-bp product including the C-tract region.

3.4.1. Radioactivity Incorporation Method

1. Prepare the master mix: 1X PCR buffer, 0.5 mM dNTP, 0.2 µM each of forward primer (C-tract F) and reverse primer (C-tract R), 1 U *Taq* DNA polymerase, 0.2 µL [α-^{32}P]dCTP, ddH$_2$O (e.g., to final volume of 20 µL).
2. Add 50 ng of genomic DNA of interest. Overlay oil and set up the PCR protocol: initial denaturing at 94°C for 3 min followed by 94°C 20 s, 55°C for 20 sec, 70°C 15 s, for 35 cycles, and final extension at 70°C for 5 min. If the samples cannot be analyzed immediately, they may be stored for up to 1 wk.
3. Dilute the PCR product with 10 times the volume of loading buffer, and after denaturing at 95°C for 5 min, load 3 µL on a denatured 6% polyacrylamide gel.

3.4.2. Primer Labeling Method

1. Mix the following reagents (for 100 PCR reactions):
   a. 5 µL of polynucleotide kinase buffer (from New England Biolabs).
   b. 1 µL of forward primer (at 100 ng/µL).
   c. 2 µL of T4 polynucleotide kinase (from New England Biolabs).
   d. 2 µL of [γ-^{32}P] ATP, 10 mCi/mL.
   e. 40 µL of ddH$_2$O.
2. Incubate at 37°C for 30 min.
3. Inactivate the polynucleotide kinase by adding 50 µL of ddH$_2$O.
4. Prepare the master mix: 1X PCR buffer, 1.5 mM dNTPs, 0.2 µM of each forward and reverse primer, *Taq* DNA polymerase 0.5 U, 1 µL of labeled forward primer (e.g., to final volume of 10 µL).
5. Add 10–50 ng of total DNA of sample of interest. Overlay oil and set up the PCR protocol: initial denaturing for 3 min followed by 94°C for 20 s, 60°C for 20 s, 72°C for 20 s, for 35 cycles, and final extension at 72°C for 5 min.
6. Add 3 µL of loading buffer, and after denaturing at 95°C for 5 min, load 3 µL on a denaturing 6% polyacrylamide gel.

4. Note

1. The Sanger method is based on the use of specific terminations of DNA chain elongation: 2′,3′-dideoxynucleoside-5′-triphosphates (ddNTP). These deoxy-nucleoside triphosphate analogs can be incorporated by a DNA polymerase into a growing DNA chain through their 5′-triphosphate group. However, because these analogs lack a hydroxyl group at the 3′-position, they cannot form phosphodiester bonds with the next incoming deoxynucleoside-5′-triphosphate (dNTP), and the chain extension terminates whenever an analog is incorporated. Thus, when a specific ddNTP is included along with the four dNTPs normally required for

DNA synthesis, the resulting extension products are a series of discrete-length DNA chains that are specifically terminated at that dideoxy residue.

References

1. Clayton, D. (1991) Replication and transcription of vertebrate mitochondrial DNA. *Annu. Rev. Cell Biol.* **7,** 453–478.
2. Cadet, J., Berger, M., Douki, T., et al. (1997) Oxidative damage to DNA: formation, measurement, and biological significance. *Rev. Physiol. Biochem. Pharmacol.* **131,** 1–87.
3. Polyak, K., Li, Y., Hong, Z., et al. (1998) Somatic mutations of the mitochondrial genome in human colorectal tumors. *Nature Genet.* **20,** 291–293.
4. Fliss, M. S., Usadel, H., Caballero, O. L., et al. (2000) Facile detection of mitochondrial DNA mutations in tumors and bodily fluid. *Science* **287,** 2017–2019.
5. Jones, J. B., Song, J. J., Hempen, P. M., et al. (2001) Detection of mitochondrial DNA mutation in pancreatic cancer offer a "Mass"-ive advantage over detection of nuclear DNA mutation. *Cancer Res.* **61,** 1299–1304.
6. Nishikawa, M., Nishiguchi, S., Shiomi, S., et al. (2001) Somatic mutation of mitochondrial DNA in cancerous and noncancerous liver tissue in individuals with hepatocellular carcinoma. *Cancer Res.* **61,** 1843–1845.
7. Horton, T. M., Petros, J. A., Heddi, A., et al. (1996) Novel mitochondrial DNA deletion found in a renal cell carcinoma. *Genes Chromosomes Cancer* **15,** 95–101.
8. Tamura, G., Nishizuka, S., Maesawa, C., et al. (1999) Mutations in mitochondrial control region DNA in gastric tumor of Japanese patients. *Eur. J. Cancer* **35,** 316–319.
9. Parfait, B., Rustin, P., Munnich, A., et al. (1998) Co-amplification of nuclear pseudogenes and assessment of heteroplasmy of mitochondrial DNA mutations. *Biochem. Biophys. Res. Commun.* **247,** 57–59.
10. Jen, J., Powell, S. M., Papadopoulos, N., et al. (1994) Molecular determinants of dysplasia in colorectal lesion. *Cancer Res.* **54,** 5523–5526.
11. Sanchez-Cespedes, M., Parrella, P., Nomoto, S., et al. (2001) Identification of a mononucleotide repeat as a major target for mitochondrial DNA mutations in human tumors. Detailed screening and patterns of alteration in lung and head and neck tumors. *Cancer Res.* **61,** 7015–7019.
12. Kang, D., Miyako, K., Kai, Y., et al. (1997) In vivo determination of replication origins of human mitochondrial DNA by ligation-mediated polymerase chain reaction. *J. Biol. Chem.* **272,** 15,275–15,279.

6

In Situ PCR of the Common Human mtDNA Deletion

Is It Related to Apoptosis?

Steven J. Zullo

1. Introduction

We have developed an *in situ* polymerase chain reaction (PCR) assay for the elucidation of the human mitochondrial DNA^{4977} deletion (common mtDNA deletion), which occurs between two 13-bp direct repeats situated 4977 bp apart *(1)*. Although the origin of the common mtDNA deletion is unknown, its presence in a muscle sample is associated with reduced energy capacity *(2)*. My colleagues and I have demonstrated in age-matched putamen and frontal gyrus of the brain that levels of the common mtDNA deletion are statistically related to medical conditions associated with chronic hypoxia *(3)*. No other factors we considered (age, sex, ethnicity, drug use, postmortem interval [PMI]) carried any statistical significance. This analysis strongly suggests that aging is not a factor (a factor investigated in many other studies with smaller sample sizes, see two paragraphs below, also **refs.** *4* and *5*), inferring that there could be genetic regulation of the common mtDNA deletion. Likewise, it has been reported that the common mtDNA deletion levels induced by γ-radiation are directly related to radiosensitivity of the cells *(6)*, again implicating regulation. Of note, a nuclear-encoded protein, a mitochondrial helicase, has recently been associated with multiple mtDNA deletions. However, it is apparently not associated with the common mtDNA deletion *(7)*.

Disclaimer: The identification of any commercial product or trade name does not imply endorsement or recommendation by the National Institute of Standards and Technology.

From: *Methods in Molecular Biology, vol. 197: Mitochondrial DNA: Methods and Protocols*
Edited by: W. C. Copeland © Humana Press Inc., Totowa, NJ

The common mtDNA deletion might also have a replicative advantage over the nondeleted molecule *(1)*. The deletion continues to possess both origins of replication (heavy and light; O_H and O_L, respectively) and is about two-thirds the size of the nondeleted mitochondrial genome. Because of the decreased size, along with the possible deletion of genetic downregulation elements *(8)*, replicative efficiency of the mtDNA[4977] deletion molecule may be increased over its nondeleted counterpart. The mtDNA deletion mutation may also metabolically compromise the mitochondria, leading to the activation of mitochondrial proliferation *(8)*. We have also noted reduced oxygen consumption in cells harboring the common mtDNA deletion *(9)*.

Whereas the minimum level of the common mtDNA deletion required to generate mitochondrial dysfunction in tissue culture is > 60% *(2)*, the mtDNA deletion levels found in brain tissue have been generally < 1% *(3,10–13)*. Mitochondrial genomes with the common mtDNA deletion are also found in a number of disorders associated with abnormal mitochondrial function *(14–17)*. The low concentration of the common mtDNA deletion in the brain led to one of two conclusions: (1) that there could be a few in each cell, suggesting there may be no physiological consequence, or (2) that the mtDNA deletions could be concentrated in a few cells, suggesting that the mtDNA deletions are really markers for a cell in trouble, dying, or dead. Indeed, physiological consequences of the common mtDNA deletion (lowered genomic DNA adduct formation associated with increased deletion levels) were shown at even the low levels found in the brain *(18)*. Thus, it could be important to determine the cellular distribution of the common mtDNA deletion in the cell.

We felt that PCR would be necessary to "amplify" any signal to detect the mtDNA deletions, and also recognized that for an *in situ* system, we would need to ensure that only PCR products of the deleted mtDNA genomes were amplified and detected. *In situ* hybridization might not be accurate, as there are two 13-bp repeats in the undeleted genomes, creating a large common area for the undeleted and deleted molecules and, thus, potentially raising the background and lowering the signal-to-noise ratio. Therefore, we judged that we could not use labeled nucleotides, which would be present in the cell, or in single strands synthesized but not involved in PCR (because the other primer is situated 5000 bases away in the nondeleted molecules), regardless of a successful PCR. We also felt that the PCR reaction needed to be as brief as possible, to reduce the chance of altering or destroying the cellular architecture during the required temperature fluxes. We settled on 127-bp amplicons, replacing the 500+-bp amplicons we previously used *(3)*. The reporter dye-quencher dye probe system (Taqman) developed by Perkin-Elmer offered a possible solution. We developed *in vitro* and *in situ* PCR primers for this system (**Table 1**).

Table 1
Oligonucleotide Primers Used in Solutional PCR Experiments and *In Situ* PCR Experiments

Name[#]	Sequence	Location[a]
HSSN8416	5'-CCT TAC ACT ATT CCT CAT CAC C-3'	8416-8437
HSAS8542	5'-tgt ggt ctt tgg agt aga aac c-3'	13519-13498 8542-8521
HSSN1307	5'-GTA CCC ACG TAA AGA CGT TAG G-3'	1307-1328
HSAS1433	5'-tac tgc taa atc cac ctt cg-3'	1433-1414
deletion probe	5'-fam-TGG CAG CCT AGC ATT AGC AGT*-tamra 3'	13461-13480 8484-8503
rRNA probe	5'-tet-CCC ATG AGG TGG CAA GAA AT-tamra 3'	1340-1359
HSAS1433amp	┌────────AT GAC GA-fitc-5' └ tt agt tta ctg cta aat cca cct tcg-3' └─dabcyl	1433-1414
HSAS8542amp	┌──────── TA CAC CA-tritc-5' └ tc gat gat gtg gtc ttt gga gta gaa acc-3' └─dabcyl	13519-13498 8542-8521

[#]Designations HSSN and HSAS represent *Homo sapiens* sense and *H. Sapiens* antisense, respectively.

[a]Location of primers relative to mtDNA sequence published by Anderson et al *(20)*. (The first range for HSAS13519amp is the Anderson numbering, the second range is of the deleted molecule). Italicized bases of amplifluor™ primers added for stem-loop creation.

*3'-T added for attachment of tamra.

Source: Modified from Melnov et al. (unpublished data) and Pogozelski et al. (in preparation)

To test our system, we needed cells that maintained the common mtDNA deletions. We found cultured lymphocytes from a boy with Pearson's marrow syndrome consistently maintained the common mtDNA deletion at around 50% of the mtDNA present in a PCR assay of isolated genomic DNA *(19;* Fischel-Ghodsian, personal communication). Interestingly, the boy's mother suffered from chronic progressive external ophthalmoplegia (CPEO) and possessed the deletion in her eye muscles. We feel that the distribution of the mtDNA deletion in this family strongly implicated genetic inheritance of an enzymatic process, with possible tissue-specific (mis-)regulation.

While the Taqman system resulted in versatile and efficient PCR detection of the deleted mtDNA molecule in isolated DNA (Melnov, et al. [unpublished data], Pogelzelski, et al., [unpublished data]), the Taqman system proved inadequate for *in situ* PCR. We found that when used *in situ*, the released bases

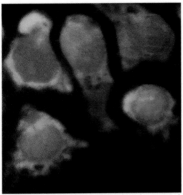

Mother with mtDNA Deletion in muscle

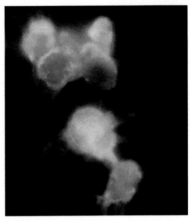

Son with mtDNA Deletion in marrow

LEGEND
GREEN Probe for total mtDNA
RED Probe for mtDNA deletion

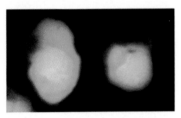

Asymptomatic Daughter

Fig. 1. Representative results from *in situ* PCR utilizing the reporter dye-quencher dye system (Taqman). Green probe for total mtDNA (HSSN1307 and HSAS1433 primers with rRNA probe primer); red probe for mtDNA deletion (HSSN8416 and HSAS8542 primers with deletion probe primer).

with the reporter fluors attached apparently leached into the cytoplasm from the mitochondria in which the PCR occurred, and mitochondrial localization was not possible (**Fig. 1**). However, we have adapted the sunrise (now called ampilfluor™) primer technology (**Table 1**), which keeps the reporter fluor attached to a PCR product that was larger than a single nucleotide and less likely to leach from the fixed mitochondria, leading to more efficient localization (**Fig. 2**). Mitochondria in cells with large numbers of deletions might degenerate, releasing the deletion PCR product (**Fig. 3A,B**).

Immunological detection of BAX and *in situ* PCR of the mtDNA deletions within the same cell depicts BAX inserted into the membranes of mitochondria not harboring the deletion (**Fig. 3a,b**). These data suggest that energetic deficits initiated by the mtDNA deletion could trigger BAX insertion into membranes of mitochondria without the deleted genome, with consequent cytochrome-*c* release and activation of the programmed cell death pathway (apoptosis; *see*

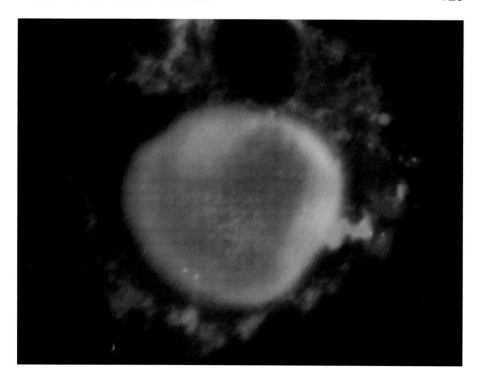

Fig. 2. Multiplex *in situ* PCR of common mtDNA deletion-containing lymphoblast with amplifuor primers (fitc green: HSSN1307 and HSAS1433amp primers) (tritc red: HSSN8416 and HSAS8542amp deletion primers).

Note 1). Indeed, the deletion has been associated with reduced oxidative phosphorylation (OXPHOS) *(22)*, and OXPHOS is essential for BAX insertion into the mitochondrial membrane of yeast *(23)*.

2. Materials

1. The primers used for these experiments are listed in **Table 1**.
2. 1X Phosphate-buffered serum (PBS) (Life Technologies, cat. no. 20012).
3. 2X SSC: 0.3 M NaCl, 0.03 M sodium citrate, pH 7.0.
4. Cytospin centrifuge. If one does not have a cytospin centrifuge, dropping the cells onto a slide is sufficient.
5. Formalin (Sigma, cat. no. HT50-1-128).
6. Anti-Bax antibody, mouse monoclonal IgG_1 (Sigma, cat. no. B8554).
7. 4X SSC/Tween-20: 100 mL 20X SSC, 400 mL H_2O, 0.5 mL Tween-20 (Sigma, cat. no. P1379)
8. AlexaFluor488-conjugated goat anti-mouse antibody (Molecular Probes, Oregon, cat. no. A11001).

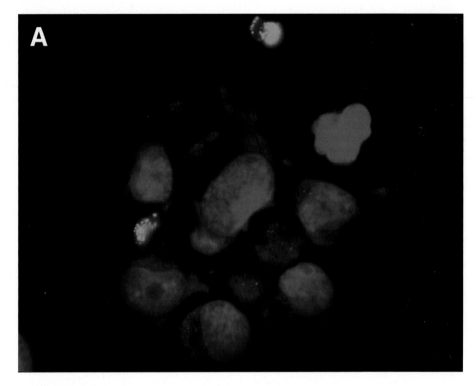

Fig. 3. Elucidation of BAX and the common deletion. (**A**) Elucidation of BAX in a culture of common mtDNA deletion-containing lymphoblasts. Note the localization indicating insertion into mitochondrial membranes (BAX channel, Alexa488 fluorescence).

  9. VectaShield with DAPI (Vector Laboratories, California, cat. no. H1200).
 10. Epifluorescent microscope.

3. Methods
3.1. In Vitro PCR Reaction

 1. Assemble a 50-µL PCR reaction containing the following:
    a. 50–200 ng genomic DNA.
    b. 0.5 µM of forward and reverse primers.
    c. 200 nM of probe.
    d. 50 mM KCl.
    e. 10 mM Tris-HCl, pH 8.3.
    f. 3.5 mM MgCl$_2$.
    g. 200 nM of dATP, dCTP, dGTP, and dTTP.
    h. 0.025 U/µL *Taq* polymerase.

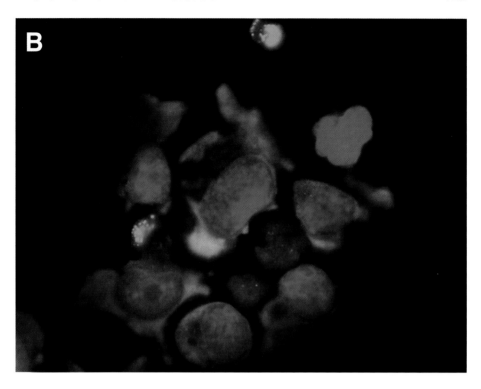

Fig 3. (**B**) Elucidation of BAX and *in situ* PCR for the common mtDNA deletion (HSSN8416 and HSAS8542amp primers, tritc fluorescence).

2. Run the PCR reaction in a PCR machine using the following cycling conditions:
   a. 95°C for 10 min.
   b. 25 cycles of 94°C for 15 s, 58°C for 15 s, 72°C for 1 min.
   c. 72°C for 5 min.
   d. 4°C and hold.

3.2. In situ *PCR Reaction*

1. Cytospin cells onto silanized slides 10 min at 112.90g (1000 rpm).
2. Fix in buffered formalin (e.g., Sigma HT50-1-128, St. Louis, MO) for 5 min.
3. Wash twice in 1X PBS, 5–10 min each.
4. Add 50 μL PCR master mix
   a. 0.5 μM of forward and reverse primers.
   b. 50 mM KCl.
   c. 10 mM Tris-HCl, pH 8.3.
   d. 3.5 mM MgCl$_2$.
   e. 200 nM of dATP, dCTP, dGTP, and dTTP.
   f. 0.025 U/μL *Taq* polymerase.

5. Cover with 22 × 22-mm² cover slip (or 24 × 24; 18 × 18 mm² needs less master mix).
6. Seal with rubber cement.
7. Run PCR on a Slide PCR machine under the following cycling conditions:
   a. 50°C for 2 min.
   b. 95°C for 10 min.
   c. 25 cycles of 94°C for 15 s, 58°C for 15 s, 72°C for 1 min.
   d. 72°C for 5 min.
   e. 4°C and hold.
8. Remove cover slip.
9. Wash once in 1X PBS, 5 min with gentle shaking (washes in Coplin jars).
10. Wash once in 2X SSC, 5 min with gentle shaking.
11. Anti-Bax antibody *(21)*, (mouse IgG1: monoclonal, B8554; Sigma, St. Louis, MO) was diluted at the working concentration according to supplied protocol in 1% BSA in 4X SSC/Tween-20.
12. Add 100 µL of antibody working solution to a 22 × 40-mm² cover slip, touch slide to cover slip and incubate in hybridization chamber for 45–60 min at 37°C.
13. Wash in 4X SSC/Tween-20, three times for 5 min each with shaking at room temperature.
14. Add 100 µL AlexaFluor488-conjugated goat anti-mouse antibody working solution to cover slip, touch slide to cover slip and incubate for 15 min at room temperature.
15. Wash in 4X SSC/Tween-20, three times for 5 min each with shaking at room temperature.
16. Wash in 2X SSC.
17. Wash 5 min in distilled H_2O with gentle shaking.
18. Dry with an ethanol series, 3 min each in 70%, 90%, and 100% ethanol.
19. Add 15 µL VectaShield with DAPI to a cover slip (22 × 22 or 24 × 24; 18 × 18 requires less), invert sample on cover slip.
20. Observe under oil on an epifluorescent microscope.

4. Note

1. These observations then led to the hypothesis that the common mtDNA deletion could be generated by an active enzymatic process to begin apoptosis, or at least to reduce the OXPHOS potential on a cellular level. At some threshold level, the cell would be unable to maintain its metabolism at a level consistent with life; then, BAX would be inserted into mitochondria with functional OXPHOS systems to commence apoptosis. The presence of the deletion in mitochondria precludes insertion of BAX into the membranes of mitochondria containing the deletions. These data suggest that the deletion might be an early event in the apoptosis process, disabling mitochondria until a cellular threshold level (of OXPHOS dysfunction) is attained—at which time, BAX insertion aids the apoptotic process via mitochondria not containing the deletion.

The distribution of the mtDNA deletion in the cultured lymphocytes ranged from isolated cellular regions to cells that are apparently saturated with the common mtDNA deletions. These observations might be consistent with a progression of the common mtDNA deletion from a single region (mitochondrion) to deletions throughout the cell. We are currently investigating the progression of these events. These studies could provide important insights into the mitochondrial involvement in apoptosis and pathophysiological processes.

References

1. Wallace, D.C. (1993) Mitochondrial diseases: genotype versus phenotype. *Trends Genet.* **5,** 9–13.
2. Hayashi, J.-I., Ohta, S., Kikuchi, A., et al. (1991) Introduction of disease-related mitochondrial DNA deletions into HeLa cells lacking mitochondrial DNA results in mitochondrial dysfunction. *Proc. Natl. Acad. Sci. USA* **88,** 10,614–10,618.
3. Merril, C. R., Zullo, S., Ghanbari, H., et al. (1996) Possible relationship between conditions associated with chronic hypoxia and brain mitochondrial DNA deletions. *Arch. Biochem. Biophys.* **326,** 172–177.
4. Merril, C. R. and Zullo, S. J. (1995) Mitochondrial genome deletions in the brain and their role in neurodegenerative diseases. *Int. Rev. Psychiatry* **7,** 385–398.
5. Zullo, S. and Merril, C. R. (1997) Deletions of the mitochondrial genome and neurodegenerative diseases, in *Principles of Neural Aging* (Dani, S. U., Hori, A., and Walter, G. F., eds.), Elsevier, Amsterdam, pp. 69–81.
6. Kubota, N., Hayashi, J., Inada, T., et al. (1997) Induction of a particular deletion in mitochondrial DNA by X rays depends on the inherent radiosensitivity of the cells. *Radiat. Res.* **148,** 395–398.
7. Spelbrink, J. N., Li, F. Y., Tiranti, V., et al. (2001) Human mitochondrial DNA deletions associated with mutations in the gene encoding Twinkle, a phage T7 gene 4-like protein localized in mitochondria. *Nature Genet.* **28,** 223–231.
8. Shoubridge, E. A., Karpati, G., and Hastings, K. E. (1990) Deletion mutants are functionally dominant over wild-type mitochondrial genomes in skeletal muscle fiber segments in mitochondrial disease. *Cell* **62,** 43–49.
9. Zullo, S. J., Parks, W. T., Wong, A., et al. (1999) The "common" human mitochondrial DNA 4977-bp deletion in cultured lymphocytes: decreased oxygen consumption and apoptotic characteristics. NIH Res. Festival. Minisymposium: Mitochondrial Disease from Bench to Bedside. Bethesda, MD.
10. Cortopassi, G. and Arnheim, N. (1990) Detection of a specific mitochondrial DNA deletion in tissues of older humans. *Nucleic Acids Res.* **18,** 6927–6933.
11. Ikebe, S., Tanaka, M., Ohno, K., et al. (1990) Increase of deleted mitochondrial DNA in the striatum in Parkinson's disease and senescence. *Biochem Biophys. Res. Commun.* **170,** 1044–1048.
12. Soong, N. W., Hinton, D. R., Cortopassi, G., et al. (1992) Mosaicism for a specific somatic mitochondrial DNA mutation in adult human brain. *Nature Genet.* **2,** 318–323.

13. Corral-Debrinski, M., Horton, T., Lott, M. T., et al. (1992) Mitochondrial DNA deletions in human brain: regional variability and increase with advanced age. *Nature Genet.* **2,** 324–329.
14. Holt, I. J., Harding, A. E., and Morgan-Hughes, J. A. (1988) Deletions of muscle mitochondrial DNA in patients with mitochondrial myopathies. *Nature* **331,** 717–719.
15. Moraes, C. T., Schon, E. A., DiMauro S., et al. (1989) Heteroplasmy of mitochondrial genomes in clonal cultures from patients with Kearns-Sayre syndrome. *Biochem. Biophys. Res. Commun.* **160,** 765–771.
16. Shoffner, J. M., Lott, M. T., Voljavec, A. S., et al. (1989) Spontaneous Kearns–Sayre/chronic external ophthalmoplegia plus syndrome associated with a mitochondrial DNA deletion: a slip-replication model and metabolic therapy. *Proc. Natl. Acad. Sci. USA* **86,** 7952–7956.
17. Schon, E. A., Rizzuto R., Moraes, C. T., et al. (1989) A direct repeat is a hotspot for large-scale deletion of human mitochondrial DNA. *Science* **21,** 346–349.
18. Zullo, S. J., Cerritos, A., and Merril, C. R. (1999) Possible relationship between conditions associated with chronic hypoxia and brain mitochondrial DNA deletions. II. Reduction of genomic 8-hydroxyguanine levels in human brain tissues containing elevated levels of the human mitochondrial DNA4977 deletion. *Arch. Biochem. Biophys.* **367,** 140–142.
19. Bernes, S. M., Bacino, C., Prezant, T. R., et al. (1993) Identical mitochondrial DNA deletion in mother with progressive external ophthalmoplegia and son with Pearson marrow–pancreas syndrome. *J. Pediatr.* **123,** 598–602.
20. Anderson, S., Bankier, A. T., Barrell, B. G., et al. (1981) Sequence and organization of the human mitochondrial genome. *Nature* **290(5806),** 457–465.
21. Hsu, Y.-T. and Youle, R. J. (1997) Nonionic detergents induce dimerization among members of the Bcl-2 family. *J. Biol. Chem.* **272,** 13,829–13,834.
22. Tang, Y., Schon, E. A., Wilichowski, E., et al. (2000) Rearrangements of human mitochondrial DNA (mtDNA): new insights into the regulation of mtDNA copy number and gene expression. *Mol. Biol. Cell* **11,** 1471–1485.
23. Harris, M. H., Vander Heiden, M. G., Kron, S. J., et al. (2000) Role of oxidative phosphorylation in Bax toxicity. *Mol. Cell. Biol.* **20,** 3590–3596.

7

Reproducible Quantitative PCR of Mitochondrial and Nuclear DNA Copy Number Using the LightCycler™

Alice Wong and Gino Cortopassi

1. Introduction

Recent developments in PCR fluorimetry have allowed for quick quantification of target molecules. Before the invention of fluorimetric quantitative PCR, researchers who wanted to quantify the amount of a gene in a sample did so painstakingly by limiting dilution, competitive polymerase chain reaction (PCR), or other methods, including high-performance liquid chromatography (HPLC), solid phase assays, dot blots, or immunoassay (1).

In the past few years, several companies have introduced products to quantify PCR. One method relies on a fluorimetric dye (either alone or conjugated to a probe) that binds double-stranded DNA. One particular method that we find easy to use and produces reproducible results is using Roche Molecular Biochemicals' LightCyler™ (2–4). The instrument and software package allows real-time measurement of PCR products (i.e., one can visualize in real time the increase in fluorescence that is the result of amplification of PCR products). In this chapter, we describe a protocol for the quantification of mitochondrial and nuclear genes in biological samples.

2. Materials
2.1. Fluorescence-Based Quantitative PCR (see Note 1)

1. Autoclaved, distilled, deionized water.
2. 10X PCR buffer: 500 mM KCl, 100 mM Tris-HCl, pH 8.3, 5 mg/mL bovine serum albumin (BSA), aliquot and store at –20°C.

From: *Methods in Molecular Biology, vol. 197: Mitochondrial DNA: Methods and Protocols*
Edited by: W. C. Copeland © Humana Press Inc., Totowa, NJ

3. 10X dNTPs: 2 m*M* dATP, 2 m*M* dCTP, 2 m*M* dTTP, and 2 m*M* dGTP, aliquot and store at –20°C.
4. DNA template; 1–100 ng for nuclear gene detection, 0.005–1.0 ng for mitochondrial gene detection.
5. 10X SYBR Green, made from 10,000X stock solution (Molecular Probes, Eugene, OR), dilute in dimethyl sulfoxide (DMSO) and mix well, aliquot, and store at –20°C (*see* **Note 2**).
6. 16 m*M* MgCl$_2$, aliquot and store at –20°C.
7. Primers (2.2 µ*M* stock solution), aliquot and store at –20°C (*see* **Note 3**).

2.2. Agarose Gel Electrophoresis

1. 1X TAE buffer: 40 m*M* Tris-acetate, pH 8.3, 1.0 m*M* EDTA, stored at room temperature.
2. Appropriate molecular weight ladder.
3. Gel loading dye: 0.25% bromophenol blue, 0.25% xylene cyanol, 30% glycerol in water, stored at 4°C.
4. 1–4% Agarose, depending on the size of the amplified fragment.

3. Methods
3.1. Preparation of PCR Buffer

Mix the following to prepare buffer for 33 reactions: 66 µL 10X buffer, 13.2 µL 10X SYBR green, 165 µL 16 m*M* MgCl$_2$, 132 µL 10X dNTPs, store at –20°C and freeze and thaw, not more than three times (*see* **Note 4**).

3.2. Preparation of Standard Curves

1. Prepare the standard curves for both nuclear and mitochondrial PCR (**Table 1**).
2. Determine the number of copies each sample will have by assuming that, in 1 ng bulk cellular DNA, there are approximately 150 haploid nuclear copies, and in many cell types, there are about 300,000 mitochondrial copies.

3.3. Preparation of PCR Master Mix

1. Calculate how many reactions will be done and add one or two extra reactions to account for pipetting error.
2. Mix the following to prepare the PCR master mix by multiplying the volume by the number of reactions: 11.4 µL PCR buffer, 1.8 µL each primer, 0.08 µL *Taq*; mix well.

3.4. Preparation of reaction

1. Add 15 µL of PCR master mix to each capillary tube.
2. Carefully add 5 µL of DNA sample to each tube (water for negative control) and cap tubes. (*See* **Note 5**.)
3. Centrifuge capillaries to accumulate liquid at the bottom and place capillaries into the carriage. Run desired PCR program. (*See* **Note 6**.)

Table 1
Theoretical mtDNA and Haploid Nuclear Targets
in a Given Amount of DNA (ng)

Total cellular DNA input (ng)	Theoretical mtDNA targets assuming 2000 mtDNA genomes/nuclear gene	No. of output molecules calculated by the LightCycler	Theoretical haploid nuclear targets	No. of output molecules calculated by the LightCycler
0.005	1,500	1,591	0.75	n.d.
0.01	3,000	2,905	1.5	n.d.
0.05	15,000	14,970	7.5	n.d.
0.1	30,000	28,980	15	n.d.
1	300,000	312,500	150	162
10	3,000,000	n.d.[a]	1,500	1,654
25	7,500,000	n.d.	3,750	4,023
50	15,000,000	n.d.	7,500	7,842
100	30,000,000	n.d.	15,000	15,346

[a]n.d.= not done; that is, PCR analysis with less than 30 or more than 300,000 template copies is usually not as accurate, efficient, or reproducible. Thus, LightCycler analysis works best in our hands with mitochondrial DNA and nuclear DNA targets at 150 < targets < 150,000 (i.e., a 1000-fold range).

3.5. Analysis of Data

Polymerase chain reaction can be analyzed using Roche's software package or by visualizing the samples on an agarose gel. Examples of amplification curves for nuclear and mitochondria genes are shown in **Fig. 1**. Primers (forward: 5′-AGCAGAGTACCTGAAACAGGAA-3′; reverse: 5′-AGCTTACCCATAGAGGAAACATAA-3′) complementary to sequences in the cystic fibrosis (CF) gene were used to amplify a 460-bp single-copy nuclear gene, using 0, 1, 10, 25, and 100 ng K562 DNA (Life Technologies, Gaithersburg, MD). PCR was performed under the following conditions: initial 30-s denaturation at 95°C followed by 40 cycles of 0 s at 95°C, 5 s at 60°C, and 13 s at 72°C. For the mitochondrial gene, ND5 primers (forward: 5′-AGGCGCTATCACCACTCTGTTCG-3′; reverse: 5′-AACCTGT-GAGGAAAGGTATTCCTG-3′) were used to amplify a 326-bp fragment using 0, 0.005, 0.01, 0.05, 0.1, and 1 ng K562 DNA. The PCR conditions were the same as nuclear gene amplification. A standard curve is made for both the nuclear and mitochondrial genes, in order to calculate the number of copies of either gene in a given amount of DNA (**Fig. 1**). The equations for the standard curves are shown as well as the correlation coefficients (r^2). This is an example

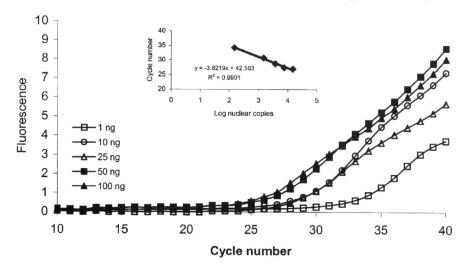

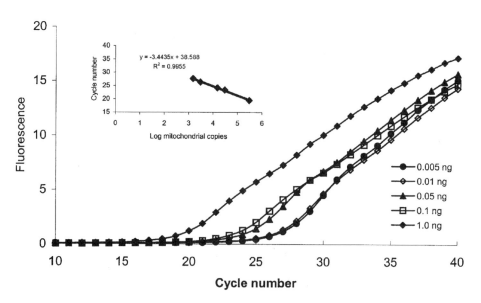

Fig. 1. Amplification and standard curves for nuclear and mitochondrial genes. Amplification curves of (**A**) nuclear cystic fibrosis gene and (**B**) ND5 gene amplified from 0–100 ng and 0–1 ng K562 DNA, respectively. The standard curves obtained from the amplifications are embedded in the graphs. The equations of the standard curves and correlation coefficients are included.

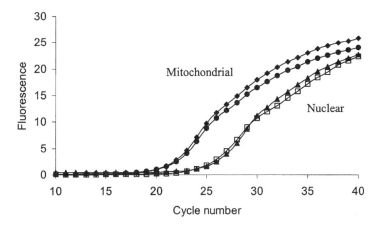

Fig. 2. Example of nuclear and mitochondrial genes amplified from a sample of DNA. Amplification curves of 10 ng DNA (nuclear, cystic fibrosis) and 0.1 ng DNA (mitochondrial, ND5).

of a "good" standard curve, whereas a relatively poor standard curve would have a correlation coefficient of less than 0.98. We do not use standard curves when the correlation coefficient is less than 0.98. The efficiency (E) is 1.83 for the nuclear CF gene amplification and 1.95 for the mitochondrial gene amplification (see **Note 7** for more on efficiency). Calculated PCR efficiencies vary depending on the primers, genomic complexity, and PCR conditions.

Amplification of a sample of DNA is shown in duplicate for nuclear and mitochondrial gene products (**Fig. 2**). Ten nanograms of total DNA was used to amplify the nuclear gene, whereas 0.1 ng total DNA was used to amplify the mitochondrial gene. Using these amounts of DNA is appropriate for this particular PCR because the number of targets in each sample falls within what we observe to be the most reproducible and convenient range of initial targets to work with (i.e., between 150 and 150,000 target molecules). Amplification of both mitochondrial and nuclear copies in the sample gives 56,286 ± 543.1 mitochondrial copies per 0.1 ng bulk DNA and 2265 ± 221.4 nuclear copies per 10 ng bulk DNA (i.e. about 2500 mitochondrial targets/nuclear targets. Different cell types may have different ratios of mitochondrial to nuclear targets in our experience.

In addition to the above quantitative analysis, PCR products can also be qualitatively analyzed on the LightCycler, by using the melting curve data, which analyzes the denaturation profile of PCR products. The melting temperature (T_m) of a PCR product is determined, of course, by its length, sequence, and the ionic makeup of the buffer. Many PCR products have sufficiently different T_m values, thus two different PCR products can be quantified

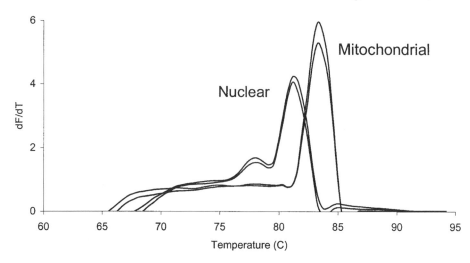

Fig. 3. Melting curves for both nuclear and mitochondrial genes. Melting curves for two nuclear (Cystic fibrosis) and mitochondrial (ND5) PCR products are shown. The melting temperatures are 81.5°C and 83.58°C, respectively.

in the same PCR without having to run the PCR samples on an agarose gel. Sample melting curves are shown for the nuclear and mitochondrial products in **Fig. 3**. The melting temperatures for these particular nuclear and mitochondrial products are 81.5°C and 83.58°C, respectively, as seen by the peak.

More traditional gel electrophoretic analysis is a good idea, either in the initial stage of primer design for the LightCycler, to confirm the specificity of the reaction, and also for troubleshooting. To analyze the PCR products the "old-fashioned" way, prepare a 1–4% agarose gel (depending on the length of PCR fragments). Remove caps from capillaries and place the capillaries upside down in an Eppendorf tube. Spin down the contents and remove 5 µL. Mix with loading dye and load onto gel. Visualize gel by staining with ethidium bromide (**Fig. 4a,b**, nuclear and mitochondrial genes, respectively).

4. Notes

1. Contamination. Good laboratory skills help prevent contamination of PCR with PCR product. Prepare reagents in a separate room from where gel electrophoresis

Fig. 4. *(see opposite page)* PCR product analysis by agarose gel electrophoresis. (**A**) nuclear gene (cystic fibrosis) products (460 bp), L-123-bp ladder, 0–100 ng K562 DNA, and A, 10 ng unknown DNA sample in duplicate. (**B**) mitochondrial gene (ND5) products (326 bp), L-123-bp ladder, 0–1 ng K562 DNA, and A, 0.1 ng unknown DNA sample in duplicate.

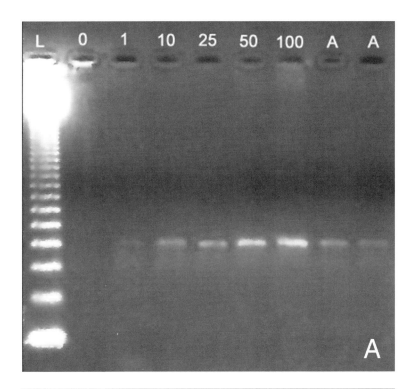

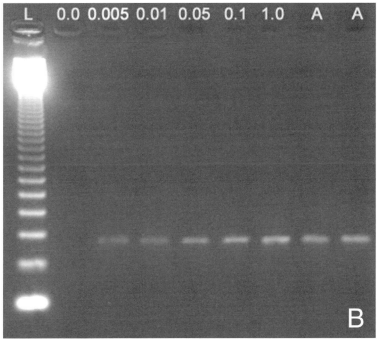

Fig. 4.

occurs. Use a laminar-flow hood or "PCR" hood to help minimize aerosol contamination (i.e., PCR product contamination of PCR setup). The use of pipet tips with filter plugs can also reduce contamination. In addition, using separate and distinct pipettors can limit contamination—one for PCR set-up, one for DNA only, and yet another for PCR products only. If contamination (positives in negative controls) occurs, decontaminate the pipet shafts with 1 N HCl or 1 N NaOH for 4 h, and remove "dirty" tubes, start with new aliquots of buffers, primers, and DNA template.

2. SYBR green lability. In our hands, freeze-thaw of SYBR green more than three times reduces its quality, thereby decreasing fluorescence intensity and giving inaccurate values for quantitative PCR. Aliquot and store SYBR green at $-20°C$ (or $-80°C$) and shield from light.

3. Primer design. The primer design is an important consideration to make when amplifying both nuclear and mitochondrial genes. This is particularly important for mitochondrial amplification because mitochondrial pseudogenes are present in the nuclear genome *(5)* and co-amplification of nuclear and mitochondrial sequences can occur *(6–8)*.

4. MgCl$_2$ optimization. The final concentration of MgCl$_2$ (4 mM) in the PCR reaction will generally amplify any gene. However, one may vary the MgCl$_2$ concentration to optimize PCR efficiency.

5. Controls. Like traditional PCR, using multiple controls is good practice. One should include a positive control (one in which your PCR should work; for example, standard curve), and run two to three negative controls (no DNA) to ensure that contamination has not occurred.

6. Speed. The use of an air-cooled thermal cycler like Roche's LightCycler greatly reduces the time it takes to complete PCR, from 2–3 h to 30–45 min. An initial denaturation can be used followed by 30–40 rounds of PCR. During each cycle, denaturation at 95°C is reduced to 0 s because the DNA will be single stranded. The annealing time can be reduced to 5–10 s (usually 30–60 s in a conventional thermal cycler), and the extension time is approximately 1 s/25 bp of PCR fragment (e.g., 10 s for a 250-bp product).

7. PCR efficiency. PCR efficiency (E) is empirically defined as $E = 10^{-1/\text{slope of standard curve}}$ and may vary in samples of different genomic complexity. Thus, if a synthetic mitochondrial DNA target is used to amplify a standard curve, we mix a known amount of bulk DNA from another species to simulate genomic complexity. We usually observe standard curve slopes (m) between -3.8 and -3.4 (i.e., $-3.8 < m < -3.4$). In our mitochondrial standard curve example the empirical $E = 1.95$. The theoretical perfect efficiency of PCR is 2 (i.e., 100% doubling of template at each cycle).

References

1. Reischl, U. and Kochanowski, B. (1995) Quantitative PCR. A survey of the present technology. *Mol. Biotechnol.* **3**, 55–71.

2. Wittwer, C. T., Herrmann, M. G., Moss, A. A., et al. (1997) Continuous fluorescence monitoring of rapid cycle DNA amplification. *Biotechniques* **22,** 130–131, 134–138.

3. Wittwer, C. T., Ririe, K. M., Andrew, R. V., et al. (1997) The LightCycler: a microvolume multisample fluorimeter with rapid temperature control. *Biotechniques* **22,** 176–181.

4. Idaho Technology, Inc. (1998) *LightCycler User's Guide.* Idaho Technology, Idaho Falls, ID.

5. Tsuzuki, T., Nomiyama, H., Setoyama, C., et al. (1983) Presence of mitochondrial-DNA-like sequences in the human nuclear DNA. *Gene* **25,** 223–229.

6. Parfait, B., Rustin, P., Munnich, A., et al. (1998) Co-amplification of nuclear pseudogenes and assessment of heteroplasmy of mitochondrial DNA mutations. *Biochem. Biophys. Res. Commun.* **247,** 57–59.

7. Wallace, D. C., Stugard, C., Murdock, D., et al. (1997) Ancient mtDNA sequences in the human nuclear genome: a potential source of errors in identifying pathogenic mutations. *Proc. Natl. Acad. Sci. USA* **94,** 14,900–14,905.

8. Hirano, M., Shtilbans, A., Mayeux, R., et al. (1997) Apparent mtDNA heteroplasmy in Alzheimer's disease patients and in normals due to PCR amplification of nucleus-embedded mtDNA pseudogenes. *Proc. Natl. Acad. Sci. USA* **94,** 14,894–14,899.

8

Yeast Nuclear Genes for mtDNA Maintenance

Françoise Foury

1. Introduction

The budding yeast *Saccharomyces cerevisiae* is a particularly suitable organism for identifying nuclear genes involved in the maintenance of the mitochondrial genome. Indeed, *S. cerevisiae* can grow and divide in the absence of respiration and, moreover, without mitochondrial DNA. In addition, the complete sequences of the nuclear and mitochondrial genomes of *S. cerevisiae* are available *(1,2)*, classical genetics has extensively been carried out, and yeast genomics is exponentially developing. Based on the Yeast Proteome Database (YPD) (http://www.proteome.com/), 464 yeast genes encode mitochondrial proteins and there are probably many more. However, our knowledge concerning the genes that are specifically involved in the maintenance of the yeast mitochondrial genome remains very incomplete. A specific feature of *S. cerevisiae* is that during vegetative growth, this yeast produces at high-frequency mutant cells that contain large deletions of mitochondrial DNA (for a review, see **ref. 3**). These cells which were initially called "petites" because they make small colonies *(4)* are also known as rho– mutants. They have an irreversible loss of respiration, do not grow on glycerol, and exhibit non-Mendelian inheritance. Cytoplasmic petites that are completely devoid of mitochondrial DNA are called rho^0. Cytoplasmic petites are to be distinguished from Pet mutants, which do not grow on glycerol and exhibit classical Mendelian inheritance *(5)*. A minority of Pet mutants are not able to maintain their mitochondrial DNA (for a review, *see* **ref. 6**). Some Pet rho^0 mutants have mutations in mitochondrial proteins directly involved in mitochondrial DNA transactions, such as the yeast mitochondrial DNA polymerase MIP1 *(7)*, the single-stranded DNA-binding protein RIM1 *(8)*, or the RNA polymerase

From: *Methods in Molecular Biology, vol. 197: Mitochondrial DNA: Methods and Protocols*
Edited by: W. C. Copeland © Humana Press Inc., Totowa, NJ

RPO41 *(9)*. Others encode cytoplasmic or nuclear proteins that control the synthesis, or the flux, of metabolites necessary for mitochondrial DNA synthesis, such as the thymidylate kinase *(10)*, thymidylate synthase *(10)*, or subunits of the ribonucleotide reductase *(11,12)*. Finally, some genes encode proteins shared by the nucleus and the mitochondrion such as the CDC9 ligase *(13,14)*. For more detailed information, we wish to refer to the remarkable review by Contamine and Picard *(6)*.

The identification of nuclear genes required for the maintenance of the mitochondrial genome has largely been based on classical genetical approaches. These are often heavy and time-consuming. However, these past years, powerful yeast mutant libraries based on random transposon insertion in the yeast genome have been developed *(15,16)*, which could be used for large-scale screening of nuclear mutants impaired in the maintenance of the mitochondrial genome.

2. Materials
2.1. Stock solutions

1. 1 *M* sodium phosphate adjusted to pH 6.5 with NaOH.
2. 1 mg Oligomycin (Sigma) dissolved in 1 mL ethanol.
3. 10 m*M* of each dATP, dCTP, dGTP, dTTP.
4. *Taq* DNA polymerase Biotools (1 unit/μL) and its (10X) buffer.
5. Phenol–chloroform–isoamylic alcohol (25:24:1) adjusted to pH 8.0 with 0.1 *M* Tris-HCl.
6. 1 *M* Tris-HCl, pH 8.0
7. 250 m*M* EDTA adjusted to pH 8.0 with NaOH
8. Oligonucleotide primers for PCR amplification of part of the mitochondrial oli1 gene: OLI1-P1, 5′ATATTATGCAATTAGTATTAGCAG (five bases upstream of the start codon of the oli1 gene) and OLI1-P2, 5′ATGTATCTTTTAAGTATGAT GCTG 5 (322 bases downstream of the start codon)
9. 10 mg/mL Freshly prepared ethidium bromide. The solution must be kept in the dark.

2.2. Strains

1. Parental strains: W303-1B (*MATalpha ade2-1 his3-15 leu2-3,115 trp1-1 ura3-1* rho⁺), its isogenic derivative of opposite mating type W303-1A (courtesy of R. Rothstein, Department of Human Genetics and Development, Columbia University) and D273-10B/A1 (*MATalpha met6* rho⁺) (courtesy of A. Tzagoloff, Department of Biological Sciences, Columbia University) (*see* **Notes 1** and **2**).
2. Strain used for cytoduction: JC7/DS400/A4 (*MATa leu1 kar1* rho⁻[DS400/A4]) *(17)*, which harbors the [DS400/A4] rho⁻ genome characterized by an oli1 resistance (oli1ᴿ) marker *(18)*.

3. Rho⁰ tester strain: IL166-6C/rho⁰ (*MATa ura1* rho⁰) *(19)*.
4. Rho⁺ tester strain : NW38-4C (*MATa his1* rho⁺ Oˢ) *(20)*.

2.3. Growth Media

1. YPD: 2% glucose, 1% yeast extract (Difco), 2% bactopeptone (Difco), and 40 mg/L adenine.
2. YPG: 3% glycerol, 1% yeast extract (Difco), 2% bactopeptone (Difco), and 40 mg/L adenine.
3. YPG-oli: 3% glycerol, 1% yeast extract (Difco), 2% bactopeptone (Difco), 25 mM sodium phosphate, pH 6.5, 40 mg/L adenine, and 3 mg/L oligomycin (*see* **Note 3**).
4. WO: 2% glucose, 0.67% yeast nitrogen base (Difco), and the required amino acids and bases.

3. Methods
3.1. EMS Mutagenesis

Large-scale screenings do not allow to distinguish easily Pet/rho⁰ mutants, which arise from rare events, from cytoplasmic petites, which are produced frequently. Therefore, it is necessary to search for conditional Pet mutants, which can maintain their mitochondrial genome at the permissive temperature (25°C) but lose it at the nonpermissive temperature (37°C). The mutants can be obtained by in vivo ethylmethanesulfonic acid (EMS) mutagenesis under conditions that give 50% cell lethality. The following steps are summarized in **Fig. 1**:

1. Strains W303-1B (or D273-10B/A1) are grown in 10 mL of liquid YPD medium to the end of the exponential phase, harvested, washed twice with 5 mL water, and resuspended in a 100-mL flask with 10 mL of a medium containing 0.12 M sodium phosphate, pH 8.0, and 2% glucose; 2% EMS are added in a ventilated hood.
2. After 90 min of incubation at 30°C with shaking, cells are collected and washed three times by centrifugation with 10 mL of ice-cold water. The cell pellet is then suspended in 10 mL YPD and incubated for 2 h at 30°C.
3. By assuming 50% cell lethality, cells are spread on YPD plates for single colonies so that approximately 200 colonies per plate are obtained with a total of 40,000 colonies for 1 mutagenesis. Plates are incubated at 25°C for 3–5 d.

3.2. Minitransposon-Based Mutant Libraries

An alternative to EMS mutagenesis consists in using insertion mutant libraries. Yeast genomic libraries, containing either the *LEU2* or *URA3* marker, have been constructed by randomly inserting a minitransposon in yeast genomic

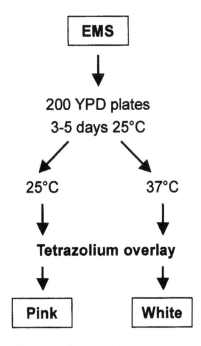

Fig. 1. Screen for conditional Pet mutants.

DNA *(15,16)*. These libraries are fully described at http://ygac.med.yale.edu and are freely available upon request. They are extremely powerful because the mutagenized gene can be recovered directly without any cloning step by the so-called "PCR-vectorette" procedure (http://genome-www.stanford.edu/group/botlab/protocols/vectorette.html). The transposon insertion site in the genome can be identified immediately by direct sequencing of a PCR-amplified yeast DNA fragment flanking the transposon. Although we have used these mutant libraries very successfully for other purposes, we have never used them to isolate Pet mutants; therefore, this approach will not be described here. We refer the reader to the websites mentioned.

3.3. Screening for Pet Mutants Accumulating rho⁻ at 37°C

Strains that have a functional respiratory chain accumulate a red compound called formazan in the presence of tetrazolium *(21)* and are pink; colonies which do not respire remain white (**Fig. 1**).

1. The mutagenized colonies grown on YPD plates (see **Subheading 3.1.**) are replicated on YPD plates.
2. The plates are incubated for 2 d at two temperatures (25°C and 37°C) and are overlayed with a medium containing 2% glucose, 50 m*M* potassium phosphate,

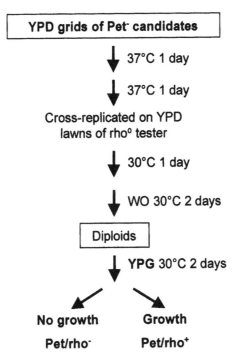

Fig. 2. Identification of Pet mutants required for mitochondrial DNA maintenance.

pH 7.0, 1.5% agar, and 0.5% tetrazolium cooled to 42°C and prepared immediately before use.

3. The plates are kept in the dark for 2–3 h, and those colonies that are pink after growth at 25°C and are white after growth at 37°C are selected as possible thermosensitive (ts) Pet mutants.

4. These ts candidates are gridded on YPD plates and their respiratory deficiency is verified by their inability to grow on YPG plates at 37°C. Purified clones from each initial Pet mutant are isolated by streaking for single colonies on YPD plates.

3.4. Cross with rho⁰ Tester Strain

The test described here (**Fig. 2**) allows one to ascertain that the glycerol growth defect at 37°C is associated with loss of mitochondrial DNA, and not simply with loss of respiration. After EMS mutagenesis, mutations in the mitochondrial DNA are rare. However, it is also necessary to verify by tetrad analysis that the ts phenotype is produced by a single nuclear mutation.

1. The Pet⁻ candidates (purified as described in **Subheading 3.3.**) are gridded on YPD plates (50 mutants per plate), and after incubation at 37°C for 24 h, they are replicated to fresh YPD plates and incubated at 37°C for another 24-h period.

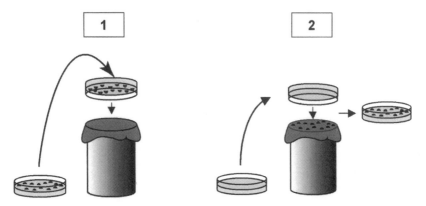

Fig. 3. Cross-replicating of a master plate on lawns of a tester strain of opposite mating type. (1) The YPD master plate of Pet mutants is replicated on a velvet; (2) the YD lawn of the tester strain is cross-replicated on the same velvet.

2. They are cross-replicated on YPD lawns of a rho[0] tester strain of opposite mating type and complementary auxotrophies (such as IL166-6C/rho[0]), as illustrated in **Fig. 3**, and incubated at 30°C for 1 d to produce diploids (**Fig. 2**).
3. Diploids are selected for their capacity to grow on WO medium without amino acids and bases by replica-plating on WO plates and incubation at 30°C for 2 d.
4. The diploid patches are replicated on YPG plates and incubated for 2 d at 30°C. No diploid growth on YPG implies that mitochondrial DNA has been lost during cell incubation at 37°C. In contrast, diploid growth on glycerol means that the Pet mutants do not lose mitochondrial DNA. Some papillae can be observed on YPG plates if loss of mitochondrial DNA has only been partial (*see* **Note 4**).
5. The Pet mutants (purified as described in **Subheading 3.3.**) are crossed with strain W303-1A on YPD medium, the diploid progeny is induced to sporulation, and tetrad analysis is performed by standard procedures *(22)* in order to verify the 2:2 segregation of the glycerol growth defect at 37°C.

3.5. Maintenance of rho⁻ Mitochondrial Genomes in Pet Mutants

Although many Pet mutants are not able to maintain a wild-type mitochondrial genome *(6)*, they can stably replicate rho⁻ mitochondrial DNA. This is the case for mutants affected in mitochondrial transcription or protein synthesis *(6)*. In contrast, mutants deficient in basic components of the mitochondrial DNA replication machinery cannot maintain rho⁻ genomes. To distinguish between these two classes of Pet mutants, the stability of a rho⁻ genome, which is stable in the wild-type strain, is simultaneously followed in the wild-type strain and Pet mutants. The stable [DS400/A4] rho⁻ genome is introduced by cytoduction in rho[0] wild-type and Pet strains (**Fig. 4**). Cytoduction is obtained

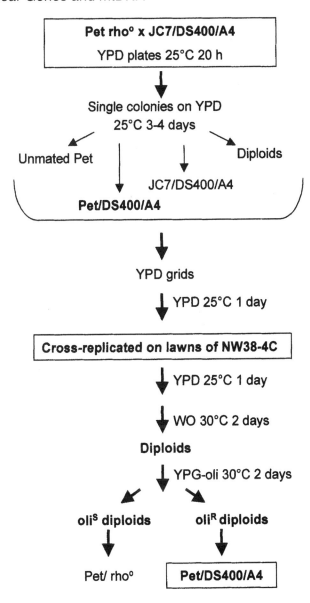

Fig. 4. Isolation of a Pet/DS400/A4 cytoductant.

by using the JC7 strain, which carries a *kar1* mutation, which prevents nuclear fusion and allows cytoplasm mixing in the zygote *(17)*.

1. Pet rho⁰ mutants are obtained by ethidium bromide treatment. The strains are inoculated at a concentration of 5×10^6 cells/mL in liquid YPD medium

containing 50 mM sodium phosphate, pH 6.5, and incubated in the dark in the presence of 50 μg/mL ethidium bromide for 24 h at 30°C. Cells are spread for single colonies on YPD plates and incubated at 30°C for 3–4 d. This treatment does not affect cell survival but elicits total loss of mitochondrial DNA. A colony is selected and and a purified clone is isolated by streaking on YPD plate for single colonies.

2. Pet rho^0 mutants and the JC7/DS400/A4 rho$^-$ strain, which bear a *kar1* mutation, are refreshed by an overnight incubation at 25°C on YPD, crossed on YPD plates, and incubated at 25°C for 20 h. For each cross, the cell patches are spread for single colonies on YPD plates. These colonies can be haploid cells of JC7/DS400/A4 or Pet/rho^0 mutants, diploids (because *kar1* is a leaky mutation), or Pet/DS400/A4 cytoductants (**Fig. 4**).

3. The colonies are gridded on YPD master plates (at least 2 plates for each cross and 50 colonies on each plate), and after 1 d at 25°C, the plates are cross-replicated on YPD lawns of the oligomycin-sensitive rho$^+$ tester strain NW38-4C. Only original Pet/rho^0 and cytoductant Pet/DS400/A4 strains can mate to NW38-4C. The diploids issued from this cross are selected for prototrophy by replica-plating on WO plates and incubation at 30°C for 2 d.

4. Diploids are replicated on YPG-oli plates. Oligomycin-resistant diploids result from the mating of strain NW38-4C to Pet/DS400/A4 cytoductants, followed by genetic recombination between rho$^+$ and [DS400/A4] rho$^-$ genomes.

5. A purified clone from each Pet/DS400/A4 mutant is isolated at 25°C by spreading for single colonies on a YPD plate and incubation for 3–4 d at 25°C. The presence of the [DS400/A4] rho$^-$ genome is verified by crossing the Pet/DS400/A4 mutant with NW38-4C, as described in **step 3** and **step 4**.

6. Quantification of the maintenance of the rho$^-$ genome is done as shown in **Fig. 5**. Cell patches of the purified Pet/DS400/A4 mutants are incubated on YPD plates at 25°C and 37°C, respectively, for 1 d, replicated on fresh YPD plates. After incubation for 1 or 2 additional days at these temperatures, they are spread for single colonies (at least 500 colonies for each Pet/DS400/A4 mutant and 100 colonies per plate) on YPD plates and incubated for 3–4 d at 25°C.

7. The YPD plates are cross-replicated on YPD lawns of the NW38-4C tester strain, and oligomycin-resistant and oligomycin-sensitive diploids are selected and identified as described in **step 3** and **step 4**. The ratio of the number of oligomycin-resistant diploids to the total number of diploids is an estimate of the fraction of Pet colonies that have retained the [DS400/A4] genome when they are incubated at 25°C or 37°C.

3.6. PCR Amplification of Mitochondrial DNA

A very sensitive and complementary alternative to this purely genetical approach is to use polymerase chain reaction (PCR) to detect the [DS400/A4] genome. However, a reliable estimation of the fraction of Pet mutants that have

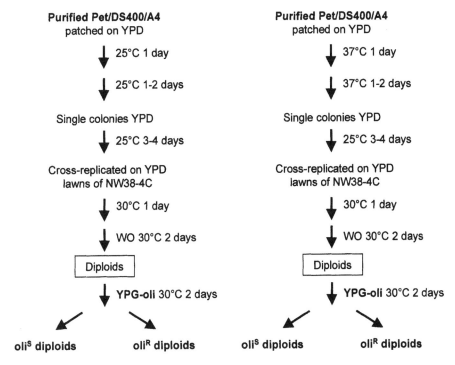

Fig. 5. Estimation of the fraction of Pet/DS400/A4 clones that retain the [DS400/A4] rho⁻ genome at 25°C and 37°C.

retained the DS400/A4 genome at 25°C or 37°C requires one to run PCR assays with at least 96 independent isolated colonies for each Pet mutant.

1. The Pet/DS400/A4 colonies isolated as described in **step 6** of **Subheading 3.5.** are grown 24 h at 25°C in test tubes containing 2 mL of YPD medium, harvested in 2-mL Eppendorf tubes by a 30-s centrifugation at the maximum speed of a microcentrifuge and washed twice with 1 mL of water.

2. The pellets are suspended in 200 μL of a buffer containing 10 mM Tris-HCl, 100 mM NaCl, 1 mM EDTA, pH 8.0. Four hundred microliters of glass beads (Sigma, acid washed, 425–600 μm) and 300 μL phenol–chloroform–isoamylic alcohol (25:24:1) are added to the suspension.

3. The yeast cells are mixed thoroughly for 2 min using a vortex mixer and the suspension is centrifuged for 5 min at room temperature at the maximum speed of a microcentrifuge.

4. The supernatant is transferred in a 1.5-mL Eppendorf tube and extracted a second time with an equal volume of phenol–chloroform. The mixture is centrifuged as in **step 3** and the supernatant is extracted twice with an equal volume of ether.

The final extract, which contains nuclear and mitochondrial DNAs, can be frozen at –20°C and kept indefinitely.

5. The PCR reactions are performed in a 0.5-mL tube containing 39 μL of water, 5 μL (10X) Biotools buffer, 1 μL of a mix containing 10 mM of each dNTP, 1 μL of OLI1-P, and OLI1-P2 (250 ng/μL) oligonucleotide primers, 2 μL DNA extract, and 1 μL *Taq* DNA polymerase (Biotools).

6. The PCR amplification is run in a PTC-100 apparatus (MJ Research, Inc.) as follows: initial denaturation of 1 min at 94°C; 30 cycles of 15 s at 94°C, 30 s at 50°C, and 45 s at 68°C; a final cycle of 5 min at 68°C.

4. Notes

1. W303 and D273 strains grow well on nonfermentable media and have good respiration. The S288c derivatives are not suitable for studying Pet mutants, as S288c exhibits defects in respiration.

2. D273-10B is cir^0 (it has no 2-μm plasmid) and thus cannot be transformed with 2-μm-based plasmids.

3. Sodium phosphate is prepared from a 1 M stock solution, pH 6.5, autoclaved separately.

4. Ethylmethane sulfonic acid can produce mild mutations in nuclear genes, giving partial loss of mitochondrial DNA even if the gene is essential for the maintenance of the mitochondrial genome. After identification of the mutated gene, the latter must be deleted to verify that mitochondrial DNA is completely lost.

References

1. Goffeau, A., Barrell, B. G., Bussey, H., et al. (1996) Life with 6000 genes. *Science* **274,** 563–567.
2. Foury, F., Roganti, T., Lecrenier, N., and Purnelle, B. (1998) The complete sequence of the mitochondrial genome of Saccharomyces cerevisiae. *FEBS Lett.* **440,** 325–331.
3. Dujon, B. (1981) Mitochondrial genetics and function, in *The Molecular Biology of the Yeast Saccharomyces: Life Cycle and Inheritance* (Strathern, J. N., Jones, E. W., and Broach, J. R., eds.), Cold Spring Harbor Laboratory Press, Cold Spring Harbor, NY, pp, 505–635.
4. Ephrussi, B., Hottinguer, H., and Tavlizki, J. (1949) Action de l'acriflavine sur les levures. I. La mutation'petite colonie'. *Ann. Inst. Pasteur* **76,** 351–367.
5. Tzagoloff, A. and Dieckmann, C. L. (1990) PET genes of *Saccharomyces cerevisiae*. *Microbiol. Rev.* **54,** 211–225.
6. Contamine, V. and Picard, M. (2000) Maintenance and integrity of the mitochondrial genome: a plethora of nuclear genes in the budding yeast. *Microbiol. Rev.* **64,** 281–315.
7. Foury, F. (1989) Cloning and sequencing of the nuclear gene MIP1 encoding the catalytic subunit of the yeast mitochondrial DNA polymerase. *J. Biol. Chem.* **264,** 20,552–20,560.

8. Van Dyck, E., Foury, F., Stillman, B., and Brill, S.J. (1992) A single-stranded DNA binding protein required for mitochondrial DNA replication in *S. cerevisiae* is homologous to *E. coli* SSB. *EMBO J.* **11,** 3421–3430.

9. Greenleaf, A. L., Kelly, J. L., and Lehman, I. R. (1986) Yeast RPO41 gene is required for transcription and maintenance of the mitochondrial genome. *Proc. Natl. Acad. Sci. USA* **83,** 3391–3394.

10. Newlon, C. S. and Fangman, W. L. (1975) Mitochondrial DNA synthesis in cell cycle mutants of *Saccharomyces cerevisiae*. *Cell* **5,** 423–428.

11. Elledge, S. J. and Davis, R. W. (1987) Identification and isolation of the gene encoding the small subunit of ribonucleotide reductase from *Saccharomyces cerevisiae*: DNA damage-inducible gene required for mitotic viability. *Mol. Cell. Biol.* **7,** 2783–2793.

12. Wang, P. J., Chabes, A., Casagrande, R., Tian, X. C., Thelander, L., and Huffaker, T. C. (1997) Rnr4p, a novel ribonucleotide reductase small subunit protein. *Mol. Cell. Biol.* **17,** 6114–6121.

13. Willer, M., Rainey, M., Pullen, T., and Stirling, C. J. (2000) The yeast CDC9 gene encodes both a nuclear and a mitochondrial form of DNA ligase I. *Curr. Biol.* **9,** 1085–1094.

14. Donahue, S. L., Corner, B. E., Bordone, L., and Campbell, C. (2001) Mitochondrial DNA ligase function in *Saccharomyces cerevisiae*. *Nucleic Acids Res.* **29,** 1582–1589.

15. Burns, N., Grimwade, B., Ross-Macdonald, P. B., Choi, E-Y., Finberg, K., Roeder, G. S., et al. (1994) Large-scale analysis of gene expression, protein localization, and gene disruption in *Saccharomyces cerevisiae*. *Genes Dev.* **8,** 1087–1105.

16. Kumar, A., Cheung, K-H., Tosches, N., et al. (2002) The TRIPLES database: a community resource for yeast molecular biology. *Nucl. Acids Res.* **30,** 73–75.

17. Conde, J. and Fink, G. R. (1976) A mutant of *Saccharomyces cerevisiae* defective for nuclear fusion. *Proc. Natl. Acad. Sci. USA* **73,** 3651–3655.

18. Macino, G. and Tzagoloff, A. (1979) Assembly of the mitochondrial membrane system: partial sequence of a mitochondrial ATPase gene in *Saccharomyces cerevisiae*. *Proc. Natl. Acad. Sci. USA* **76,** 131–135.

19. Genga, A., Bianchi, L., and Foury, F. (1986) A nuclear mutant of *Saccharomyces cerevisiae* deficient in mitochondrial DNA replication and polymerase activity. *J. Biol. Chem.* **261,** 9328–9332.

20. Foury, A. and Goffeau, A. Genetic control of enhanced mutability of mitochondrial DNA and gamma-ray sensitivity in Saccharomyces cerevisiae. *Proc. Natl. Acad. Sci. USA* **76,** 6529–6533.

21. Ogur, M. and John, R. S. (1956) A differential and diagnostic plating method for population studies of respiration deficiency in yeast. *J. Bacteriol.* **72,** 500.

22. Sherman, F. and Hicks, J. (1994) Micromanipulation and dissection of asci, in *Guide to Yeast Genetics and Molecular Biology*, Methods in Enzymology Vol. 194 (Guthrie, C. and Fink, G. R., eds.), Academic, New York, pp. 21–37.

9

Measuring mtDNA Mutation Rates in *Saccharomyces cerevisiae* Using the mtArg8 Assay

Micheline K. Strand and William C. Copeland

1. Introduction

Mitochondrial mutation rates have historically been more refractory to genetic analysis than nuclear mutation rates. This is the result of the lack of auxotrophic genes in the mitochondria, the multiple (50–100) copies of mitochondrial genomes per cell *(1)*, the random segregation of mitochondrial organelles during meiotic and mitotic cell division, and the apparent unequal replication of mitochondria *(2)*.

Although large mitochondrial deletions are relatively common and are readily detectable by polymerase chain reaction (PCR) or gel electrophoresis *(3–7)*, these techniques are based on analysis of one sample at a time. Identification of new genes involved in mitochondrial DNA replication, DNA repair, biogenesis, segregation, or regulation of copy number requires the ability to rapidly screen large numbers of samples and determine mutation rates. Nuclear mutation rates are typically determined by selection for mutations in nutrition- or antibiotic-resistance genes. No auxotrophic genes exist naturally in mitochondria. Erythromycin can be used to select mitochondrial mutations; however, resistance to erythromycin in *Saccharomyces cerevisiae* has been shown to be the result of a single C to G transversion at nucleotide 3993 of the large ribosomal RNA gene *(8)*. Erythromycin is a recessive mutation; thus, homoplasmy of the mutation must occur for resistance. Mutations that do not survive to become homoplasmic will not be counted. Using erythromycin to screen for mitochondrial mutators will favor mutators with reduced mitochondrial copy number and will be less likely to pick up mutators with increased copy number.

From: *Methods in Molecular Biology, vol. 197: Mitochondrial DNA: Methods and Protocols*
Edited by: W. C. Copeland © Humana Press Inc., Totowa, NJ

Strains with high mitochondrial mutation rates are also likely to form mutations that lower the efficiency of oxidative phosphorylation. Cells respond to insufficient ATP by increasing mitochondrial copy number. Thus, mitochondrial mutators may have more mitochondrial genomes, confounding mutation rate determinations that are based on a recessive mutation becoming homoplasmic. We have, in fact, observed precisely this phenomenon with the Mcg 1 mutation *(9)*. A second genetic screening technique is to measure rates of petite colony formation. This is a nonselective assay; nevertheless, this approach has been used successful by a number of investigators *(10,11)*. A third approach is to estimate mutation rates based on polymorphism frequency. Although this approach has value in evolutionary studies, it is not suited for isolating mutants with altered mitochondrial properties.

The desire to insert more tractable genes into the mitochondrial genome in order to measure a wider variety of mutations and to be able to measure mitochondrial mutation rates more easily has been hindered by the differences in codon usage between mitochondria and the nucleus and the physical difficulties in transforming mitochondria. However, some groups have had success in transforming mitochondria, and in 1996, Fox's group recoded the nuclear Arg8 gene using the mitochondrial genetic code and inserted this gene into the Cox3 gene in the mitochondrial genome *(12)*. This construct enables genetic selection of revertants using media lacking arginine.

We describe here our protocol for measuring mitochondrial mutation rates using a +1 insertion at the 1592-bp (base pair) position in the mtarg8 gene *(9)*. This mtarg8 strain could be used to measure a variety of types of mutations; the Petes lab has inserted various microsatellite repeats into this same strain to measure mitochondrial microsatellite stability rates *(13)*.

The mtarg8$^-$ derivative of TF235 (TF236) was a spontaneous arg8 mutant isolated by Fox. We sequenced this mutant and determined that it had three mutations in the mtArg8 gene (**Fig. 1**). We also sequenced Arg$^+$ reversions of this strain and determined that reversion of one of these mutations gives rise to the Arg$^+$ phenotype. Consequently, this assay measures one base deletion at nucleotide 1592 of the mtArg8 gene. Thus, the reversion (i.e., a −1 frameshift) of the mitochondrial mutation at position 1592 results in an Arg$^+$ phenotype. The sequence of the mtarg8 gene in TF236 and the Arg$^+$ reversions are shown in **Fig. 1**.

The mtarg8 reversion assay described in this chapter offers a number of advantages to the field. At the simplest level, it provides an additional selectable point mutation in a different gene than the erythromycin assay. However, the mtArg8 reversion is a dominant mutation; consequently, not all genomes must be mutated in order for the cell to be able to form a colony. Because the erythromycin mutation is recessive, that assay favors mutants with reduced

A published mtArg8:
700 aaa gct ttt caa gta aca aca tat...
     Lys Ala Phe Gln Val Thr Thr Tyr...

1583 gta tca aat tat gta tta gat aca att gct gat gaa...
      Val Ser Asn Tyr Val Leu Asp Thr Ile Ala Asp Glu...

B TF236 (arg⁻)
700 aaa gct ttt caa gta aca a**T**a tat...
     Lys Ala Phe Gln Val Thr **Ile** Tyr...

1583 gta tca aat **A**ta tgt att aga ta**T** aat tgc tga tga a...
      Val Ser Asn **Met Cys Ile Arg Tyr Asn Cys Trp Trp**...

Fig. 1. (**A**) The DNA sequence of the mtarg8 mutant (TF236) around the sites of reversions; (**B**) the DNA sequence of the Arg⁺ revertants. Sequence changes are shown in bold.

mitochondrial copy number. This assay, on the other hand, favors mutants with increased copy number. Finally, the TF236 strain has no oxidative phosphorylation (as a result of the partial deletion of the mtCox3 gene). Consequently, the effects of the mutation on oxidative phosphorylation are not a factor. We have demonstrated the utility of this assay by using this method to isolate a number of genes that affect mitochondrial mutation rates not isolated by investigators using other assays (*9*). We have used this assay to screen the effect of nuclear gene deletions or mutations in genes of unknown function to assess their role in mitochondrial DNA metabolism.

2. Materials

All reagents were obtained from Difco (Detroit, MI), unless otherwise noted.

1. Yeast strain: TF236 genotype is MATα *ino1*::*His3 arg8*::*hisG pet9 (op1) ura3-52 lys2 cox3*::*mtArg8* and was a generous gift from Tom Fox (Cornell University). (*See* **Note 1**.)
2. YPD plates: 20 g dextrose, 20 g bacto-peptone, 10 g bacto-yeast extract, 20 g bacto-agar (ICN Biomedicals, Aurora, OH), double distilled (dd) H_2O to 1 L, mix well, autoclave for 25 min, cool to 50°C, pour into 100-mm round Petri plates.
3. SD-arg plates: 20 g dextrose, 1.7 g yeast nitrogen base without amino acids, 5 g $(NH_4)_2SO_4$, 20 g bacto-agar, 800 mL ddH_2O. Mix well, autoclave for 25 min, cool to 50°C, add 200 mL of filter-sterilized 5X arg dropout mix (300 mg L-isoleucine, 200 mg L-adenine, 200 mg L-methionine, 2 g L-threonine, 200 mg L-uracil, 1 g L-leucine, 1.58 g L-valine, 300 mg L-lysine HCl, 500 mg L-phenylalanine, 300 mg L-tyrosine, 200 mg L-tryptophan, 200 mg L-histidine, ddH_2O to 1 L.

Store at 4°C in the dark. All amino acids were obtained from Sigma-Aldrich (Boston, MA).

3. Methods (*see* Notes 2–4)
3.1. Growth of the arg8 Mutation Strain

Streak for individual TF236 colonies on fresh YPD plates. After 2 d, pick individual colonies into 200 μL ddH$_2$O. Very slow growing strains may need to be grown for 3 d. We pick no more than four at a time to limit the amount of time that the cells spend in water. Vortex well (20 s) and remove 2 μL for further dilution. Plate 99% of the colony on the SD-arg plate. Dilute the other 1% appropriately and plate on a YPD plate to determine colony size. Dilute onto YPD so that 200–400 colonies grow on the YPD plate. The exact dilution will depend on the growth rate of the strain—we generally plate 2×10^{-5} of the colony per YPD plate. Incubate plates in a 30°C incubator. If the incubator has a fan, place open bottles of sterile water in the incubator to keep humidity levels up.

3.2. Scoring

The YPD plates were scored after 3 d. SD-arg plates were scored after 3 wk. Our observation has been that Arg$^+$ prototrophs form continuously during the 3 wk. If the assay could be continued beyond 3 wk, there were most likely be even more prototrophs; however, in our hands, the plates became too dry after 3 wk to support further growth.

3.3. Sequencing of Revertants

We sequenced revertants using an ABI 377 and followed the manufacturer's instructions.

3.4. Mutation Rate Calculations

We determined the mutation rate using the method of Luria and Delbruck *(14)* if the majority of SD-arg plates have no colonies, and the method of Lea and Coulson *(15)* if the majority of plates have > 1 colony. Other methods have been used by other investigators. Although many labs measure mutation frequencies instead of mutation rates and numerous investigators use the two terms interchangeably, measuring mutation frequencies ignores the impact of early mutational events and can, consequently, give misleading results. (*See* **Notes 5–6**.)

4. Notes

1. The *cox3* mutation in the TE236 strain renders the mitochondria in this strain nonfunctional, but the *pet9* mutation is lethal in the absence of mitochondria.

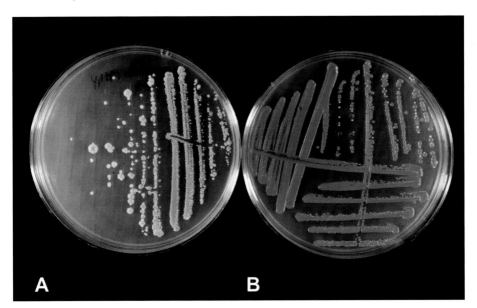

Fig. 2. Growth of a wild-type (YPH925) and mutant (YPH925mcg1) on YPD plates: (A) the mcg1 mutant, showing an increased frequency of petite colony formation as well as an increased frequency of petite sectors within grande colonies, (B) the wild-type, showing a normal level of petite colony formation for this strain. This strain contains the ade2-101 mutation, which results in the red pigment in cells with functioning mitochondria. The mcg1 mutant has a 50-fold increase in mitochondrial mutation rates and a 3-fold increase in mitochondrial DNA per cell *(9).*

Therefore, although the mitochondria are not functioning in this strain, the strain cannot lose its mitochondria.

2. The assay can be used for screening nuclear genes that affect the mitochondrial mutation rate by random mutagenesis of the nucleus with methyl-methanesulfonate, *N*-methyl-*N'*-nitro-*N*-nitrosoguanidine, *N*-methyl-*N*-nitrosourea, or other mutagens.

3. We have also used this assay to screen nuclear gene deletion of genes with unknown function to test for their role in mitochondrial DNA metabolism. Disruption plasmids can be obtained from the Yale Genome Analysis Center (http://ycmi.med.yale.edu/YGAC/home.html). These disruption plasmids are used to make gene disruptions in the TF236 strain. All disruptions should be verified by PCR and multiple disruptions should be assayed to avoid potentially misleading effects of secondary mutations (which may run as high as 10%).

4. This assay may also be used to screen for toxins that preferentially target mitochondrial DNA, such as ethidium bromide or dideoxynuclosides.

5. Once a mitochondrial mutator gene is identified, the effects on nuclear mutation rates should also be examined. We assayed nuclear mutation rates by disrupting

the gene in a strain that contained lys2-801 and ade2-101 mutations. The nuclear mutation rates were measured by counting the reversions of these two loci on SD plates lacking either lysine or adenine and by calculating mutation rates as described above in **Subheading 3.4.**. The lys2-801 and ade2-101 loci were chosen for convenience. A variety of nuclear mutation assays exist, and other assays will be easier if using other strains.

6. As this assay is based on reversion rates of a mutation in the arginine synthesis pathway, apparent mutators must be checked for alterations in arginine transport. This can be done by measuring growth rates in media with very low concentrations of arginine. We have also used petite colony and petite sector formation as a second assay to ensure that the mutation affects mitochondria and not the arginine pathway. This is easily done visually in a strain that contains an ade2 or an ade3 mutation, as petite colonies or petite sectors are white (**Fig. 2**).

References

1. Grimes, G. W., Mahler, H. R., and Perlman, R. S. (1974) Nuclear gene dosage effects on mitochondrial mass and DNA. *J. Cell Biol.* **61,** 565–574.
2. Davis, A. F. and Clayton, D. A. (1996) In situ localization of mitochondrial DNA replication in intact mammalian cells. *J. Cell Biol.* **135,** 883–893.
3. Ayala-Torres, S., Chen, Y., Svoboda, T., Rosenblatt, J., and Van Houten, B. (2000) Analysis of gene-specific DNA damage and repair using quantitative polymerase chain reaction. *Methods* **22,** 135–147.
4. Chen, T. J., Boles, R. G., and Wong, L. J. (1999) Detection of mitochondrial DNA mutations by temporal temperature gradient gel electrophoresis. *Clin. Chem.* **45,** 1162–1167.
5. Khrapko, K., Coller, H. A., Andre, P. C., Li, X. C., Hanekamp, J. S., and Thilly, W. G. (1997) Mitochondrial mutational spectra in human cells and tissues. *Proc. Natl. Acad. Sci. USA* **94,** 13,798–13,803.
6. Muniappan, B. P. and Thilly, W. G. (1999) Application of constant denaturant capillary electrophoresis (CDCE) to mutation detection in humans. *Genet. Anal.* **14,** 221–227.
7. Salazar, J. J. and Van Houten, B. (1997) Preferential mitochondrial DNA injury caused by glucose oxidase as a steady generator of hydrogen peroxide in human fibroblasts. *Mutat. Res.* **385,** 139–149.
8. Sor, F. and Fukuhara, H. (1984) Erythromycin and spiramycin resistance mutations of yeast mitochondria: nature of the rib2 locus in the large ribosomal RNA gene. *Nucleic Acids Res.* **12,** 8313–8318.
9. Strand, M. K., Stuart, G., Longley, M. J., Dominick, O. C., and Copeland, W. C. (2002) Identification of a kinase gene and characterization of its role in mitochondrial mutagenesis, submitted.
10. Senbongi, H., Ling, F., and Shibata, T. (1999) A mutation in a mitochondrial ABC transporter results in mitochondrial dysfunction through oxidative damage of mitochondrial DNA. *Mol. Gen. Genet.* **262,** 426–436.

11. Backer, J. and Foury, F. (1985) Repair properties in yeast mitochondrial DNA mutators. *Curr. Genet.* **10,** 7–13.
12. Steele, D. F., Butler, C. A., and Fox, T. D. (1996) Expression of a recoded nuclear gene inserted into yeast mitochondrial DNA is limited by mRNA-specific translational activation. *Proc. Natl. Acad. Sci. USA* **93,** 5253–5257.
13. Sia, E. A., Butler, C. A., Dominska, M., Greenwell, P., Fox, T. D., and Petes, T. D. (2000) Analysis of microsatellite mutations in the mitochondrial DNA of Saccharomyces cerevisiae. *Proc. Natl. Acad. Sci. USA* **97,** 250–255.
14. Luria, S. E. and Delbruck, M. (1943) Mutations of bacteria from virus sensitivity to virus resistance. *Genetics* **28,** 491–511.
15. Lea, D. E. and Coulson, C. A. (1949) The distribution of the number of mutants in bacterial populations. *J. Genet.* **49,** 264–285.

10

Measuring Oxidative mtDNA Damage and Repair Using Quantitative PCR

Janine H. Santos, Bhaskar S. Mandavilli, and Bennett Van Houten

1. Introduction

The human mitochondrial genome was completely sequenced in 1981 by Anderson and co-workers (1) and consists of a closed circular supercoiled DNA molecule of 16,569 base pairs. Mammalian cells characteristically contain a few hundred to several thousand mitochondria, each with 2–10 copies of the genome. The mitochondrial genome encodes 13 polypeptides, 22 transfer RNAs (tRNAs), and 2 rRNA. The 13 polypeptides encoded by the mitochondrial DNA (mtDNA) are essential subunits of the electron transport chain (ETC) and ATP synthase; cells lacking mtDNA are completely dependent on glycolysis for survival (2).

During oxidative phosphorylation, oxygen is reduced to water in a four-electron reduction at cytochrome oxidase (complex IV) in the mitochondrial ETC. More than 90% of the body's oxygen is consumed by the ETC (3). It has been estimated that about 1–3% of the oxygen consumed in the mitochondria is released as superoxide and hydrogen peroxide (H_2O_2) at complexes I and III, making mitochondria a major source of endogenous reactive oxygen species (ROS) (4,5). These ROS can damage macromolecules in mitochondria, including proteins, lipids, RNA, and DNA.

In the last two decades, several studies have suggested that mtDNA is more susceptible than nuclear DNA to genotoxic agents, most notably ROS (6). In this context, numerous studies applied high-performance liquid chromatography (HPLC)/eletrochemical detection to measure the levels of 8-oxodeoxyguanosine (8-oxodG), a common byproduct of oxidative stress, both in mitochondrial and nuclear DNA (nDNA). Early reports indicated that the mtDNA had higher

From: *Methods in Molecular Biology, vol. 197: Mitochondrial DNA: Methods and Protocols*
Edited by: W. C. Copeland © Humana Press Inc., Totowa, NJ

amounts of oxidative damage than the nuclear genome *(7)*. However, it was found that the lower yields of mtDNA were more likely to be oxidized during the purification and processing steps. This would, in turn, lead to unrealistically high estimates of 8-oxodG, which were, indeed, an artifact of DNA isolation *(8,9)*.

Mitochondrial DNA was also reported to be methylated three to seven times more extensively than the nDNA after treatment with *N*-methyl-*N*-nitrosourea (MNU), *N*-nitrosodimethylamine (NDMA), and *N*-methyl-*N*-nitro-nitrosoguanidine (MNNG) (reviewed in **ref. 6**). It is interesting to note that the same group was able to show that many DNA-damaging agents bind mtDNA preferentially, including polyaromatic hydrocarbons (PAH), nitrosamines (NA) and cisplatin (reviewed in **ref. 6**).

Damage to the mtDNA, if not repaired, could lead to mutations during replications. More importantly, it could have further implications in cell physiology and, ultimately, in human health. In this context, several studies showed that mutations in mtDNA are associated with a number of hereditary human diseases, including Kearns–Sayre syndrome, Leber's hereditary optic neuropathy, Pearson's syndrome, and chronic progressive external opthal-moplegia *(10)*. mtDNA point mutations and/or deletions were also associated with the carcinogenic process, having been found to occur in some tumors *(11,12)* as well as in aging tissues *(13)*.

The first documented study of mammalian mtDNA repair was performed by Clayton and colleagues in 1974 *(14)*, who studied the loss of ultraviolet (UV)-induced cyclobutane dimers (CPD) from covalently closed supercoiled mtDNA, using a modified T4-endonuclease-sensitive assay. Their finding that CPD were not removed from mtDNA has lead, unfortunately, to the idea that mitochon-dria do not undergo DNA repair. Although it is true that mammalian mitochondria do not have the capacity for nucleotide excision repair, probably because of the lack of mitochondrial transport of the complex machinery to mediate this process, mitochondria have a robust base excision repair system *(6,15)*.

The detection of gene-specific damage and repair has been studied in nuclear and mitochondrial DNA by the use of Southern analysis *(16,17)*. This method involves the generation of lesion-specific strand breaks within a restriction fragment of genomic DNA, using a damage-specific endonuclease such as T4 endonuclease V at the site of CPD, formamidopyrimidine (FAPY) glycosylase at sites of oxidized purines, and endonuclease III at sites of oxidized pyrimidines or *Escherichia coli* UvrABC at a variety of lesions. The frequency of the lesions is measured as endonuclease-sensitive sites after hybridization with the radiolabeled probe. Wilson and co-workers were the first group to use this approach to study mtDNA damage and repair after the alkylating agent streptozotocin in a rat insulinoma cell line *(18)*. They observed significant repair within the first 8 h after treatment and 70% removal of the mtDNA damage by 24 h. Using a similar approach in the same cell line, they *(19)* found that a 1-h treatment with alloxan (5 m*M*), which generates superoxide and hydrogen peroxide by redox cycling, caused significant mtDNA damage, which was fully repaired within 4 h.

1.1. Development of the QPCR Assay

Southern blot techniques used for estimation of gene-specific damage and repair has several limitations. The method requires knowledge of the restriction sites flanking the damaged site, the use of large quantities of DNA, and incision of DNA lesions with a specific endonuclease. A quest for developing alternative methods to measure gene-specific damage and repair led our laboratory to develop a quantitative polymerase chain reaction (QPCR) *(20)*. The basis of the QPCR assay is that lesions present in the DNA block the progression of any thermostable polymerase on the template, resulting in the decrease of DNA amplification in the damaged template when compared to the undamaged DNA *(20)*. The PCR actually measures the fraction of undamaged template molecules, which decreases with increasing amount of lesions. Gene-specific damage is measured as loss of template amplification, and DNA repair activity is measured as the restoration of the amplification signal *(20)*. Analysis of long PCR products allowed increased sensitivity of the assay *(21,22)*. We have successfully used this approach to follow damage and repair of a wide variety of DNA lesions in mtDNA from yeast, rodent and human cells and tissues (reviewed in **ref. 23**).

1.2. mtDNA Is More Prone to Oxidative DNA Damage

Hydrogen peroxide (H_2O_2) was found to induce two to three times more damage to mtDNA than nuclear targets in human fibroblasts *(24)*. These lesions were repaired within 1.5 h after a 15-min treatment with 200 μ*M* H_2O_2. However, a 60-min exposure to 200 μ*M* hydrogen peroxide led to persistent mitochondrial DNA damage even though the nuclear damage was repaired within 1.5 h. Treatment with glucose oxidase (GO), a steady generator of hydrogen peroxide, resulted in large amounts of mtDNA injury with little or no nuclear damage *(25)*. Human fibroblasts treated with 6 mU GO/mL of media generated about 10 μ*M* H_2O_2 during a 15-min treatment and produced approx 1.0 lesion/mitochondrial genome. This damage was repaired within 4 h after the treatment *(25)*. This finding of increased mtDNA injury after an oxidant has been repeated in a large number of cell types from both human and rodent cells *(26–29)*. It is interesting to note that the antiapoptotic protein Bcl-2, although not capable of blocking ROS-induced mitochondrial damage, leads to increased rates of mtDNA repair *(27)*. Even yeast cells suffer significantly more mtDNA lesions than nuclear damage following oxidative stress (*see* Fig. 3;

Chen and Santos, unpublished observation). These observations have lead to two working hypotheses: (1) increased divalent metal ions in mitochondria produce more mtDNA lesions after ROS damage; (2) damaged mitochondria can generate ROS and lead to a vicious cycle of damage that, even in the presence an active repair DNA system, results in persistent mtDNA damage.

The QPCR technique is a sensitive and reliable assay for the detection of gene-specific damage and repair. Current detection limits are on the order of 1 lesion/10^5 nucleotides from as little as 5 ng of total genomic DNA. Successful application of this technique relies on the careful DNA extraction and quantitation, followed by the establishment of optimal conditions for amplification. A broad view of the steps currently in use by our group to perform this assay can be observed in **Fig. 1**. The following section will deal with the materials necessary for carrying out QPCR.

2. Materials

2.1. DNA Samples

For QPCR of long genomic targets, DNA of high molecular weight must be isolated to achieve efficient amplification reactions. Unless recommended by the manufacturer, all the components of the following kits are kept at room temperature.

1. QIAGEN® Genomic Tip and Genomic DNA Buffer Set (QIAGEN, cat. nos. 10323 and 19060, respectively), for mammalian DNA extractions.
2. MasterPure™ Yeast DNA Purification Kit, available from Epicentre (cat. no. MPY80200), for yeast DNA isolations.

It is important to emphasize that we have found that these kits give the highest reproducibility and show less variation from lot to lot. Moreover, we observed that the DNA purified is of high stability, yielding comparable results over a long period of time.

2.2. DNA Quantitation and PCR Analysis

For both DNA quantitation and the analysis of the PCR products (*see* **Subheading 3.3.**), we make use of the PicoGreen® dsDNA Quantitation Kit (Molecular Probes; cat. no. P-7589). This kit comprises the PicoGreen reagent (1 mL of PicoGreen solution in dimethyl sulfoxide [DMSO]), 20X TE buffer (25 mL of 200 mM Tris-HCl, 20 mM EDTA, pH 7.5) and lambda [λ] standard DNA [1 mL of 100 µg/mL in TE]. The PicoGreen solution should be aliquoted into 50 µL volumes, must be kept at –20°C, and protected from light. It should be thawed just prior to use. When kept properly, the solution is quite stable. Aliquots can be thawed and refrozen several times without any loss of activity. The TE buffer is usually diluted in sterile water (giving rise to a working

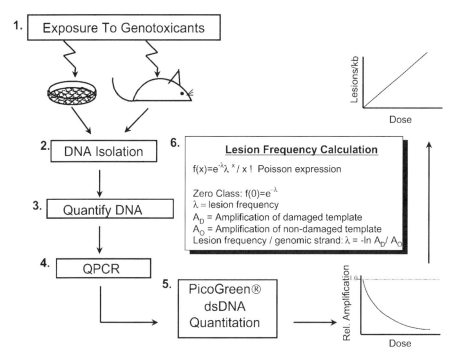

Fig. 1. Overview of the steps necessary to carry out the QPCR gene-specific repair assay. Adapted from **ref. 23**, with permission from Academic Press.

solution of 1X TE) and is kept at room temperature. The λ DNA as well as the DNA samples to be quantitated are kept at 4°C. A 96-well microtiter plate is also necessary.

2.3. QPCR Reagents

All the stock reagents are stored at –20°C, whereas vials that are currently in use are kept at 4°C. We routinely use the GeneAmp XL PCR Kit (PerKin-Elmer/Roche) for QPCR reactions. The kit includes the following:

1. *rTth* DNA polymerase XL (400 units; 2 units/μL).
2. 3.3X XL PCR buffer.
3. 25 mM Mg(OAc)$_2$.
4. dNTPs are purchased from Pharmacia (cat. no. 27-2035-01) and a solution of 10 mM of total dNTPs is prepared, using 2.5 mM of each nucleotide mixed in sterile and deionized water. To minimize degradation, aliquots of 100 μL are made and kept at –80°C.
5. Primer stocks are maintained at –20°C. In our laboratory, we routinely dilute the primers in sterile deionized water and store as small aliquots (10 μM). The vials of the diluted stocks are kept at 4°C.

It is important to note that the polymerase used is not infinitely thermostable and, for greater efficiency, the enzyme should not be put under unnecessary denaturing steps. In this context, the addition of BSA (bovine serum albumin, 100 ng/µL final concentration) in the PCR mix increases the resistance of the polymerase *(23)*.

The main concern regarding the QPCR is DNA cross-contamination. Therefore, care should be taken when handling all of the samples and reagents before and after the PCR. In this regard, the use of disposable gloves throughout the procedure is strongly recommended, given that it helps to avoid the introduction of nucleases or other contaminants that can cause either degradation of the template or inhibition of the polymerase during cycling. We have found it necessary to have distinct workstations, in separate rooms, to complete different steps of the procedure. For example, DNA isolation takes place at a separate area than where PCR is set up, and DNA quantitation and post-PCR analysis must be carried out in a different isolated area. Moreover, the setup is performed in a hood equipped with ultraviolet (UV) light, where, prior to use, all the micropipets, racks, tubes, tips, and so forth must be decontaminated (UV lamps are turned on for 1 min prior to PCR arrangement). We recommend that all materials used for the PCR setup be of exclusive use for the assay and that they be kept always inside the hood.

The next section will deal with the methods necessary for the successful realization of quantitative PCR of long amplification products.

3. Methods

3.1. DNA Extraction

As already stressed, caution must be taken to avoid any kind of contamination of the DNA samples used for QPCR. Thus, it is recommended that decontamination of equipment take place prior to usage and also that centrifuges, baths, and so forth are not for common use for manipulation of samples from different organisms. For example, in our laboratory, yeast and mammalian isolations are performed with distinct apparatus and in different rooms.

For the DNA extraction itself, various kits are available commercially. We use the DNA extraction kits from QIAGEN and Epicentre, which, in our hands, give rise to templates of high quality and yield. This is important because our group previously demonstrated that the integrity of DNA template is essential for the reliable amplification of long PCR targets *(21)*. In both cases, protocols are followed as suggested by the manufacturers. It is worth noting that if isolations are not going to take place on the same day of the experiments, the samples should be kept at –80°C.

3.2. Quantitation of the DNA Template

A crucial step of quantitative PCR is the concentration of the DNA sample *(23)*. In fact, the accuracy of the assay relies on initial template quantity because all of the samples must have exactly the same amount of DNA. Therefore, following DNA extraction, precise quantitation is required.

Great improvement regarding DNA quantitation was achieved in our laboratory after adoption of the PicoGreen dsDNA Quantitation Kit (Molecular Probes), which measures the concentration of double-stranded DNA (dsDNA) solutions using a fluorometer or a fluorescence microplate reader. The underlying mechanism is that the reagent is an unsymmetrical cyanine dye. Free dye is essentially nonfluorescent and exhibits >1000-fold fluorescence enhancement upon binding to dsDNA. The assay displays a linear correlation between dsDNA quantity and fluorescence, being extremely sensitive (detection range extending from 25 pg/mL to 1 µg/mL). PicoGreen has several advantages, such as (1) the high affinity of the dye for dsDNA over RNA, DNA single stranded, and oligonucleotides, (2) the lack of base selectivity, which make the results not compromised by proteins, nucleotides, and other common sample contaminants *(30)*, (3) the rapidity of analysis—15 min total for analysis using a microtiter fluorometer plate reader, (4) the manipulation of a nonhazardous material, (5) the inexpensive price.

Using the above-mentioned kit, quantitation is performed in two different steps. The first one, called prequantitation, gives a rough estimate of the amount of DNA in each sample. The final quantitation gives the exact amount of DNA needed to make a 3-ng/µL solution.

The protocol for quantitation consists of the following:

1. Dilution of λ DNA (in 1X TE buffer), yielding different concentrations to generate a standard curve. We suggest 1.25, 2.5, 5.0, 7.5, 10, and 20 ng/µL.
2. Use 10 µL of λ DNA as well as 10 µL of water as background control for the standard curve.
3. For prequantitation, pipet 2 µL of the sample DNA in the microtiter plate already containing 90 µL of buffer.
4. Prepare a solution containing the PicoGreen reagent (5 µL reagent per mL of 1X TE), preferentially in another room. This solution is mixed and 100 µL are pipetted into each well containing the DNA samples.
5. Incubate for 10 min, at room temperature, in the dark (the plate can be covered with foil paper).
6. Read fluorescence with the following parameters (in our laboratory we use the FL600 Microplate Fluorescence Reader form Bio-Tek®): excitation and emission wavelengths of 485 nm and 530 nm, respectively; sensitivity limit of 75, and shaking of the plate set at level 3 for 20 s.

The DNA concentration of the samples is calculated by means of average and slope obtained from the λ DNA standard curve, using a Microsoft Excel spread sheet. After the first estimation, the DNA samples are diluted, in TE buffer, to 10 ng/μL from the original DNA stock. This is the solution that goes through the final quantitation, which is done similarly. Note that for this last step, 10 μL of the 10-ng/μL stock solution should be loaded into the wells. This last estimation will allow calculations for a 3-ng/μL DNA sample, which is the recommended concentration of DNA to perform the QPCR assay *(23)*.

3.3. Quantitation of PCR Products

The PicoGreen kit has not only proven efficient in regard to template quantitation but also to PCR products analysis. In fact, the accuracy of the data obtained with this assay is comparable to those accomplished with ^{32}P-radiolabeled nucleotides (Y. Chen, unpublished observation).

In our laboratory, analysis of PCR products is performed similarly as for the DNA quantitation step. After the last PCR cycle is finished, 10 μL of the PCR products (aliquoted in a separate room from where the reaction is set up and run) are added to 90 μL of 1X TE buffer, which is mixed with 100 μL of the diluted PicoGreen reagent (as in **Subheading 3.2.**). These solutions are incubated in the dark, at room temperature, for 10 min and the fluorescence emitted is read (*see* **Subheading 3.2.**). For data analysis, see **Subheading 3.4.4.**

3.4. QPCR

3.4.1. Primer Selection

It is well established that various parameters influence the performance of the QPCR assay *(23,31)*. In this context, the appropriate primer selection is an important issue to be considered. In general, the oligonucleotides should comprise 20–24 bases in length with a G+C content of approx 50% and a T_M of approx 68°C. As usual, the selected primers should be evaluated for secondary structures using an appropriate software because it is known that formation of artifacts, such as primer-dimers, can compete against the QPCR reaction *(23)*. We have purchased oligonucleotide primers from several vendors and find that the primers work well unpurified. **Table 1** shows the sequences of the oligonucleotides currently in use in our laboratory to amplify human, mouse, rat, and yeast target genes.

3.4.2. PCR Reaction

Once the primers are selected, finding the optimal reaction condition is the next step. Different target genes (from different species) usually require distinct conditions. Our laboratory has established optimal concentrations of reagents

Table 1
Gene Targets and Primer Pairs for QPCR

Human Primers
17.7-kb fragment from the 5'flanking region near the beta-globin gene, accession number, J00179

44329	5'-TTG AGA CGC ATG AGA CGT GCA G-3'	Sense
62007	5'-GCA CTG GCT TAG GAG TTG GAC T-3'	Antisense
48510	5'-CGA GTA AGA GAC CAT TGT GGC AG-3'	Sense

Note: Using 48510 and 62007, we get a robust 13.5kb fragment which is of adequate length to give sufficient sensitivity.

16.2-kb mitochondria fragment, accession number, J01415

15149	5'-TGA GGC CAA ATA TCA TTC TGA GGG GC-3'	Sense
14841	5'-TTT CAT CAT GCG GAG ATG TTG GAT GG-3'	Antisense
14620	5'-CCC CAC AAA CCC CAT TAC TAA ACC CA-3'	Sense
5999	5'- TCT AAG CCT CCT TAT TCG AGC CGA-3'	Sense

Note: Using primers 14841 and 5999, we get a fragment of 8.9 kb.

10.4-kb fragment encompassing exons 2–5 of the hprt gene, accession number, J00205

14577	5'-TGG GAT TAC ACG TGT GAA CCA ACC-3'	Sense
24997	5'-GCT CTA CCC TCT CCT CTA CCG TCC-3'	Antisense

12.2-kb region of the DNA polymerase gene beta, accession number, L11607

2372	5'-CAT GTC ACC ACT GGA CTC TGC AC-3'	Sense
3927	5'-CCT GGA GTA GGA ACA AAA ATT GCT G-3'	Antisense

Mouse Primers
8.7-kb fragment of the β-globin gene, accession number, X14061

21582	5'-TTG AGA CTG TGA TTG GCA ATG CCT-3'	Sense
30345	5'-CCT TTA ATG CCC ATC CCG GAC T-3'	Antisense

6.5-kb fragment of the DNA polymerase beta, accession number, AA79582

MBFor1	5'-TAT CTC TCT TCC TCT TCA CTT CTC CCC TGG-3'	Sense
MBEX1B	5'-CGT GAT GCC GCC GTT GAG GGT CTC CTG-3'	Antisense

10-kb mitochondria fragment

3278	5'-GCC AGC CTG ACC CAT AGC CAT AAT AT-3'	Sense
13337	5'-GAG AGA TTT TAT GGG TGT AAT GCG G-3'	Antisense

117-bp mitochondria fragment

13597	5'-CCC AGC TAC TAC CAT CAT TCA AGT-3'	Sense
13688	5'-GAT GGT TTG GGA GAT TGG TTG ATG T-3'	Antisense

Rat Primers
12.5-kb fragment from the clusterin (TRPM-2) gene, accession number M64733

5781	5'-AGA CGG GTG AGA CAG CTG CAC CTT TTC-3'	Sense
18314	5'-CGA GAG CAT CAA GTG CAG GCA TTA GAG-3'	Antisense

13.4-kb mitochondria fragment

13559	5'-AAA ATC CCC GCA AAC AAT GAC CAC CC-3'	Sense
10633	5'-GGC AAT TAA GAG TGG GAT GGA GCC AA-3'	Antisense

235-bp mitochondria fragment

14678	5'-CCT CCC ATT CAT TAT CGC CGC CCT TGC-3'	Sense
14885	5'-GTC TGG GTC TCC TAG TAG GTC TGG GAA-3'	Antisense

(continued)

Table 1 *(continued)*

Yeast (Sc) Primers (23a)		
16.4-kb region encompassing genes Rad14, PFK2, and HFA1 on chromosome XIII		
666791 5′-TAG TAG GGC TAA CGA CGG TGA TC-3′		Sense
683201 5′-CGC TAA AAT CCC GTG TAT CCC TTG-3′		Antisense
9.3-kb region encompassing genes PFK2, and HFA1 in chromosome XIII		
673854 5′-CAA AGA ACC GTC ACC ACC ACA AA-3′		Sense
683201 5′-CGC TAA AAT CCC GTG TAT CCC TTG-3′		Antisense
6.9-kb mitochondria fragment in COX1 gene		
13999 5′-GTG CGT ATA TTT CGT TGA TGC GT-3′		Sense
20948 5′-GTC ACC ACC TCC TGC TAC TTC AA-3′		Antisense
14297 5′-TTC ACA CTG CCT GTG CTA TCT AA-3′		Antisense

Adapted from **ref. *23***, with permission from Academic press.

to amplify the genes of our interest. Using the Perkin-Elmer kit mentioned earlier, the PCR reactions are prepared as follows:

1. 15 ng of DNA.
2. 1X Buffer.
3. 100 ng/μL of BSA.
4. 200 μM of dNTPs.
5. 20 pmol of each primer.
6. 1.3 mM of Mg^{2+}.
7. Water to complete a total volume of 45 μL.
8. Begin PCR by a "hot start:" in which 1 U of the enzyme (diluted in water: 0.5 μL of polymerase in 5 μL total) is added to each PCR reaction tube, at 75°C (*see* **Note 3**).

It is important to emphasize that primers and magnesium concentrations are subject to variation (*see* **Notes 1–5**). Another important feature is the sequence in which the components are added into the PCR tube. The DNA template should always be added first (remember to include a control tube without any template), followed by the PCR mix and, finally, the enzyme.

3.4.3. Cycle Number and Thermal Parameters

The usefulness of the QPCR assay for the detection of DNA damage requires that amplification yields be directly proportional to the starting amount of template. These conditions must be met by keeping the PCR in the exponential phase. It is important, therefore, to perform cycle tests to determine quantitative conditions for the gene of interest *(20)*. Thus, once the amplification rate of nondamaged DNA is within the exponential range, always run a 50% control and a no template control. The 50% control is made by diluting the nondamaged sample in TE buffer to make a new solution with 1.5 ng of DNA per microliter. This control should give a 50% reduction of the amplification signal (*see* **Note 3**). The nontemplate control will detect contamination with spurious DNA or PCR products.

Another concern when carrying out QPCR is finding the optimal thermal conditions for amplification. As mentioned earlier, QPCR in our laboratory is routinely performed using a hot start, which is accomplished by bringing the reaction mixture (containing buffer components, dNTPs, template, and primers) to 75°C prior to the addition of enzyme and subsequent cycling. This approach produces cleaner PCR products because it prevents nonspecific annealing of primers to each other, as well as to the template, before enzyme addition. Keep in mind that the annealing temperature of the primers determines how stable the primers hybridize to the DNA template. Thus, it is important to check this parameter with suitable software beforehand, although, sometimes, empirical evaluation is necessary. **Table 2** and **Fig. 2** show the most favorable conditions for human, rodent, and yeast amplifications currently used in our laboratory.

3.4.4. Data Analysis

Analysis of data in our laboratory, obtained by quantitation of the PCR products with the PicoGreen kit, is routinely done using a Microsoft Excel spreadsheet. After fluorescence readings for all the samples are taken and subtracted from the blank (the PCR tube, which had no template), relative amplification is calculated. This can be accomplished by dividing fluorescence values of the treated samples by the control (nontreated sample). These results are then used to determine the lesion frequency per fragment at a particular dose, such that lesions/strand (average for both strands) at dose $D = -\ln (A_D/A_C)$. This equation is based on the "zero class" of a Poisson expression. Note that a Poisson distribution requires an assumption that DNA lesions are randomly distributed, (more details can be found in **ref. *23***).

3.4.5. Normalization to mtDNA Copy Number

Attention must be paid when interpreting data regarding amplification from mitochondrial genome. Because it is known that this organelle can vary in its DNA copy number, one has to keep in mind that variation in amplification may be observed (e.g., in samples from distinct areas of a specific organ). This fluctuation can either arise from different levels of lesions within the sample or from variation in the number of copies of the mitochondrial genome. To cope with this latter problem, we recommend amplifying a short fragment of the gene under study, usually no longer than 300 bp (base pairs). The underlying idea is that the short fragment will reflect undamaged DNA as a result of the low probability of introducing lesions in small segments. Therefore, variations in the degree of amplification are assumed to be the result of fluctuations in

Table 2
PCR Conditions for Human, Rodent, and Yeast Targets

Target	Primer set	Mg^{2+} conc. (mM)	T_m (°C)	Cycle no.	PCR profile
Human					
Large mito	15149/14841	1.3	64	26	B
Large mito	5999/14841	1.3	64	19	B
Small mito	14620/14841	1.3	60	19	C
hprt	14577/24997	1.3	64	29	B
β-globin	48510/62007	1.3	64	27	B
β-pol	3927/2372	1.3	64	26	B
Mouse					
Large mito	3278/13337	1.0	65	20	B
Small mito	13688/13597	1.1	60	18	C
β-globin	21582/30345	1.1	65	25	B
β-pol	MBFor1/MBEX1B	1.1	65	25	B
Rat					
Large mito	10633/13559	1.1	65	18	B
Small mito	14678/14885	1.1	60	18	C
Clusterin	5781/18314	1.1	65	28	B
Yeast					
Nuclear 9.3 kb	673854/683201	1.3	64	20	B
Nuclear 16.4 kb	666791/683201	1.3	64	26	A
Mito	13999/20948	1.3	64	17	B
Small mito	13999/14297	2.5	60	19	C

Adapted from **ref. 23**, with permission from Academic Press.

gene copy number, and the results from small mitochondrial fragments can be used to normalize the data obtained for the large products (7–15 kb).

3.5. Application of the QPCR Assay for the Analysis of mtDNA Damage

Although there is a robust amount of data showing the reliability of QPCR, criticism still exists regarding the sensitivity of the assay because it is known that only specific types of DNA lesions block the progression of the thermostable polymerase. Our group has been addressing these questions for the past 5 yr and the data so far obtained not only shows that the QPCR assay is a tool able to identify damage and/or repair induced by a wide variety of agents, but also that it is trustworthy for the detection of lesions in both nuclear and mitochondrial genes (*24,25,29,32–34*).

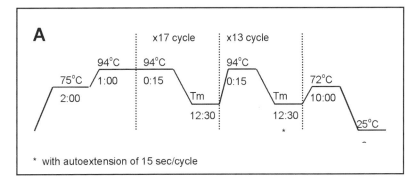

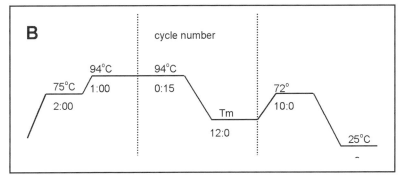

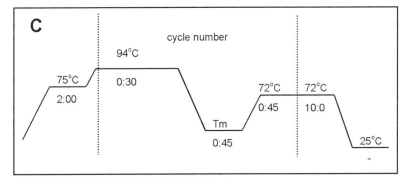

Fig. 2. Cycle parameters used to perform QPCR. Each reaction is initiated by the addition of *rTth* polymerase during a hot start at 75°C. Dashed lines indicate cycle phase (note that time periods are not drawn to scale). Adapted from **ref. 23**, with permission from Academic Press.

The effects of hydrogen peroxide (H_2O_2) in different model systems have long been of interest of our laboratory. As mentioned previously, Yakes and Van Houten *(24)* were able to observe that SV40-transformed human fibroblasts when challenged with this genotoxicant show damage both in mitochondrial

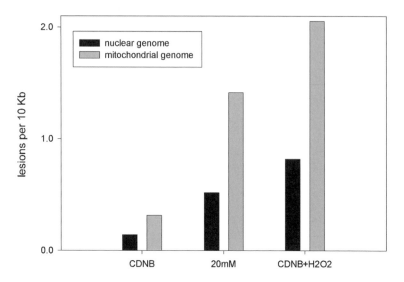

Fig. 3. 1×10⁷ cells/mL of the yeast wild type strain ale 1000 were inoculated in YPD broth. Eight hours after inoculation, cells were harvested and resuspended in PBS. Treatments included exposure for 30 min to either 1 mM of CNDB or 20 mM of hydrogen peroxide alone, or a combined exposure in which CDNB was administered first, followed by 30 min to H_2O_2. Lesions (per 10 kb), observed in both nuclear and mitochondrial genomes, are plotted as a function of treatments (data of mitochondria normalized for mitochondria copy number).

and nuclear genomes. Moreover, the damage is two to three times higher and persists longer in mtDNA in relation to nuclear DNA. We have recently been examining the production of mtDNA damage after reactive oxygen species both in cells from the budding yeast *Saccharomyces cerevisiae* and fibroblasts expressing telomerase. Treatment of a wild-type yeast strain (ale 1000) with 20 mM of H_2O_2 for 30 min gives rise to damage in both genomes, with the mitochondrial genome showing, at least, threefold more damage than a nuclear gene. It is noteworthy that this effect can be further enhanced with a pretreatment with 1-chloro-2,4-dinitrobenzene (CDNB), a glutathione inhibitor (**Fig. 3**).

We have recently examined the production of hydrogen peroxide-induced nuclear and mitochondrial DNA damage in telomerase expressing fibroblasts (a generous gift from Dr. Carl Barret). As can be seen in **Fig. 4A**, the nuclear genome of these cells is not affected by the incubation of H_2O_2 (200 µM), whereas the mitochondrial DNA is severely damaged (**Fig. 4B**). When compared to our previous results with human SV40-transformed fibroblasts *(24)*,

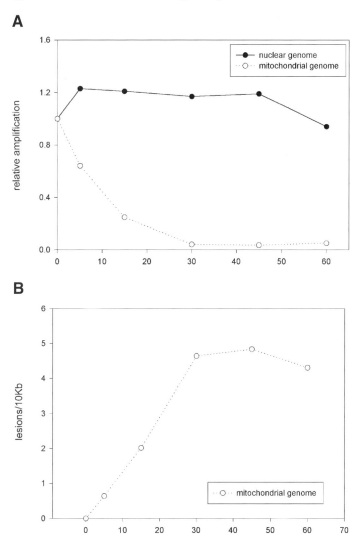

Fig. 4. Time-course of hydrogen peroxide-induced damage in human fibroblasts expressing telomerase. 10^5 cells were plated per milliliter of media, in duplicates, 16 h before treatments. Cells were treated for 1 h with 200 μM of H_2O_2 and harvested at 5, 15, 30, 45, and 60 min. DNA was extracted and quantified, and PCR reactions were made in duplicates. **(A)** Relative amplification observed for both nuclear and mitochondrial genomes. **(B)** Amount of lesions per 10 kb detected in the mitochondrial genome (with normalization of mitochondria copy number). Note the lack of damage in the nuclear gene.

these data suggest that SV40 transformation of fibroblasts may make nuclear DNA more susceptible to damage following hydrogen peroxide damage.

4. Notes

1. Primers. The same batch of primers when used over a long period of time (several months) can give rise to lower amplification, it is, therefore, valuable to make new dilutions from time to time. Always protect primer stocks from unnecessary temperature fluctuation and contamination. If frozen, primer stocks should be completely thawed prior to use.
2. Magnesium. The optimal concentration is determined for each set of primers and template. The *rTth* polymerase is extremely sensitive to magnesium; therefore, we advise that amplification of the fragment of interest should be performed using varying quantities of Mg^{2+}, starting from 0.9 and going up with 0.1 increments.
3. Quantitative aspect of amplification. During each set of amplifications, we routinely amplify a control sample in which only 50% of the template is added to the QPCR. Depending on the DNA quality and the products being amplified, relative amplification ranging from 40% to 60% is considered acceptable. Any experiments that are outside this range are not satisfactory and the entire set of reactions is discarded. It may be necessary to reoptimize the PCR by varying the number of cycles to establish a linear response to increasing template concentrations from 1.25 to 30 ng (*see* **ref. *31***).
4. Enzyme concentration. Increasing amounts of the thermostable polymerase beyond 2.5 U per reaction can increase the production of nonspecific amplification products;
5. dNTPs. A higher mutation frequency for the enzyme as well as a reduction in magnesium concentration has been observed when dNTPs exceed 200 μM.

References

1. Anderson, S., Bankier, A. T., Barrell, B. G., deBruijn, M. H. L., Coulsen, A. R., Drouin, J., et al. (1981) Sequence and organization of the human mitochondrial genome. *Nature* **290,** 457–465.
2. Desjardins, P., deMuys, J. M., and Morais, R. (1986) An established avian fibroblast cell line without mitochondrial DNA. *Somatic Cell Mol. Genet.* **12,** 133–139.
3. Boveris, A., Oshino, N., and Chance, B. (1972) The cellular production of hydrogen peroxide. *Biochem. J.* **128,** 617–630.
4. Turrens, J. F. and Boveris, A. (1980) Generation of superoxide anion by the NADH dehydrogenase of bovine heart mitochondria. *Biochem. J.* **191,** 421–427.
5. Wei, Y. H., Scholes, C. P., and King, T. E. (1981) Ubisemiquinone radicals from the cytochrome b-c1 complex of mitochondrial electron transport chain: demonstration of QP-S radical formation. *Biochem. Biophys. Res. Commun.* **99(4),** 1411–1419.
6. Sawyer, D. E. and Van Houten, B. (1999) Repair of DNA damage in mitochondria. *Mutat. Res.* **434,** 161–176.

7. Richter, C., Park, J. W., and Ames, B. N. (1988) Normal oxidative damage to mitochondrial and nuclear DNA is extensive. *Proc. Natl. Acad. Sci. USA* **85**, 6465–6467.

8. Helbock, H. J., Beckman, K. B., and Ames, B. N. (1999) 8-Hydroxydeoxyguanosine and 8-hydroxyguanine as biomarkers of oxidative DNA damage. *Methods Enzymol.* **300,** 156–166.

9. Helbock, H. J., Beckman, K. B., Shigenaga, M. K., Walter, P. B., Woodall, A. A., Yeo, H. C., et al. (1998) DNA oxidation matters: the HPLC-electrochemical detection assay of 8-oxo-deoxyguanosine and 8-oxo-guanine. *Proc. Natl. Acad. Sci. USA* **95(1),** 288–293.

10. Wallace, D. C. (1992) Mitochondrial DNA mutations and neuromuscular diseases. *Trends Genet.* **5,** 9–13.

11. Fliss, M. S., Usadel, H., Caballero, O. L., Wu, L., Buta, M. R., Eleff, S. M., et al. (2000) Facile detection of mitochondrial DNA mutations in tumors and bodily fluids. *Science* **287(5460),** 2017–2019.

12. Cortopassi, G. A., Shibata, D., Soong, N. W., and Arnheim, A. (1992) A pattern of accumulation of a somatic deletion of mitochondrial DNA in aging human tissues. *Proc. Natl. Acad. Sci. USA* **89,** 7370–7374.

13. Polyak, K., Li, Y., Zhu, H., Lengauer, C., Wilson, J. K., Markowitz, S. D., et al. (1998) Somatic mutations of the mitochondrial genome in human colorectal tumours. *Nature Genet.* **20(3),** 291–293.

14. Clayton, D. A., Doda, J. N., and Freidberg, E. C. (1974) The absence of pyrimidine dimer repair mechanism in mammalian mitochondria. *Proc. Natl. Acad. Sci. USA* **71,** 2777–2781.

15. Van Houten, B. and Friedberg, E. C. (1999) Mitochondrial DNA damage and repair. *Mutat. Res.* **434,** 133–254 (special issue).

16. Croteau, D. L., Stierum, R. H., and Bohr, V. A. (1999) Mitochondrial DNA repair pathways. *Mutat. Res.* **434,** 137–148.

17. Le Doux, S. P., Driggers, W. J., Hollensworht, B. S., and Wilson, G. L. (1999) Repair of alkylation and oxidative damage in mitochondrial DNA. *Mutat. Res.* **434,** 149–159.

18. Pettepher, C. C., LeDoux, S. P., Bohr, V. A., and Wilson, G. L. (1991) Repair of alkali-labile sites within the mitochondrial DNA of RINr 38 cells after exposure to the nitrosourea streptozotocin. *J. Biol. Chem.* **266,** 3113–3117.

19. Driggers, W. J., LeDoux, S. P., and Wilson, G. L. (1993) Repair of oxidative damage within the mitochondrial DNA of RINr 38 cells. *J. Biol. Chem.* **268,** 22,042–22,045.

20. Kalinowski, D., Illenye, S., and Van Houten, B. (1992) Analysis of DNA damage and repair in murine leukemia L1210 cells using a quantitative polymerase chain reaction assay. *Nucleic Acids Res.* **20,** 3485–3494.

21. Cheng, S., Chen, Y., Monforte, J. A., Higuchi, R., and Van Houten, B. (1995) Template integrity is essential for PCR amplification of 20- to 30-kb sequences from genomic DNA. *PCR Methods Applic.* **4,** 294–298.

22. Yakes, F. M., Chen, Y., and Van Houten, B. (1996) PCR-based assays for the detection and quantitation of DNA damage and repair, in *Technologies for Detec-*

tion of DNA Damage and Mutations (Pfeifer, G. P., ed.), Plenum, New York, pp. 171–184.

23. Ayala-Torres, S., Chen, Y., Svoboda, T., Rosenblatt, J., and Van Houten, B. (2000) Analysis of gene-specific DNA damage and repair using quantitative PCR, in *Methods. A Companion to Methods in Enzymology* (Doetsch, P. W., ed.), Academic, New York, **22,** 135–147.

23a. Mewes, H. W., Hani, J., Pfeiffer, F., and Frishman, D. (1998) MIPS: a database for protein sequences and complete genomes. *Nucleic Acids Res.* **26,** 33–37.

24. Yakes, F. M. and Van Houten, B. (1997) Mitochondrial DNA damage is more extensive and persists longer than nuclear DNA damage in human cells following oxidative stress. *Proc. Natl. Acad. Sci. USA* **94,** 514–519.

25. Salazar, J. J. and Van Houten, B. (1997) Preferential mitochondrial DNA injury caused by glucose oxidase as a steady generator for hydrogen peroxide in human fibroblasts. *Mutat. Res.* **385(2),** 139–149.

26. Ballinger, S. W., Van Houten, B., Coklin, C. A., Jin, A., and Godley, B. (1999) Hydrogen peroxide causes significant mitochondrial DNA damage in human RPE cells. *Exp. Eye Res.* **68(6),** 765–772.

27. Deng, G., Su, J. H., Ivins, K. J., Van Houten, B., and Cottman, C. (1999) Bcl-2 facilitates recovery from DNA damage after oxidative stress. *Exp. Neurol.* **159,** 309–318.

28. Ballinger, S. W., Patterson, C., Yan, C. N., Doan, R., Burow, D. L., Young, C. G., et al. (2000) *Circ. Res.* **86(9),** 960–966.

29. Mandavilli, B. S., Syed, F. A., and Van Houten, B. (2000) DNA damage in brain mitochondria caused by aging and MPTP treatment. *Brain Res.* **885(1),** 45–52.

30. Singer, V. L., Jones, L. J., Yue, S. T., and Haugland, R. P. (1997) Characterization of PicoGreen reagent and development of a fluorescence-based solution assay for double-stranded DNA quantitation. *Anal. Biochem.* **249(2),** 228–238.

31. Van Houten, B., Cheng, S., and Chen, Y. (2000) Measuring DNA damage and repair in human genes using quantitative amplification of long targets from nanogram quantities of DNA. *Mutat. Res.* **460(2),** 81–94.

32. Chandrasekhar, D. and Van Houten, B. (1994) High resolution mapping of UV-induced photoproducts in the E. coli lacI gene: inefficient repair in the nontranscribed strand correlates with high mutation frequency. *J. Mol. Biol.* **238,** 319–322.

33. Chen, K. H., Srivastava, D. K., Yakes, F. M., Singhal, R. K., Rawson, T. Y. , Sobol, R. W., et al. (1998) Up-regulation of base excision repair correlates with enhanced protection against a DNA damaging agent in mouse cell lines. *Nucleic Acid Res.* **26(8),** 2001–2007.

34. Horton, J. K., Roy, G., Piper, J. T., Van Houten, B., Awashi, Y. C., Mitra, S., et al. (1999) Characterization of chlorambucil-resistant human ovarian carcinoma cell line overexpressing glutathione s-transferase μ. *Biochem. Pharmacol.* **58(4),** 693–702.

11

Migration of mtDNA into the Nucleus

Mary K. Thorsness, Karen H. White, and Peter E. Thorsness

1. Introduction
1.1. Background

The endosymbiotic hypothesis proposes that mitochondria were once free-living organisms that colonized another cell. Millions of years of evolutionary pressure created the present-day situation in which the mitochondrion is a semiautonomous organelle that is fully integrated into virtually every aspect of intermediary metabolism *(1)*. A process central to this integration of mitochondrial and cellular activities has been the transfer of a large number of genes from mitochondria to the nucleus *(2)*. This transfer of genetic material has altered the structure and composition of the nuclear genome and has been, perhaps, the major evolutionary influence on eucaryotic energy metabolism.

Two types of sequence derived from mitochondrial DNA (mtDNA) can be identified in the nucleus. The first are nuclear DNA sequences encoding mitochondrial gene products that were transferred from mitochondria during evolution and are no longer present in mitochondria. The second are nuclear DNA sequences that are copies of existing mtDNA. Virtually every nuclear genome contains such sequences and the transfer of mtDNA to the nucleus clearly remains an ongoing process *(2)*. Characterization of the cellular events and environmental factors that control the escape, migration, and integration of mtDNA into the nuclear genome has led to the identification of proteins important in the maintenance and function of the mitochondrial compartment and genome *(3–5)*. Additionally, mtDNA escape has attracted attention because of its potential impact on disease processes through the integrative activation or disruption of genetic loci in the nucleus, analogous to the pathological affect of viral DNA integration *(6–9)*.

From: *Methods in Molecular Biology, vol. 197: Mitochondrial DNA: Methods and Protocols*
Edited by: W. C. Copeland © Humana Press Inc., Totowa, NJ

An assay was developed to experimentally investigate the escape and transfer of mtDNA to the nucleus in the yeast *Saccharomyces cerevisiae*. Biolistic transformation of mitochondria and subsequent identification of appropriately recombinant mitochondrial genomes placed a genetic tag in mitochondria that could be followed upon its transfer to the nucleus *(10)*. Several practical considerations of mitochondrial genetics and molecular methodologies limit this approach to this yeast. First, *S. cerevisiae* is a petite positive organism—one that can survive either with mutated mtDNA or in the complete absence of mtDNA. Second, mitochondrial genetic markers can be followed during the mating of haploid yeast, and recombinants can be easily identified and recovered because of the high frequency of genetic recombination in yeast mitochondria. Finally, because there are multiple copies of the mitochondrial genome, the loss of one or more copies to the nucleus does not significantly impact the ability of the cell or the mitochondrial compartment to function.

1.2. Overview of the mtDNA Escape Assay

Yeast mtDNA was tagged with nuclear genes that serve as selectable phenotypic markers in the nucleus. The yeast nuclear genes *TRP1* and *LYS2* were integrated into yeast mtDNA upstream of the *COX2* locus. *TRP1* encodes the enzyme phosphoribosylanthranilate isomerase that is necessary for tryptophan biosynthesis *(11)*, and *LYS2* encodes the enzyme aminoadipate–semialdehyde dehydrogenase that is necessary for lysine biosynthesis *(12)*. Associated with both of these genetic markers is *ARS1*, a well-characterized nuclear autonomous replicating sequence. The assay strain for mtDNA escape must contain a tagged mitochondrial genome as well as a corresponding mutant allele of the *TRP1* and/or *LYS2* nuclear loci. These yeast behave phenotypically as tryptophan or lysine auxotrophs because the mitochondrial *TRP1* and *LYS2* loci are not functional. The appearance of mtDNA-derived *TRP1* or *LYS2* sequences in the nucleus, however, complements the nuclear mutation, creating tryptophan or lysine prototrophs in a background of auxotrophs (**Fig. 1**). The frequency of mtDNA escape to the nucleus is reflected by the frequency of tryptophan or lysine prototropy. This frequency is characteristic of the genetic background of the cell and the conditions under which the cells are grown.

The mtDNA escape assay is typically carried out on Petri plates. Cells are plated on media that contains the appropriate nutrients: tryptophan or lysine depending on the mtDNA marker, as well as all other required nutritional supplements. After a period of growth, the yeast are replica-plated to media lacking either tryptophan or lysine in order to assay for the escape of the mitochondrially encoded nutritional marker gene. The appearance of tryptophan or lysine prototrophic colonies in a background of nongrowing auxotrophic

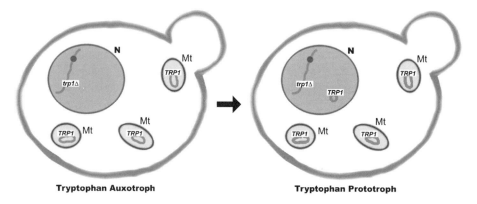

Tryptophan Auxotroph   **Tryptophan Prototroph**

Fig. 1. Schematic diagram of mtDNA escape in yeast. On the left, the tryptophan auxotroph contains the *trp1-Δ1* allele in the nucleus and the wild-type *TRP1* allele in the mitochondria. Because the mitochondrially encoded *TRP1* is not expressed, this yeast requires tryptophan for growth. On the right, a copy of the mitochondrially encoded *TRP1* gene has escaped mitochondria and migrated to the nucleus. Expression of *TRP1* in the nucleus generates a tryptophan prototroph.

cells is monitored over the course of several days (**Fig. 2**). The nuclei of prototrophs contain DNA sequences corresponding to the nutritional marker in association with various amounts of authentic mtDNA sequences *(13)*. The escaped mtDNA is mitotically unstable in over 99% of independent isolates, indicating that in most circumstances, the mtDNA is not integrated into the chromosome and, presumably, replicates in the nucleus utilizing the *ARS1* sequences associated with the nutritional marker.

2. Materials

2.1. Strains

Table1 describes yeast strains that can be used to assay mtDNA escape. There are two essential features of these yeast strains. First, the strains contain the appropriate nuclear mutation paired with the corresponding wild-type copy of the gene integrated into the mitochondrial genome; that is, the *trp1-Δ1* nuclear allele is coupled with the *TRP1* mitochondrial allele, and the *lys2* nuclear allele is coupled with the *LYS2* mitochondrial allele. Second, the strains PTY33, PTY44, PTY52, PTY133, PTY144, and PTY179 are isogenic. Thus, relative rates of mtDNA escape can be compared in these strains. The process of mtDNA escape is very sensitive to genetic background, and comparison of relative escape rates in strains that are not isogenic pairs is not valid. The effect of any given mutation on mtDNA escape can be assayed in these strains using

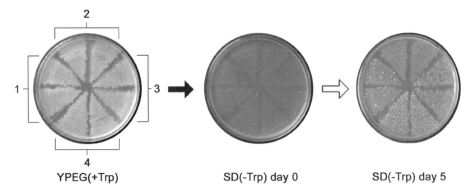

Fig. 2. Replica plate assay for mtDNA escape. Cells are streaked to lawns in a pie-plate configuration on media containing tryptophan (YPEG [+Trp]). The lawns are replica-plated to media lacking tryptophan (SD [-Trp]) to score mtDNA escape. At d 0, a light lawn of cells is present on the plate, but no growth is observed. By d 5, individual Trp+ colonies are evident in the background of nongrowing Trp⁻ cells. **(1)** PTY44 wild type [rho+ *TRP1*] exhibits a low level of mtDNA escape; **(2)** PTY144 wild type [rho+, *LYS2*] does not exhibit mtDNA escape because the mitochondrial genome is marked by *LYS2*, whereas the assay scored for escape of the *TRP1* marker; **(3,4)** [rho+, *TRP1*] yeast with nuclear mutations that exhibit increased rates of mtDNA escape.

current allele-replacement techniques or by backcrossing a mutant strain five or more times to the wild-type escape strain. Alternatively, it is possible to transfer mitochondrial genomes between yeast strains using a karyogamy deficient strain *(14,15)*. PTY200 and PTY201 can be used to transfer a *TRP1-* or *LYS2*-tagged mitochondrial genome into a different strain background, allowing mtDNA escape to be assayed in alternative genetic backgrounds. PTY52 and PTY179 carry a disrupted allele of the nuclear gene *YME1*, *yme1-Δ1::URA3*, and can be used as a control for an elevated level of mtDNA escape.

2.2. Media

Yeast cells are grown on the following media:

1. Complete glucose medium (YPD) containing 2% glucose, 2% Bacto-peptone, 1% yeast extract, supplemented with adenine and tryptophan at 40 mg/L.
2. Complete ethanol–glycerol medium (YPEG) containing 3% glycerol, 3% ethanol, 2% Bacto-peptone, 1% yeast extract, supplemented with adenine and tryptophan at 40 mg/L.
3. Synthetic glucose medium (SD) containing 2% glucose, 6.7 g/L yeast nitrogen base without amino acids (Difco, Detroit, MI) and supplemented with the

Table 1
Yeast Strains

Strain	ATCC#[a]	Genotype[b]
PTY33	MYA-1747	*MATa ura3-52 ade2 leu2-3,112 trp-Δ1* [*rho⁺, TRP1*]
PTY44	MYA-1748	*MATα ura3-52 lys2 leu2-3,112 trp-Δ1* [*rho⁺, TRP1*]
PTY52	MYA-1749	*MATα ura3-52 lys2 leu2-3,112 trp-Δ1 yme1-Δ1*::URA3 [*rho⁺, TRP1*]
PTY133	MYA-1750	*MATa ura3-52 ade2 lys2 leu2-3,112 trp-Δ1* [*rho⁺, LYS2*]
PTY144	MYA-1751	*MATα ura3-52 lys2 leu2-3,112 trp-Δ1 his3::hisg* [*rho⁺, LYS2*]
PTY179	MYA-1752	*MATα ura3-52 lys2 leu2-3,112 trp-Δ1 his3::hisg yme1-Δ1::URA3* [*rho⁺, LYS2*]
PTY200	MYA-1753	*MATa kar1-1 ura3-52 ade2* [*rho⁺, TRP1*]
PTY201	MYA-1754	*MATa kar1-1 ura3-52 ade2* [*rho⁺, LYS2*]

[a] All strains are deposited in the American Type Culture Collection (10801 University Blvd. Manassas, VA 20110-2209; tel: 703-365-2882).
[b] The mitochondrial genome is bracketed.

appropriate nutrients. Uracil is added at 40 mg/L, adenine at 40 mg/L, tryptophan at 40 mg/L, lysine at 60 mg/L, histidine at 20 mg/L, and leucine at 100 mg/L.

4. Synthetic galactose medium (SG) containing 2% galactose, 6.7g /L yeast nitrogen base without amino acids, supplemented with the appropriate nutrients as described in **item 3**.

Bacto-agar is added at 20 g/L for solid media. mtDNA escape is assayed on solid synthetic media containing either glucose (SD) or galactose (SG) and lacking either tryptophan or lysine, depending on the mitochondrial marker gene being assayed. For plate-to-plate comparisons, best results are obtained using selectable media from the same batch.

3. Methods

3.1. Replica Plate Assay for Determining Relative Rates of mtDNA Escape

To compare relative rates of mtDNA escape, lawns of test strains are grown on solid media and then replica-plated to synthetic media lacking the nutrient corresponding to the nutritional marker present in mtDNA. Growth on this selective "escape plate" is indicative of mtDNA escape events.

Because mtDNA escape is very sensitive to genetic background, isogenic strains must be used for comparisons of escape rates between strains or treatments. Additionally, it is important to include a wild-type strain on each plate to serve as a baseline for mtDNA escape. PTY52 can also be included as a control that shows a high rate of mtDNA escape resulting from the *yme1-Δ1*::*URA3* mutation. Comparisons of mtDNA escape in related strains should be made on the same Petri plate, as there is plate-to-plate variability of the assay. This variability is the result of differences in the number of cells transferred, in the hydration state of the solid media, and in the consistency of the replica medium. Relative rates of mtDNA escape are easily determined using this plate assay as long as appropriate controls are present on each Petri plate.

1. Streak test strains on media supplemented with tryptophan. One-quarter to one-sixteenth of a plate is patched in a "pie-plate" configuration with test strains (**Fig. 2**). Grow strains under conditions that select for the highest number of respiring cells, selecting for plasmids if necessary. Growth on YPEG is preferred if selection is not necessary (*see* **Note 1**). Care should be taken to streak cells to a lawn rather than to single colonies in order to maximize reproducibility and to better assay low rates of mtDNA escape (*see* **Note 2**).
2. Incubate plates overnight to 5 d, until growth is confluent. There should be a solid lawn of cells (**Fig. 2**), but results are more reproducible if cells are not extensively overgrown (*see* **Note 3**).
3. Replica to synthetic media lacking the nutrient corresponding to the mitochondrial marker gene (**Fig. 2**). Yeast strains containing the *TRP1* mitochondrial marker should be replica-plated to media lacking tryptophan, whereas those containing the *LYS2* mitochondrial marker should be replica-plated to media lacking lysine (*see* **Note 4**).
4. Incubate at optimum growth temperature until colonies are visible on control sectors of the plate, usually 4–5 d (**Fig. 2**).
5. Score the number of escape colonies in test strains and compare to control sectors (*see* **Note 5**).

3.2. Quantitative Determination of mtDNA Escape Rate

Although the plate assay of relative mtDNA escape rates is in many ways the most useful measure of this cellular event, there are circumstances that require a more quantitative measurement of mtDNA escape. In such cases, the statistical analysis first described by Luria and Delbrück *(16)* to analyze the rate at which mutants arise in a population of cells can be applied. This assay is, however, fairly cumbersome. Unless there is a need for quantitative comparison, the plate assay described in **Subheading 3.1.**) provides a simple method for comparing relative rates of mtDNA escape in different yeast strains. For the quantitative assay, mtDNA escape must be compared in multiple identical cultures. We typically assay 10 identical cultures. The total number of cells and the number of cells prototrophic for the nutrient corresponding to the marker integrated in the mtDNA is determined for each culture, and these numbers are used to calculate a rate of mtDNA escape.

1. Plate yeast cells to isolated single colonies on solid media. Each single colony is effectively an "identical culture" (*see* **Note 6**).
2. Harvest 10 colonies by cutting a plug of media containing the cells from a plate and suspending the cells in 10 mL of sterile water. Vortex vigorously to suspend the cells.
3. Determine the total number of cells in the colony by diluting the resulting suspension 1:200 in water (5 µL of cells in 995 µL of water) and plating 5 µL of this dilution on complete glucose media (YPD) in triplicate.
4. Incubate plates for 3 d and determine the average number of colonies on the three plates. The total number of cells in the single colony culture is the average number of cells on the three plates multiplied by 40,000, typically between 1×10^6 and 1×10^7.
5. Concurrently determine the average number of prototrophic colonies in the single colony culture by plating cells in triplicate from the 10-mL suspension to synthetic media lacking tryptophan or lysine, depending on the mitochondrial marker being used. The volume of culture to plate must be determined empirically, as differences in the rate of mtDNA escape between strains can lead to vastly different numbers of prototrophs in a given culture. Typically, 5 µL, 50 µL, 100 µL, and 200 µL of culture is plated on selective media.
6. Incubate plates for 3 d and determine the average number of prototrophic colonies on the three plates. The total number of prototrophs in the 10-mL suspension is calculated by multiplying the average number of prototrophic colonies by the appropriate dilution factor.
7. Calculate the rate of mtDNA escape using the equation derived by Luria and Delbrück **(16)**: $r = aN_t \ln(N_t Ca)$, where r is the observed average number of prototrophs per culture, C is the number of single colony cultures examined (10 in this example), N_t is the total number of cells in a culture, and a is the number of escape events per time unit. If a is multiplied by $\ln 2$, the time unit is a cell division **(13)**. This rate is probably an underestimate of the true rate of mtDNA escape, as the rate analysis assumes faithful inheritance of the genetic change. However, the ARS-based plasmids bearing the nutritional maker are often transmitted to as few as 5% of the progeny.

3.3. Assay of mtDNA Escape in Nongrowing Cells

Certain treatments of nongrowing yeast cells induce mtDNA escape **(13)**. The effect of these "stress" treatments can be assayed in a manner similar to the quantitative assay described earlier.

1. Prepare a single culture of yeast and divide into control and test fractions.
2. Treat test fractions as desired.

3. Determine the number of viable cells and the number of prototrophs in control and test fractions as described in **Subheading 3.2.**

4. Compare the relative number of prototrophic colonies in control and treated fractions to assess the affect of treatment on mtDNA escape.

4. Notes

1. Frequently, the absence of observable mtDNA escape is the result of the conversion of cells from rho$^+$ to rho$^-$ or rho^0. If the *TRP1* or *LYS2* marker is lost from the mitochondrial genome in this conversion, escape cannot be observed. Growth on YPEG selects for functional mitochondria and thus assures that assayed cells contain mtDNA. If, however, the test strains carry plasmids, it is important to select for those plasmids because they may be rapidly lost during cell division. When plasmids are present, streak strains on SD or SG rather than YPEG. Use of synthetic ethanol/glycerol medium to concurrently select for plasmids and for respiration is not recommended because of very poor growth of strains on this medium. Additionally, mutations that affect the stability of the mitochondrial genome will alter the apparent rate of mtDNA escape as a result of loss of the mitochondrial marker gene. Because many mutations that affect mtDNA escape rate alter mtDNA stability, it is important to determine the relative number of rho$^+$ and rho$^-$/rho^0 yeast in strains in which mtDNA escape rate is compared.

2. Escape of mtDNA can be assayed in single colonies, but it is much less reproducible than from a lawn of cells. In particular, single-colony replicas frequently exhibit an apparently high rate of mtDNA escape because the colony was "founded" by a cell that had previously experienced a mtDNA escape event. Repeated colony purification and mtDNA escape assays can distinguish isolates that have a true high rate of mtDNA escape from those in which the rate is high because of an early escape event. This approach can be used for screening escape on tetrad dissection plates or during mutant screens; however, such screens are typically very difficult to score.

3. All sectors should contain lawns of similar density. If comparisons are to be made between strains with significantly different growth rates, slower-growing strains should be streaked before faster-growing ones.

4. Transfer of too many cells to the escape plate inhibits the growth of prototrophs, reducing the apparent mtDNA escape rate. To minimize this problem, make replicas from one velvet to several escape plates. Later replicas will transfer fewer cells, which may allow a more accurate determination of relative escape rates. Alternatively, excess cells can be removed from the first velvet by blotting with a second velvet. The first velvet is then used to transfer cells to the escape plate. Additionally, velvets with a very short nap tend to give a cleaner result on the escape plate. If a plasmid is present in the tested strains, maintain selection for the plasmid on the escape assay plate.

5. Plate-to-plate differences can significantly affect observed mtDNA escape rates. These differences are minimized by using plates from the same batch, velvets

from the same lot, as well as by including a wild-type control on each plate to control for plate-to-plate differences.

6. Identical cultures for the quantitative assay of mtDNA escape can, alternatively, be grown in liquid medium. A small inoculum, ideally a single cell, is used to inoculate multiple 1- to 2-mL cultures, which are grown to the same optical density and analyzed as described.

Acknowledgments

The work described here was supported by NIH grant GM47397. We thank Justin White for figure preparation.

References

1. Lang, B. F., Gray, M. W., and Burger, G. (1999) Mitochondrial genome evolution and the origin of eukaryotes. *Annu. Rev. Genet.* **33,** 351–397.

2. Thorsness, P. E. and Weber, E. (1996) Escape and migration of nucleic acids between chloroplasts, mitochondria, and the nucleus. *Int. Rev. Cytol.* **165,** 207–234.

3. Campbell, C. L. and Thorsness, P. E. (1998) Escape of mitochondrial DNA to the nucleus in *yme1* yeast is mediated by vacuolar-dependent turnover of abnormal mitochondrial compartments. *J. Cell Sci.* **111,** 2455–2464.

4. Hanekamp, T. and Thorsness, P. E. (1996) Inactivation of *YME2/RNA12*, which encodes an integral inner mitochondrial membrane protein, causes increased escape of DNA from mitochondria to the nucleus in *Saccharomyces cerevisisae*. *Mol. Cell. Biol.* **16,** 2764–2771.

5. Thorsness, P. E., White, K. H,. and Fox, T. D. (1993) Inactivation of *YME1*, a gene coding a member of the *SEC18, PAS1, CDC48* family of putative ATPases, causes increased escape of DNA from mitochondria in *Saccharomyces cerevisiae*. *Mol. Cell. Biol.* **13,** 5418–5426.

6. Shay, J. W. and Werbin, H. (1992) New evidence for the insertion of mitochondrial DNA into the human genome: significance for cancer and aging. *Mutat. Res.* **275,** 227–235.

7. Hadler, H. I. (1989) Comment: mitochondrial genes and cancer. *FEBS Lett.* **256,** 230–232.

8. Richter, C. (1988) Do mitochondrial DNA fragments promote cancer and aging? *FEBS Lett.* **241,** 1–5.

9. Liang, B. C. (1996) Evidence for association of mitochondrial DNA sequence amplification and nuclear localization in human low-grade gliomas. *Mutat. Res.* **354,** 27–33.

10. Thorsness, P. E. and Fox, T. D. (1993) Nuclear mutations in *Saccharomyces cerevisiae* that affect the escape of DNA from mitochondria to the nucleus. *Genetics* **134,** 21–28.

11. Braus, G. H., Luger, K., Paravicini, G., et al. (1988) The role of the TRP1 gene in yeast tryptophan biosynthesis. *J. Biol. Chem.* **263,** 7868–7875.

12. Bhattacharjee, J. K. (1985) Alpha-aminoadipate pathway for the biosynthesis of lysine in lower eukaryotes. *Crit. Rev. Microbiol.* **12,** 131–151.
13. Thorsness, P. E. and Fox, T. D. (1990) Escape of DNA from mitochondria to the nucleus in *Saccharomyces cerevisiae. Nature* **346,** 376–379.
14. Berlin, V., Brill, J. A., Trueheart, J., Boeke, J. D., and Fink, G. R. (1991) Genetic screens and selections for cell and nuclear fusion mutants. *Methods Enzymol.* **194,** 774–792.
15. Rose, M. D. and Fink, G. R. (1987) KAR1, a gene required for function of both intranuclear and extranuclear microtubules in yeast. *Cell* **48,** 1047–1060.
16. Luria, S. E. and Delbrück, M. (1943) Mutations of bacteria from virus sensitivity to virus resistance. *Genetics* **28,** 491–511.

12

2D Gel Electrophoresis of mtDNA

Heather E. Lorimer

1. Introduction

The neutral/neutral two-dimensional (2D) agarose gel technique has proved to be a powerful tool for analyzing DNA structure. It was initially adapted and developed by Brewer and Fangman to localize and define bidirectional DNA replication origins on plasmids in yeast *(1)*. Since then, neutral/neutral 2D gels have been used to analyze chromosomal replication origins, the efficiency of replication origin use, direction of replication fork movement, and barriers to DNA replication fork progression (reviewed in **ref. 2**). In recent years, this technique has also been used to analyze rolling circle replication in T4 phage *(3)* and topologically complex forms of intact plasmid DNA *(4)*.

Analyzing DNA structures by neutral/neutral 2D gels in organellar DNA presents some additional technical difficulties as well as well as opportunities. Some mitochondrial genomes are small enough to examine in their native uncut state by fairly conventional 2D gels. On the other hand, some organellar genomes are quite large, and their replication mechanisms are often substantially different from the classic bidirectional replication seen in chromosomal DNA. Some, such as the mitochondrial DNA (mtDNA) in the yeast *Saccharomyces cerevisiae*, exhibit very active recombination and a plethora of unusual DNA structures. Two-dimensional gels are a means by which some of these complex structures can be separated and observed. At this time, 2D gels have been used to examine recombination and a variety of forms of organellar DNA, including recombination intermediates, D-loops, rolling circles, and bidirectional replication forks in plants, yeast, and mammals *(5–8)*.

The first goal of analyzing mtDNA structure is to make sure that all nucleic acids are isolated. Many DNA isolation methods are aimed at isolating nuclear

From: *Methods in Molecular Biology, vol. 197: Mitochondrial DNA: Methods and Protocols*
Edited by: W. C. Copeland © Humana Press Inc., Totowa, NJ

DNA and may involve techniques that separate the nucleus from other cellular components. This will obviously diminish yields of mitochondrial DNA. Other techniques aim to purify a mitochondrial fraction from other cellular components. Mitochondria are dynamic and variable organelles. Thus, the methods that purify mitochondria tend to yield a subset of the total mitochondria. Similarly, methods that purify mtDNA by cesium chloride gradients on the basis of different A/T contents of mitochondrial DNA isolate a specific fraction of mtDNA, potentially missing supercoiled forms and complex structures along with single-stranded regions. Therefore, the method of DNA isolation used here is a modification of a small-scale preparation of yeast total cellular DNA, often called a "smash and grab" *(9)*.

The conditions under which the gels are run should be modified according to the size of the DNA being analyzed. The percent agarose in the gels can be lowered for larger DNA molecules or raised for smaller DNA molecules. In general, the first dimension is run very slowly, in a low-percentage agarose gel, in the absence of ethidium bromide to separate molecules primarily on the basis of size. The DNA-containing lane from the first dimension is rotated 90° counterclockwise and the second-dimension gel is poured around it. The second dimension is run at a much higher voltage in a much higher-percentage agarose, in the presence of ethidium bromide. Ethidium bromide is an intercalating dye that makes DNA molecules more rigid, amplifying any changes mobility resulting from the shape of the molecule.

The method presented here is designed to analyze whole genomes of small size or small repeat size (under 5 kbp [kilobase pair], such as are found in ρ^- mutants of the yeast *S. cerevisiae* or to examine restriction fragments between 1 and 5 kbp.

The last steps of 2D gel analysis of mtDNA from yeast requires very even and complete transfer of a variety of DNA molecules onto a membrane, followed by a very sensitive probing system. A modified version of Southern blotting *(10)* is presented in the chapter. Several different systems to make mtDNA-specific probes can be used. In general, I have only achieved good results using high-activity ^{32}P-labeled probes, but fluorescein-labeled probes have been reported that were sensitive enough to detect low-abundance replication intermediates from blots of 2D gels *(11)*.

2. Materials

2.1. Total Cellular DNA Isolation

1. 0.2 *M* EDTA (*see* **Notes 1** and **2**).
2. Acid-washed glass beads: For 1 kg of 425- to 600-μm glass beads (the author uses Sigma G-9268), place in 2-L beaker, cover with concentrated HCl, incubate

at room temperature (in a fume hood) for 10 min, and carefully rinse with deionized water until the water in the beaker is the same pH as the water going into the beaker. Spread beads in a clean glass tray; bake until dry. (*See* **Note 3.**)

3. Lysis buffer: 2% Triton X-100, 1% sodium dodecyl sulfate (SDS), 100 mM NaCl, 10 mM Tris-HCl, pH 8.0, 10 mM EDTA. Can be autoclaved. Store refrigerated.
4. PCIA: 50% equilibrated phenol *(12)*, 48% chloroform, 2% isoamyl alcohol.
5. 10X TE: 100 mM Tris-HCl, pH 8.0, 10 mM EDTA.
6. 3M Sodium acetate, pH 5.2 (*see* **Note 1**).
7. 100% Ethanol.
8. 75% Ethanol, 25% H_2O at 0°C.

2.2. Agarose Gel Electrophoresis

1. TBE: 85 mM Tris base, 89 mM boric acid, 2.5 mM EDTA, pH 8.0, (*see* **Note 1**).
2. Low electroendosmosis (EEO) agarose.
3. Ethidium bromide: 10 mg/mL in H_2O store at 4°C protected from light.
4. Submarine gel electrophoresis box with ports for buffer circulation that holds a minimum of a 15-cm × 15-cm gel (*see* **Note 4**).
5. Comb with 0.5-cm-wide teeth.
5. Power supply sufficient to supply a minimum of 500 V and 300 mA.
6. Ultraviolet (UV) light box (*see* **Note 5**).
7. Fluorescent ruler.
6. Peristaltic pump.

2.3. Southern Blot

1. BBI: 0.25 N HCl.
2. BBII: 0.5 M NaOH, 1 M NaCl.
3. BBIII 0.5 M Tris-HCl, pH 7.5, 3 M NaCl.
4. 10X SSC: 1.5 M NaCl, 0.15 M sodium citrate, pH 7.0, (*see* **Note 6**).
5. Tray and 1- to 2-in.-thick sponge, cut to fit (*see* **Note 7**).
6. DNA binding membrane such as Boehringer Mannheim positively charged nylon membrane.
7. 3MM filter paper.
8. Paper towels (*see* **Note 8**).

3. Methods
3.1. Isolating Total Cellular DNA from Yeast

1. Grow yeast in 200 mL of liquid media (*see* **Notes 9** and **10**). While yeast is growing in liquid culture, add 10 mL of 0.2 M EDTA to two centrifuge bottles, freeze overnight (*see* **Note 11**).
2. Take 200 mL fresh liquid culture of yeast cells and transfer to a centrifuge bottles. Spin for 5 min at 4000g, and remove supernatant.

3. Suspend pellet in 20 mL ice-cold 10X TE and transfer to 50-mL conical tubes. Centrifuge at 2000*g* for 5 min at 4°C and discard supernatant. Pellets may be stored at –20°C until used.

4. Add 3 mL acid-washed 425- to 600-µm glass beads, 2 mL cold lysis buffer, and 2 mL PCIA. (*See* **Note 12**.)

5. Vortex on high speed for 30 s, then place on ice for 30 s until pellet is no longer visible at bottom of tube, continue to vortex for 10 more 30-s intervals. (*see* **Note 13**.)

6. Centrifuge for 5 min as in **step 2**. You should see a layer of beads and PCIA on the bottom, an opaque white interface in the middle, and a clear aqueous layer on the top. Transfer the top (aqueous) layer to a 15-mL conical tube (*see* **Note 14**).

7. Add 2 mL of 10X TE to original tube, vortex for 30 s, and centrifuge for 5 min on high speed in microcentrifuge. Transfer top (aqueous) layer to the tube in **step 6**.

8. Add 4 mL PCIA to combined aqueous sample. Vortex for 30 s, centrifuge for 5 min, and transfer top (aqueous) layer to new 15-mL tube.

9. Repeat **step 8** until the interface is clear (*see* **Note 15**).

10. Transfer the aqueous layer to a high-speed centrifuge tube. Add 1/10 volume (400 µL) 3 *M* sodium acetate, and 2.5 volumes (10 mL) ethanol. Mix gently; a stringy white precipitate should become apparent.

11. Chill sample on ice or in a –20°C freezer for 30 min (*see* **Note 16**). Samples may be stored at this stage in the freezer indefinitely.

12. Centrifuge the sample at 20,000*g* for 20 min.

13. Discard supernatant; wash pellet gently with ice-cold 75% ethanol.

14. Air-dry pellet. Suspend in 1X TE to 1 mg/mL concentration (*see* **Note 17**).

3.2. First Dimension: Size Separation

1. Prepare a 0.4% agarose gel in TBE (*see* **Notes 18** and **19**).

2. Load 1–5 µg total cellular DNA per lane and a size marker in another lane (*see* **Note 20**).

3. Separate the samples by electrophoresis for 20 h at 0.7 V/cm (*see* **Notes 21** and **22**).

3. Stain the gel in TBE containing 300 ng/mL ethidium bromide (*see* **Note 23**).

4. Visualize the gel briefly on a UV light box to mark where the lane will be cut measuring with a fluorescent ruler (*see* **Note 24**).

5. Cut out the lane(s) containing cellular DNA, being careful to retain the well and cut straight and close to the sample. The gel slice may be moved on a thin, flexible plastic ruler.

3.3. Second Dimension: Shape Separation

1. Chill TBE with 300 ng/mL ethidium bromide sufficient to run the second dimension to 4°C.

2. Prepare 1.2% molten agarose in TBE with 300 ng/mL ethidium bromide.

3. Rotate the gel slice from above 90° counterclockwise, placing it near the top of the gel tray (**Fig. 1**).
4. Seal the slice to the gel tray with drops of the molten 1.2% agarose solution (*see* **Note 25**).
5. Pour the rest of the 1.2% agarose around the slice.
6. When the gel has solidified, place it in the electrophoresis apparatus with the cold TBE.
7. Attach the pump to circulate the TBE from the positive electrode to the negative electrode (*see* **Note 26**).
8. Run the gel at 5 V/cm for 3 h at 4°C (*see* **Note 27**).

3.4. Southern Blot

1. Place gel face-down in a tray.
2. Add 200 mL BI, rock gently for 15 min, pour off solution, repeat (*see* **Note 28**).
3. Add 200 mL BII, rock gently for 15 min, repeat (*see* **Note 29**).
4. Add 400 mL BIII, rock gently for 30 min (*see* **Note 30**).
5. Cut a 1- to 2-in.-thick foam sponge to lie flat and cover the bottom of a 2- to 3-in.-deep glass tray. Saturate the sponge completely with 10X SSC and pour additional SSC in the tray to about 1 cm below the top of the sponge. (*See* **Note 31**.)
6. Place a piece of 3MM paper cut slightly larger than the gel on the top of the sponge. (**Fig. 2**).
7. Place gel face-down on paper.
8. Cut a piece of positively charged membrane to the size of your gel; when the membrane is dry, it is possible to mark the date and any other relevant information lightly in pencil on a corner of the membrane. (*See* **Note 32**.)
9. Wet membrane with H_2O, then in the BIII or 10X SSC
9. Place wet membrane face-down on gel,
10. Place a sheet of 3MM paper wetted with 10X SSC on top of the membrane.
11. Place parafilm or plastic wrap from the edge of the gel to the edge of the tray to cover exposed areas of the sponge.
12. Stack about 5 in. of paper towels on top of 3MM paper.
13. Place a flat weight such as a sheet of glass with a book or bottle on top to apply even pressure to the stack. Let sit overnight.
14. Remove and dispose of paper towels and 3MM paper. Crosslink the membrane face up in UV crosslinker.
15. Hybridize to a high-activity mtDNA-specific probe of choice.

4. Notes

1. Many of the solutions are made according to methods previously published (*12*).
2. All solutions used should be sterile and nuclease-free.
3. Glass beads may be purchased already acid washed. This is necessary because DNA is acidic and will adhere to glass that in not acid washed or siliconized to

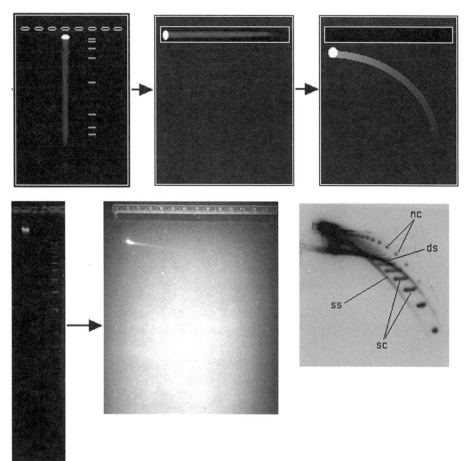

Fig. 1. The top panels show a diagram of the steps of neutral/neutral 2D gel electrophoresis of DNA. Top left panel shows total cellular DNA and a marker run in the first dimension. Top center panel shows the lane of cellular DNA excised, rotated 90° counterclockwise, with the agarose for the second dimension poured around it. Top right panel shows the DNA after the second-dimension electrophoresis is complete. The bottom panels show an actual 2D gel of total, uncut, cellular DNA in steps. Bottom left panel is after the first dimension; bottom center panel is after the second dimension. Bottom right panel is an autoradiogram of the gel after Southern blotting, probed for a ρ^- mtDNA. The autoradiogram clearly shows a continuum from very small to very large double-stranded (ds) and single-stranded (ss) DNA, as well as nicked circular (nc) and supercoiled circular discrete forms ranging from monomeric circles though multimers of greater than 10 repeats in length. Note that the larger supercoiled circles apparently exhibit a variety of amounts of supercoiling revealed by horizontal arrays of dots.

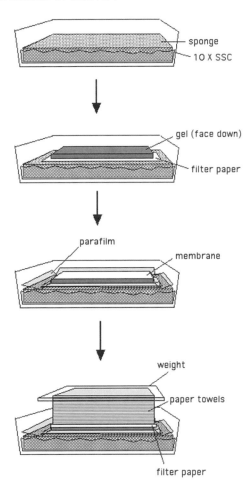

Fig. 2. Steps in the Southern blot assembly in order from top to bottom.

block ionic interactions between the glass and the DNA. Similarly, never pick up solutions containing DNA with a glass pipet.

4. Typically, we use box with a gel tray that is 20 cm long for the first dimension, and a box with a gel tray that is 20 cm wide and 20 cm long for the second dimension.

5. The UV light box needs to be large enough to illuminate the whole gel. Ideally, it will have a low setting for preparative work to minimize DNA nicking from UV energy.

6. The 20X SSC may be used in place of 10X SSC, but it is prone to precipitating during storage.

7. Soft foam sponges provide an even surface and are easy to cut to size.

8. The author prefers inexpensive single-fold natural brown paper towels that come wrapped in convenient 5-in.-high stacks.

9. This protocol is designed to produce a substantial amount of DNA that can be analyzed a number of times in different ways.

10. More or less cells may be grown depending on the final concentration. For cultures that are allowed to reach a stationary growth state, 100 mL may be more than sufficient. For early log-phase cultures, 500 mL may be more appropriate.

11. Throughout the procedure, until the DNA is precipitated, EDTA is critical to reduce the activity of any contaminating nucleases.

12. The efficiency of breaking open the cells can be increased by adding more beads.

13. Excessive vortexing, pipetting, freezing and thawing, and reprecipitating DNA can shear DNA. In particular, large, complex, and/or partially single-stranded forms are sensitive to shearing. Shearing can be reduced by erring on the side of using more beads in the initial breaking of the cells. A few microliters from the aqueous layer may be examined by light microscopy for the percent of broken cells, visible as nonrefractile "ghosts" when compared to the refractile intact cells.

14. If the volume of beads exceeds the interface of the organic and aqueous, more PCIA and lysis buffer may be added before centrifugation.

15. A common technical mistake is to fail to continue to extract until the interface is clear. As long as a white or cloudy interface is present, contaminating proteins may be present with the DNA. These contaminants may affect the moblility of the DNA in the gel and/or may lead to nicking of the DNA, resulting in loss of some forms.

16. There is evidence that freezing DNA in ethanol to enhance precipitation may be unnecessary; the author remains unconvinced of that however.

17. DNA concentration may be determined by visual examination after running on a mini-gel with a standard of known concentration, or the sample may be treated with RNase, extracted, reprecipitated, dissolved, and analyzed by UV spectrophotometry (as described in **ref. *12***).

18. This procedure was designed to analyze all of the native forms of a ρ^- genome 1 kb in repeat length This genome exists in sizes ranging form monomeric circles to multimeric forms running at the limit of mobility. For restriction fragment digest of DNA aimed at examining specific regions for replication intermediates, a higher-percent agarose may be used in the first dimension, as very large forms of DNA will not be expected.

19. For best results, prepare enough agarose melted in TBE to be able to pour the gel to be about 12 mm thick. A fast and easy way to prepare the molten agarose–TBE mix is to microwave on high power in a flask capable of holding four times the volume used, for intervals of 1 min, swirling the solution in the flask in between, until all the agarose is dissolved.

20. It is highly recommended to separate samples from each other by leaving an empty lane between them to avoid cross-contamination.

21. Volts/centimeter may be measured with a hand-held voltmeter by placing the contacts a measured distance apart in the TBE buffer in the gel box with the power on. The volts measured can be divided by the distance (in cm). Alternately, the distance between the electrodes in the box can be measured and that distance used to multiply the desired volts/centimeter to arrive at total volts to set the power supply on. For instance, if the electrode are 30 cm apart in a gel box, and the desired voltage is 0.7 V/cm ($0.7 \times 30 = 21$ V).

22. Set the power supply to deliver constant voltage.

23. Stain the gel by rocking gently in a container large enough to hold the gel in its tray. The 0.4% agarose gels are very fragile and can be easily broken when attempting to remove them from their trays and then moving them to a UV light box.

24. At this point the integrity of the DNA preparation can be monitored by the presence of a defined band of DNA at the top of the lane, indicating large chromosomal DNA that migrated at the upper limit of mobility (*see* **Fig. 1**, lower left). Be aware that exposure to UV light can nick the DNA, so minimize the time that the gel is exposed to UV light. Nicking can be revealed after exposure of the Southern blot by absence of supercoiled DNA and presence of nicked circular forms when supercoils were expected (*see* **Fig. 1**, lower right).

25. Use a Pasteur pipet to drip molten agarose thickly all the way around the gel slice. Let the drops harden before pouring the rest of the agarose.

26. Ethidium bromide will migrate opposite to the direction of the DNA in the gel in the electric field. Unless the buffer from the upper chamber is constantly pumped back to the lower chamber, the ethidium bromide concentration will change, causing irregularities in the DNA migration in the second dimension.

27. It is essential to refrigerate the gel to keep it cool and to keep the milliamps low under the high voltage and high gel concentration in the second dimension. Running the gel in a refrigerated cold room is ideal.

28. The acid BBI solution removes some purine bases from the DNA strands. Do not leave the gel for longer than the recommended times in this solution.

29. The basic BBII solution denatures the DNA strands and can destroy them if the gel is left too long in this solution.

30. This is a neutralizing solution; the gel can sit longer than 30 min in this solution if necessary.

31. Make sure the sponge is evenly and thoroughly saturated. Air bubbles in the sponge result in areas where the DNA does not transfer well to the membrane.

32. Do not touch the membrane with bare hands.

References

1. Brewer, B. J. and Fangman, W. L. (1987) The localization of replication origins on ARS plasmids in *S. cerevisiae. Cell* **51,** 463–471.

2. Friedman, K. L. and Brewer, B. J. (1995) The analysis of replication intermediates by two-dimensional agarose gel electrophoresis. *Methods Enzymol.* **262,** 613–627.

3. Boulanger, K. G., Mirzayan, C., Kreuzer, H. E., Alberts, B. A., and Kreuzer, K. N. (1996) Two-dimensional gel analysis of rolling circle replication in the presence and absence of bacteriophage T4 primase. *Nucleic Acids Res.* **24,** 2166–2175.

4. Martín-Parras, L., Lucas, I., Martínez-Robles, M. L., Hernández, P., Krimer, D. B., Hyrien, O., et al. (1998) Topological complexity of different populations of pBR322 as visualized by two-dimensional agarose gel electrophoresis. *Nucleic Acids Res.* **26,** 3424–3432.

5. Lockshon, D., Zweifel, S. G., Freeman-Cook, L. L., Lorimer, H. E., Brewer, B. J., and Fangman, W. L., (1995) A role for recombination junctions in the segregation of mitochondrial DNA in yeast. *Cell* **81,** 947–955.

6. MacAlpine, D. M., Perlman, P. S., and Butow, R. N. (2000) The numbers of individual mitochondrial DNA molecules and mitochondrial DNA nucleoids in yeast are co-regulated by the general amino acid control pathway. *EMBO* **19,** 767–775.

7. Kunnimalaiyaan, M. and Neilsen, B. L. (1997) Fine mapping of replication origins (*ori*A and *ori*B) in *Nicotiana tabacum* chloroplast DNA. *Nucleic Acids Res.* **25,** 3681–3686.

8. Holt, I. J., Lorimer H. E., and Jacobs, H. T. (2000) Coupled leading- and lagging-strand synthesis of mammalian mitochondrial DNA. *Cell* **100,** 515–524.

9. Hoffman, C. S. and Winston, F. (1987) A ten-minute DNA preparation from yeast effectively releases autonomous plasmids for transformation of *Escherichia coli. Gene* **57,** 267–272.

10. Srutkowska, S., Caspi, R., Gabig, M., and Wegrzyn, G. (1999) Detection of DNA replication intermediates after two-dimensional agarose gel electrophoresis using a fluorescein-labeled probe. *Anal. Biochem.* **269,** 221–222.

11. Southern, E. M. (1975) Detection of specific sequences among DNA fragments separated by gel electrophoresis. *J. Mol. Biol.* **98,** 503–517.

12. Sambrook, J., Fritsch, E. F., and Maniatis, T. (1982) *Molecular cloning. A laboratory Manual.* 2nd Ed. Cold Spring Harbor Laboratory Press, Cold Spring Harbor, NY.

II

METHODS FOR THE ANALYSIS OF PROTEINS INVOLVED IN mtDNA TRANSACTIONS

13

Purification of Mitochondria for Enzymes Involved in Nucleic Acid Transactions

Daniel F. Bogenhagen

1. Introduction

Ever since the demonstration that mitochondria contain DNA (mtDNA) and a complete apparatus for its replication and expression there has been an interest in identifying the enzymes involved in these processes. This problem is compounded by the fact that similar DNA transactions also occur in the nucleus, where vastly larger quantities of enzymes deal with 1000 times as much DNA. Thus, it has been a consistent challenge to the field to purify a distinct mitochondrial enzyme activity and prove that it is a *bona fide* mitochondrial enzyme. When mtDNA genomes were sequenced in the 1980s and open reading frames assigned to metabolic enzymes, it became clear that *all* proteins involved in mtDNA replication, transcription, and repair are products of nuclear genes that must be imported into mitochondria. Our laboratory has had a long-standing interest in these mitochondrial enzymes. When we first began to approach this problem, the best estimates in the literature were that the mtRNA polymerase had a molecular weight estimated at 66–68 kDa in rat liver *(1)* or 46 kDa in *Xenopus* oocytes *(2)*; similarly, some early estimates of the size of mtDNA polymerase suggested that it was a homo-tetramer of 47-kDa subunits *(3,4)*. Both the mtRNA polymerase and the cata-lytic subunit of DNA pol γ are now known to be approximately 125–140 kDa in various organisms *(5–12)*. The possibility that proteolysis contributed to the early underestimate of the molecular masses of these proteins underscores the need to work quickly and to use a complete set of protease inhibitors during purification of these enzymes. In recent years, the effort to characterize the mitochondrial complement of DNA metabolic enzymes has benefited

From: *Methods in Molecular Biology, vol. 197: Mitochondrial DNA: Methods and Protocols*
Edited by: W. C. Copeland © Humana Press Inc., Totowa, NJ

enormously from the expansion of genetic databases that provide a gold standard in quality control to assure that purified proteins have the properties predicted by primary sequence information.

The difficulty of preparing large quantities of clean mitochondria is a limiting factor in efforts to purify mitochondrial proteins. One productive strategy has been to identify a biological source for starting material that is enriched in mitochondria or relatively free of nuclear contamination. For instance, Hübscher et al. *(13)* used rat brain synaptosomes as a starting material for purification of DNA pol γ. Other labs have used *Drosophila* embryos *(11)* or *Xenopus* oocytes *(8,14–17)* as sources enriched in mitochondria stored for distribution to cells during early embryonic development. The advantage of embryos and oocytes as a starting material is evident in the case of *Xenopus*, where a single mature oocyte with a diameter of 1.4 mm contains as much mtDNA as 100,000 somatic cells, so that approximately 95% of the DNA in the cell is mtDNA. The large oocyte nucleus, also known as a germinal vesicle, is physically disrupted by homogenization, so that nuclear polymerases and other activities stored for early development are solubilized and easily washed away by differential centrifugation. A single *Xenopus* female contains 8–15 g of ovary tissue, which yields approximately as much DNA pol γ as 4–6 L of cultured HeLa cells. We typically prepare mitochondria from 30 ovaries at a time, yielding 0.7–1 g of soluble mitochondrial protein in approximately 3 h. We have found *Xenopus* oocytes to be a convenient source for preparation of DNA pol γ, mtRNA polymerase, mitochondrial transcription factor A (mtTFA), mitochondrial single-stranded DNA binding protein (mtSSB) and mtDNA ligase. All of these have been cloned and found to closely resemble their human counterparts, as shown in **Table 1**.

Most recent methods for purification of mitochondria are based on the pioneering efforts of Schneider and Hogeboom *(19)* using differential centrifugation for the preparation of a mitochondrial fraction. When cells are disrupted by homogenization, the mitochondrial reticulum is fragmented into bacterial-sized vesicles that are larger and denser than microsomal and plasma membrane vesicles, but smaller than nuclei. These mitochondrial fragments are prepared after centrifugation at low speed (1500*g*) to remove nuclei, unbroken cells, and other cell debris followed by centrifugation at 10,000*g* for 10–15 min. As long branching mitochondria are fragmented, there is some unavoidable loss of mitochondrial material (*see* **Note 1**). The method of Schneider and Hogeboom *(19)* formed the basis of protocols developed by Bogenhagen and Clayton *(20)* for purification of mtDNA. These workers followed the recovery of mtDNA that was selectively labeled with radioactive thymidine in TK-cultured cells. Because nuclear DNA is not radioactively labeled in these cells, the labeled mtDNA provides a marker for the mitochondrial fraction. This study showed

Table 1
**Comparison of Cloned Human and *Xenopus* Proteins Involved
in mtDNA Maintenance**

Protein	Human #AA [accession]	% Identity	Xenopus #AA [accession]
DNA Pol γ-A	1239 [G02750]	63.3	1200 [S68258]
DNA Pol γ-B	485 [AF177201]	49.1	463 [AF124606]
mtRNA Pol	1230 [AAB58255]	46.2	1235 [AAF19376]
mtTFA	247 [M62810]	33.6	283 [U35728]
mtSSB	148 [Q04837]	75.2	148 [JC6172]
mtDNA Ligase III[a]	1009 [X84740]	66.0	988 [AF393654]

[a]Human mtDNA ligase appears to be a translation product of the DNA ligase III gene initiated at a Met residue upstream from that reported by Wei et al. *(18)* as suggested *(17)*. The *Xenopus laevis* sequence was reported by Perez-Jannotti et al. *(33)*. For both organisms, the ligase IIIα variant with a BRCT domain is used in the calculations.

that differential centrifugation alone provided the highest yield of mtDNA, but that this procedure was not sufficient to prepare a mitochondrial fraction free of adherent contaminating nuclear DNA. The purity of mitochondria was significantly improved by using sucrose step gradients to remove most contaminating nucleic acids. Similar systematic studies of cell-fractionation procedures have been undertaken to study the cellular distribution of metabolic proteins. A number of publications have used Percoll gradients as an alternative density medium for purification of mitochondria *(21)*. The comprehensive review by Pederson et al. *(22)* is an excellent source of general information on mitochondrial purification techniques.

Sucrose or Percoll gradients are relatively low-capacity methods for purification of mitochondria that are difficult to perform on the scale necessary to purify proteins involved in mtDNA metabolism. Moreover, the use of gradient procedures to purify mitochondria does not eliminate contamination from other cell compartments. Sucrose gradient purified mitochondria are still associated with significant amounts of contaminating endoplasmic reticulum (ER) fragments. Two-dimensional (2D) gel electrophoretic analysis of proteins in a mitochondrial fraction purified by sucrose gradient sedimentation reveals persistent contamination by ER and peroxisomal proteins (Bogenhagen and

Kobayashi, unpublished data). Similar levels of contamination have been observed by others (23,24). This is likely a consequence of the tight biological interactions between mitochondria and ER at specialized junctions (25,26). Ribosomal proteins and cytoplasmic rRNAs also contaminate mitochondrial preparations based on the association of ribosomes with the outer mitochondrial membrane (27). However, very little evidence has been presented for adventitious association of nuclear proteins with sucrose gradient purified mitochondria. In particular, we have observed no significant amanitin-sensitive RNA polymerase or aphidicolin-sensitive DNA polymerase in our mitochondrial preparations. Although it is not uncommon for nuclear proteins to leach out through nuclear pores during cell fractionation, these proteins do not typically associate selectively with mitochondria and are largely removed by washing. Thus, in approaching the problem of purifying mitochondrial polymerases and accessory factors for studies of replication, transcription, and DNA repair, the researcher is always faced with the dilemma of either using rapid, higher-capacity methods or of taking pains to prepare highly purified mitochondria. Because it is always difficult to obtain sufficient quantities of mitochondria as starting material, it is recommended that mitochondria be prepared rapidly using only differential centrifugation in the early stages of characterizing a putative novel mitochondrial protein. Analytical-scale preparations using more highly purified mitochondria should be performed later to confirm that the activities purified are enriched in the mitochondrial fraction. Ultimately, the mitochondrial location can be documented by cloning the gene and demonstrating that it contains a mitochondrial targeting signal and by preparing antibodies to confirm that the antigen is located in mitochondria.

In this chapter, a protocol for purification of mitochondria from Xenopus oocytes is presented. Two variations of the procedure are provided for the use of sucrose step gradients to increase the purity of the mitochondrial fraction and for preparation of mitochondria from cultured cells. The final volume of the mitochondrial lysate should be adjusted so that the protein concentration is 5–10 mg/mL. This lysate can be stored frozen for several months, thawed and applied directly to a DEAE–Sephacel column at 0.25 M KCl as described (14,15) to remove endogenous nucleic acids and some proteins [including a substantial fraction of mtTFA (16)]. DNA pol γ, mtRNA polymerase, mtTFB, mtDNA ligase, mitochondrial AP endonuclease, mitochondrial type I topoisomerase, and mtSSB are found in the flow-through fraction of the DEAE column run at 0.25 M KCl. This flow-through fraction is dialyzed for 2–3 h to reduce the salt concentration to less than 60 mM KCl and loaded on a column of S-Sepharose. All of the activities noted earlier bind to S-Sepharose and elute as the column is washed with a gradient of KCl increasing from 50 to 750 mM, as shown in **Fig. 1**. Each of these activities has been further purified

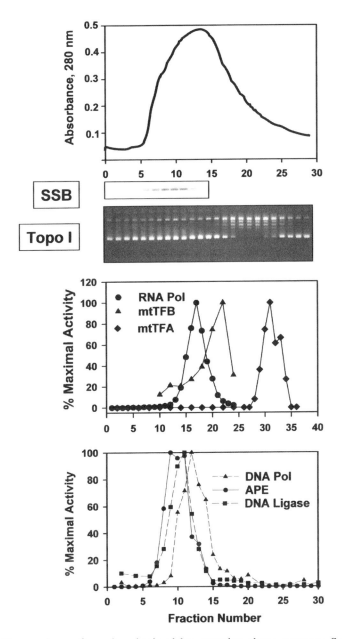

Fig. 1. Comparison of results obtained by assaying three separate S-Sepharose column eluants for a variety of *Xenopus* oocyte mitochondrial proteins. The upper panel shows a typical absorbance profile for mitochondrial proteins eluted by a 50- to 750-m*M* KCl gradient. This column was probed for mtSSB by immunoblotting, yielding a peak of protein in fraction 9, and was assayed for type I topoisomerase by a standard relaxation assay, with a peak of activity in fraction 22. The center panel shows the elution positions for proteins involved in transcription of mtDNA *(8)*. The lower panel shows the elution positions for proteins involved in base excision repair of mtDNA *(17)*. The overall protein profile and order of elution of these activities are very consistent from one preparation to another, although the exact fraction numbers vary slightly from one preparation to another.

by a combination of gel filtration, hydrophobic interaction, ion exchange and affinity chromatography procedures as described in original publications cited in this review. For some of these proteins, additional details or alternate procedures are available in other chapters in this volume.

2. Materials

2.1. Stock Solutions

All glassware and implements should be autoclaved. All solutions should be autoclaved or filter sterilized and stored on ice prior to use.

1. 0.5X SSC (1X SSC is 0.15 M NaCl, 15 mM Na citrate).
2. DHE buffer: 2 mM dithiothreitol (DTT), 30 mM HEPES, pH 8.0, 1 mM EDTA.
3. Protease inhibitor stock solutions: 0.4 M benzamidine–HCl in H_2O, 200X; 200 mM phenylmethylsulfonyl fluoride (PMSF) in isopropanol, 1000X; 5 mg/mL aprotinin in 50% glycerol, 10 mM HEPES, pH 7.5, 1000X; 5 mg/mL leupeptin in H_2O, 1000X; 1 mM pepstatin in methanol, 1000X; 2 mM E-64 in H_2O, 1000X; 5 mg/mL antipain in H_2O, 1000X.
4. Unsupplemented 2.5X MSH buffer (*see* **Note 2**). 1X buffer: 210 mM mannitol, 70 mM sucrose, 20 mM HEPES, pH 8.0, 2 mM EDTA. Filter-sterilized stocks may be stored indefinitely at 4°C.
5. MSH buffer (prepare immediately before use): dilute 2.5X MSH to 1X and add the following ingredients to the indicated final concentrations: 2 mM DTT, 2 mM benzamidine–HCl, 0.2 mM PMSF, 1 μM pepstatin, 5 μg/mL leupeptin.
6. Mitochondrial lysis buffer. This is a 1.25X buffer that will be diluted with KCl and TX-100 to reach final concentrations of 30 mM Tris-HCl, pH 8.4, 10% glycerol, 2 mM DTT, 1 mM EDTA, 2 mM benzamidine–HCl, 0.5 mM PMSF, 5 μg/mL aprotinin, 5 μg/mL antipain, 5 μg/mL leupeptin, 1 μM pepstatin, 2 μM E-64. Combine 3 mL of 1 M Tris-HCl, pH 8.4, 21.5 mL of 80% glycerol, 0.5 mL of 0.5 M DTT, 0.5 mL of 0.25 M EDTA, pH 7.5, 0.5 mL of benzamidine–HCl, 0.4 mL of 200 mM PMSF in isopropanol, 0.1 mL of 5 mg/mL aprotinin, 0.1 mL of 5 mg/mL antipain, 0.1 ml of 5 mg/mL leupeptin, 0.1 mL of 1 mM pepstatin in methanol, 0.1 mL of 2 mM E-64 with H_2O in a final volume of 80 mL.

2.2. Sucrose Step-Gradient Solutions

1. 1 M Sucrose in DHE buffer, supplemented with protease inhibitors (*see* **Subheading 2.1., items 2** and **3**).
2. 1.5 M Sucrose in DHE buffer, supplemented with protease inhibitors (*see* **Subheading 2.1., items 2** and **3**).

2.3. Solutions for Cultured Cells

1. Isotonic TDE buffer: 135 mM NaCl, 5 mM KCl, 25 mM Tris-HCl, pH 7.6, 0.7 mM Na_2HPO_4, 0.5 mM β-mercaptoethanol, 0.2 mM PMSF.

2. Buffer A: 10 mM HEPES, pH 8.0, 1.5 mM MgCl$_2$, 10 mM KCl, 1 mM DTT, supplemented with protease inhibitors PMSF, pepstatin, and leupeptin.

3. Methods

3.1. Preparation of a Mitochondrial Lysate from Xenopus Oocytes

The following is based on 300–400 g of ovarian tissue from 30–35 frogs. All animal research must be conducted with appropriate approval from the Institutional Animal Care and Use Committee. Keep the sample at 0–4°C throughout this procedure and work as quickly as possible to minimize the opportunity for proteolysis. **Subheading 3.2.** describes a modification of this method in which sucrose step-gradient sedimentation is used in place of one round of differential sedimentation to increase the purity of mitochondria. **Subheading 3.3.** describes a modification for preparation of a mitochondrial fraction from cultured cells.

1. Remove ovarian tissue from anesthetized, ice-chilled *X. laevis* females and hold in 0.5X SSC on ice until dissection is completed. (*See* **Note 3**.)
2. Rinse in 0.5X SSC until free of blood; this will require 1–2 L of 0.5X SSC.
3. Weigh ovaries quickly by picking up clumps of tissue with forceps and transferring to a tared beaker with 100 mL of MSH buffer. Use this as a rinse into MSH; pour off and replace with fresh MSH buffer.
4. Homogenize in two stages in a total volume of 400 mL of tissue plus MSH buffer.

   First: Start tissue disruption in a chilled Waring blender. Run for 10–20 s on setting 3–4. Transfer the homogenate back to a container on ice.

   Second: After the tissue has been crudely disrupted by the Waring blender, repeat the homogenization with three to five strokes of a tight-fitting motor-driven Teflon pestle in a 50-mL glass homogenizer. On this scale, processing the entire homogenate will require several rounds of homogenization. (*See* **Note 4**.)
5. Centrifuge at 1000*g* for 10 min in 250-mL centrifuge bottles in a GSA rotor to remove debris and yolk platelets. Filter supernatant through 0.6-mm Nytex gauze to remove fibrous material. Repeat this low-speed spin.
6. Transfer the supernatant to sterile 50-mL polycarbonate tubes and centrifuge at 20,000*g* for 15 min. The indicated scale will require 12 tubes.
7. Scrape off and discard the yellow lipid layer. Decant supernatant to waste or process for purification of cytoplasmic proteins. (*See* **Note 5**.)
8. Resuspend the light colored (brownish) upper layer of the pellet containing mitochondria in fresh MSH buffer while leaving behind most of the denser black layer (pigment and yolk). (*See* **Note 6**). For this and all subsequent washes, use approximately 30 mL per 50-mL tube. Distribute the resuspended mitochondria to six clean polycarbonate tubes and spin at 20,000*g* for 15 min as in **step 6**.

9. Resuspend the mitochondrial pellet in MSH+Na (100 mL of MSH + 2 mL of 5 *M* NaCl), again leaving most of the denser pigment behind. Transfer to four clean polycarbonate tubes and repeat centrifugation as in **step 6**.
10. Resuspend mitochondria in MSH, distribute to two centrifuge tubes, and repeat centrifugation.
11. Resuspend each tube in a total of 40 mL of lysis buffer to permit lysis in two batches of 50 mL each. Transfer 40 mL of resuspended mitochondria into a Dounce homogenizer. Then, add 4.1 mL of 3 *M* KCl for a final concentration of 0.25 *M* and 6.4 mL of 10% Triton X-100 to 1.25%. The final volume of this mix is 50 mL. Homogenize with the "A" pestle for six to eight strokes. Avoid frothing. Repeat this step with the other batch of resuspended mitochondria.
12. Transfer the 100 mL of mitochondrial lysate to eight Beckman polyallomer quick seal tubes for the Ti 70.1 rotor (cat. no. 342413). Seal the tubes and centrifuge for 90 min at 230,000*g* (50,000 rpm) in the Ti 70.1 rotor. Carefully remove the supernatant and leave the gelatinous pellet (*see* **Note 7**). The cleared supernatant should have a final volume of 90–94 mL and should contain approximately 90% of the initial lysate protein. The cleared lysate typically has a protein concentration of 7–10 mg/mL.
13. Freeze the lysate by immersion in liquid N_2 and store at –80°C.

3.2. Modification for Sucrose Gradient Purification of Mitochondria

This procedure is intended to replace one round of differential centrifugation to obtain a higher level of purity of mitochondrial proteins. Following the first sedimentation, mitochondria are resuspended in MSH buffer and layered on discontinuous sucrose gradients (*see* **Note 6**).

1. In advance, prepare sucrose step gradient in SW28 tubes by layering 15 mL of 1 *M* sucrose over a 10-mL pad of 1.5 *M* sucrose. All buffers should contain 2 m*M* EDTA, 2 m*M* DTT, 20 m*M* HEPES, pH 8, and protease inhibitors. Store preformed step gradients at 4°C, preferably not longer than 1 h.
2. Prepare the initial mitochondrial pellet fraction through **step 8** of **Subheading 3.1.** Instead of repelleting mitochondria, carefully layer the resuspended mito-chondrial suspension over the sucrose step gradient directly. Adjust the volume of MSH used to resuspend mitochondria to fill the appropriate number of step gradients, each of which can accommodate 8–10 mL of solution.
3. Centrifuge for 30 min at 92,600*g* at 4°C. Mitochondria sediment to the 1/1.5 *M* sucrose interface.
4. Remove upper layers by suction, leaving the mitochondrial layer undisturbed. Then, collect the mitochondrial layer with a sterile Pasteur pipet.
5. Dilute the mitochondrial suspension with an equal volume of cold DHE buffer or with 3 volumes of MSH buffer to reduce the sucrose concentration and

resediment at 26,000g (15 krpm) in a Sorvall SS34 rotor or Beckman J₁
resuming the preparation in **step 10** of **Subheading 3.1.**

3.3. Modification for Purification of Mitochondria from Cultured Cells

The following is a procedure for preparation of mitochondria that we have used for both human HeLa and *Xenopus* A6 cells grown in culture and for mouse liver mitochondria. The procedure is adapted from the Dignam and Roeder *(28)* procedure for preparation of nuclear extracts for the purification of transcription factors and from Bogenhagen and Clayton *(20)*. In contrast to *Xenopus* oocytes, smaller somatic cells are difficult to disrupt in a glass Dounce homogenizer unless they are swollen in hypotonic buffer. The volume of buffer to be used is critical because excessive dilution of mitochondria in hypotonic buffer can lead to lysis of the organelles. It is important to maintain a fixed ratio of cell-swelling buffer A to the volume of packed cells (PCV) to prevent exposure of mitochondria to very low ionic strength. It is also critically important to add 2.5X MSH quickly to stabilize mitochondria. Following adjustment of the solution to 1X MSH, nuclei are removed by two rounds of low-speed centrifugation and mitochondria are pelleted and further purified by sucrose gradient sedimentation as in **Subheading 3.2.** Note that other protocols are available using alternative methods for cell disruption *(29)*.

1. Sediment cells grown in suspension culture in 500-mL centrifuge tubes by centrifugation for 10 min at 1000g.
2. Resuspend in isotonic TDE buffer (phosphate-buffered saline [PBS] can be substituted, but should be modified to contain 2 mM mercaptoethanol and 0.2 mm PMSF), transfer to 50-mL conical centrifuge tubes (e.g., Falcon 2098) and centrifuge at 1000g for 5 min. Use gradations on the tube to estimate the PCV.
3. Resuspend in 5 PCV of buffer A and repellet cells by centrifugation for 5 min at 1000g.
4. Add 2 PCV of buffer A and resuspend. Allow 5 min for the cells to swell and confirm swelling by inspecting a sample of the cells in the microscope. Homogenize with 8–12 strokes of a tight-fitting Dounce homogenizer. Check for complete disruption of cells and release of nuclei by microscopy (*see* **Note 8**).
5. When over 90% of cells are disrupted, add 2 PCV of 2.5X MSH to stabilize mitochondria.
6. Pellet nuclei in 50-mL conical centrifuge tubes at 1600g in a swinging bucket rotor (e.g., Sorvall RT-6000). Decant the supernatant into a clean conical tube and repeat this spin.
7. Transfer the postnuclear supernatant to a 50-mL polycarbonate tube and centrifuge at 15,000g to pellet mitochondria.

8. Resuspend the mitochondria in MSH buffer and continue the preparation with sucrose step-gradient fractionation at **step 2** of **Subheading 3.2.**, followed by additional sedimentation and lysis as in **Subheading 3.1.**

4. Notes

1. We have consistently found that centrifugation at 20,000g for 15 min provides a higher overall yield of mitochondrial proteins, presumably by capturing smaller fragments broken off mitochondria during homogenization.
2. Mannitol and sucrose do not readily enter intact mitochondria and thus stabilize the mitochondrial membranes against osmotic rupture. The unusual 2.5X concentration reflects the limited solubility of mannitol. This basic buffer is supplemented with protease inhibitors and DTT immediately before use.
3. With practice, a group of 3 workers can harvest ovaries from 30 frogs in approximately 15–20 min.
4. Wear padded protective gloves while handling the glass homogenizer to avoid heating the sample and to protect against glass breakage.
5. Methods for purification of type I and II topoisomerase and RNA polymerase III from the postmitochondrial supernatant have been published *(30–32)*.
6. At this point, resuspended mitochondria can be layered onto sucrose step gradients. Care must be taken not to overload gradients to the point that a heavy pad of mitochondria develops at the 1.0/1.5 M sucrose interface.
7. This centrifugation for 1.5 h at a clearing factor of 71 is intended to remove mtDNA and 55S mitochondrial ribosomes. For ovary mitochondria, there is a central pigment layer under the pellet. A turbid layer at the interface between the tightly packed pellet material and the supernatant is a sign of ineffective lysis due to insufficient Triton X-100.
8. If cell disruption is incomplete, it is likely that cells did not swell adequately. The addition of more hypotonic buffer A may be helpful, but this will require addition of a correspondingly greater volume of 2.5X MSH after homogenization to attain a final concentration of 1X MSH.

Acknowledgments

Research in the author's laboratory has been supported by grants GM29681 and ES03068 from the NIH. This manuscript is dedicated to the memory of Dr. Suzan S. Cairns, a former graduate student who began research in my laboratory on *Xenopus* mtDNA.

References

1. Reid, B. and Parsons, P. (1971) Partial purification of mitochondrial RNA polymerase from rat liver. *Proc. Natl. Acad. Sci. USA* **68,** 2830–2834.
2. Wu, G. and Dawid, I. (1972) Purification and properties of mitochondrial deoxyribonucleic acid dependent ribonucleic acid polymerase from ovaries of Xenopus laevis. *Biochemistry* **11,** 3589–3595.

3. Yamaguchi, M., Matsukage, A., and Takahashi, T. (1980) Chick embryo DNA polymerase γ: purification and structural analysis of nearly homogeneous enzyme. *J. Biol. Chem.* **255,** 7002–7009.

4. Matsukage, A., Tanabe, K., Yamaguchi, M., Taguchi, Y., Nishizawa, M., Takahashi, T., et al. (1981) Identification of a polypeptide component of mouse myeloma DNA polymerase γ. *Biochim. Biophys. Acta* **655,** 269–277.

5. Foury, F. (1989) Cloning and sequencing of the nuclear gene MIP 1 encoding the catalytic subunit of the yeast mitochondrial DNA polymerase. *J. Biol. Chem.* **264,** 20,552–20,560.

6. Masters, B. S., Stohl, L. L., and Clayton, D. A. (1987) Yeast mitochondrial RNA polymerase is homologous to those encoded by bacteriophages T3 and T7. *Cell* **51,** 89–99.

7. Ropp, P. A. and Copeland, W. C. (1996) Cloning and characterization of the human mitochondrial DNA polymerase, DNA polymerase γ. *Genomics* **36,** 449–458.

8. Bogenhagen, D. F. (1996) Interaction of mtTFB and mtRNA polymerase at core promoters for transcription of *Xenopus laevis* mtDNA. *J. Biol. Chem.* **271,** 12,036–12,041.

9. Carrodeguas, J., Kobayashi, R., Lim, S., Copeland, W., and Bogenhagen, D. (1999) The accessory subunit of *X. laevis* mitochondrial DNA polymerase γ increases processivity of the catalytic subunit of human DNA polymerase γ and is related to class II amino acyl tRNA synthetases. *Mol. Cell. Biol.* **19,** 4039–4046.

10. Tiranti, V., Savoia, A., Forti, F., D'Apolito, M.-F., Centra, M., Rocchi, M., et al. (1997) Identification of the gene encoding the human mitochondrial RNA polymerase (h-mtRPOL) by cyberscreening of the Expressed Sequence Tags database. *Hum. Mol. Genet.* **6,** 615–625.

11. Wernette, C. M. and Kaguni, L. S. (1986) A mitochondrial DNA polymerase from embryos of *Drosophila* melanogaster. Purification, subunit structure, and partial characterization. *J. Biol. Chem.* **261** 14,764–14,770.

12. Ye, F., Carrodeguas, J. A., and Bogenhagen, D. F. (1996) The γ subfamily of DNA polymerases: cloning of a developmentally regulated cDNA encoding *Xenopus laevis* mitochondrial DNA polymerase γ. *Nucleic Acids Res.* **24,** 1481–1488.

13. Hübscher, U., Kunzle, C., and Spadari, S. (1977) Identity of DNA polymerase γ from synaptosomal mitochondria and rat-brain nuclei. *Eur. J. Biochem.* **81,** 249–258.

14. Insdorf, N. F. and Bogenhagen, D. F. (1989) DNA polymerase γ from *Xenopus laevis* I. The identification of a high molecular weight catalytic subunit by a novel DNA polymerase photolabeling procedure. *J. Biol. Chem.* **264,** 21,491–21,497.

15. Bogenhagen, D. F. and Insdorf, N. F. (1988) Purification of *Xenopus laevis* mitochondrial RNA polymerase and identification of a dissociable factor required for specific transcription. *Mol. Cell. Biol.* **8,** 2910–2916.

16. Antoshechkin, I. and Bogenhagen, D. F. (1995) Distinct roles for two purified factors in transcription of *Xenopus* mitochondrial DNA. *Mol. Cell. Biol.* **15,** 7032–7042.

17. Pinz, K. and Bogenhagen, D. (1998) Efficient repair of abasic sites in DNA by mitochondrial enzymes. *Mol. Cell. Biol.* **18,** 1257–1265.

18. Wei, Y.-F., Robins, P., Carter, K., Caldecott, K., Pappin, D. J. C., Yu, G.-L., et al. (1995) Molecular cloning and expression of human cDNAs encoding a novel DNA ligase IV and DNA ligase III, an enzyme active in DNA repair and recombination. *Mol. Cell. Biol.* **15,** 3206–3216.

19. Schneider, W. C. and Hogeboom, G. H. (1950) Intracellular distribution of enzymes: further studies on distribution of cytochrome c in rat liver homogenates. *J. Biol. Chem.* **183,** 123–128.

20. Bogenhagen, D. and Clayton, D. A. (1974) The number of mitochondrial deoxyribonucleic acid genomes in mouse L and human HeLa cells. *J. Biol. Chem.* **249,** 7991–7995.

21. Storrie, B. and Madden, E. (1990) Isolation of subcellular organelles. *Methods Enzymol.* **182,** 203–225.

22. Pederson, P., Greenawalt, J., Reynafarje, B., Hullihen, J., Decker, G., Soper, J., et al. (1978) Preparation and characterization of mitochondria and submitochondrial particles of rat liver and liver-derived tissues. *Methods Cell Biol.* **20,** 411–481.

23. Rabilloud, T., Kieffer, S., Procaccio, V., Louwagie, M., Courchesne, P., Patterson, S., et al. (1998) Two-dimensional electrophoresis of human placental mitochondria and protein identification by mass spectrometry: toward a human mitochondrial proteome. *Electrophoresis* **19,** 1006–1014.

24. Patterson, S., Spahr, C., Daugas, E., Susin, S., Irinopoulou, T., Koehler, C., et al. (2000) Mass spectrometric identification of proteins released from mitochondria undergoing permeability transition. *Cell Death Differ.* **7,** 137–144.

25. Rutter, G. and Rizzuto, R. (2000) Regulation of mitochondrial metabolism by ER Ca^{2+} release: an intimate connection. *TIBS* **25,** 215–221.

26. Manella, C. (2000) Introduction: our changing views of mitochondria. *J. Bioenerg. Biomembr.* **32,** 1–4.

27. Kellems, R. E., Allison, V. F., and Butow, R. A. (1975) Cytoplasmic type 80S ribosomes associated with yeast mitochondria. IV. Attachment of ribosomes to the outer membrane of isolated mitochondria. *J. Cell Biol.* **65,** 1–14.

28. Dignam, J. D., Martin, P. L., Shastry, B. S., and Roeder, R. G. (1983) Eukaryotic gene transcription with purified components. *Methods Enzymol.* **101,** 582–598.

29. Adachi, S., Gottleib, R., and Babior, B. (1998) Lack of release of cytochrome C from mitochondria into cytosol early in the course of Fas-mediated apoptosis of Jurkat cells. *J. Biol. Chem.* **273,** 19,892–19,894.

30. Cozzarelli, N. R., Gerrard, S. P., Schlissel, M., Brown, D. D., and Bogenhagen, D. F. (1983) Purified RNA polymerase III accurately and efficiently terminates transcription of 5S RNA genes. *Cell* **34,** 829–835.

31. Luke, M. and Bogenhagen, D. F. (1989) Quantitation of Type II topoisomerase in oocytes and eggs of *Xenopus laevis*. *Dev. Biol.* **136,** 459–468.

32. Richard, R. E. and Bogenhagen, D. F. (1989) A high molecular weight topoisomerase I from *Xenopus laevis* ovaries. *J. Biol. Chem.* **264,** 4704–4709.

33. Perez-Jannotti, R. M., Klein, S. M., and Bogenhagen, D. F. (2001) Two forms of mitochondrial DNA ligase III are produced in *Xenopus laevis* oocytes. *J. Biol. Chem.* **276,** 48,978–48,987.

14

Purification of Mitochondrial Uracil–DNA Glycosylase Using Ugi–Sepharose Affinity Chromatography

Samuel E. Bennett, Mary Jane Shroyer, Jung-Suk Sung, and Dale W. Mosbaugh

1. Introduction

Affinity chromatography is a powerful technique that allows purification of proteins based on biospecific interactions using an affinity absorbent to selectively bind a target protein *(1,2)*. Separation is achieved through a highly specific, but reversible, interaction with a complementary binding substrate (ligand) that is immobilized on a solid support (matrix). This purification approach has been developed to incorporate protein–protein interactions into the design of the affinity matrix *(3–5)*. Affinity chromatography procedures can achieve purification levels on the order of several thousand-fold in a single separation step. The method presented here utilizes the bacteriophage PBS2 uracil–DNA glycosylase inhibitor (Ugi) protein as an immobilized ligand on Sepharose 4B beads for purification of mammalian mitochondrial uracil–DNA glycosylase.

Uracil–DNA glycosylase plays an important biological role in mutation avoidance by initiating the uracil-mediated DNA base excision repair pathway *(6–8)*. The enzyme acts to specifically remove uracil from DNA through cleavage of the *N*-glycosylic bond linking the base to the deoxyribose phosphate DNA backbone *(9)*. *Escherichia coli* uracil–DNA glycosylase (Ung) has been extensively characterized as a monofunctional, single polypeptide enzyme with a deduced molecular weight of 25,563 *(6,10)*. Remarkable conservation of amino acid sequence and tertiary structure have been detected in uracil–DNA glycosylase obtained from a wide variety of organisms ranging from *E. coli* to humans *(11–13)*. Distinct mitochondrial and nuclear forms of the enzyme have

From: *Methods in Molecular Biology, vol. 197: Mitochondrial DNA: Methods and Protocols*
Edited by: W. C. Copeland © Humana Press Inc., Totowa, NJ

been purified and partially characterized *(14–16)*. Using a six-step procedure, uracil–DNA glycosylase from isolated rat liver mitochondria was purified to apparent homogeneity, and two distinct mitochondrial forms of the enzyme were first observed *(15)*. Recently, it has been shown that both human and mouse mitochondrial (UNG1) and nuclear (UNG2) uracil–DNA glycosylase are produced from the same *UNG* gene by alternative transcriptional start sites and differential splicing of mRNA *(17,18)*. Consequently, the C-terminal 269 amino acids of the two proteins are identical, including the DNA binding and catalytic domain *(12,17)*. In humans, the unique N-terminal portion (35 amino acids) of the UNG1 precursor protein (304 amino acids) contains a mitochondrial localization sequence that appears to be cleaved upon entry into the mitochondria by the mitochondrial processing peptidase *(17,19)*. Recent reports have confirmed the presence of two mitochondrial UNG1 species and provided evidence that proteolytic processing is involved in their formation *(20)*. Although many of the biochemical properties of uracil–DNA glycosylase have been elucidated, significant biological questions concerning the native form(s) of the mitochondrial enzyme still remain. The development of an affinity chromatography technique to facilitate purification of mitochondrial uracil–DNA glycosylase should be useful in extending our understanding of the role that this enzyme plays in mitochondrial base excision repair and other DNA metabolic functions.

The bacteriophage PBS2 Ugi protein exhibits several properties that make it ideally suited as an affinity ligand for purifying various uracil–DNA glyco-sylases, including the mitochondrial species. Ugi is a small (9474 molecular weight), monomeric, heat stable, acidic ($pI = 4.2$) protein of 84 amino acids that specifically binds and inactivates uracil–DNA glycosylase from diverse biologi-cal sources *(6,12,21,22)*. However, interaction with other DNA glycosylases *(23,24)* or general DNA metabolizing enzymes *(22)* has not been established. Thus, the Ugi interaction with uracil–DNA glycosylase appears to be highly specific. The molecular mechanism for Ugi inhibition of Ung has been eluci-dated in detail using both biochemical and structural approaches. Ugi utilizes DNA mimicry that targets the highly conserved uracil–DNA glycosylase DNA binding domain to form a tightly bound noncovalent Ung•Ugi complex with 1:1 stoichiometry *(12,21)*. Stable binding of Ugi is accomplished by a two-step reaction characterized by an initial reversible association with Ung followed by transformation to a "locked" complex that is essentially irreversible under physiological conditions *(25)*. A mutant form of the inhibitor protein containing an amino acid substitution of Leu for Glu at residue 28 (Ugi E28L) has been produced that exhibits near (88%) wild-type Ugi activity but demonstrates reversible binding to Ung *(26)*. The high-affinity interactions between both Ugi and Ugi E28L with Ung are desirable in designing an affinity chromatography strategy because they typically facilitate highly specific binding to the target protein. However, extremely tight binding may also complicate the process of removing the enzyme from the affinity matrix and may necessitate elution of the protein using extreme conditions (e.g., pH 2.5), which can reduce recovery of the enzyme in an active form, as has been previously observed *(19)*.

In this chapter, we describe the protocols for immobilization of Ugi protein onto Sepharose 4B resin, determination of the binding capacity of the Ugi– Sepharose matrix, preparation of a mitochondrial extract for uracil–DNA glycosylase isolation, and subsequent purification of the enzyme using Ugi– Sepharose affinity chromatography. The method may be used to facilitate (1) the timely purification of mitochondrial uracil–DNA glycosylase, (2) the isolation of polymorphic variants, (3) the purification of heterogeneous forms that arise by posttranscriptional and/or posttranslational modification, and (4) the isolation of multiprotein complexes involving accessory factors.

2. Materials
2.1. Preparation of Ugi–Sepharose

1. Coupling buffer: 0.1 M NaHCO$_3$, pH 8.3, 0.5 M NaCl.
2. Blocking buffer: 0.1 M Tris-HCl, pH 8.0.
3. Low pH buffer: 100 mM glycine, pH 2.5, 1 M NaCl.
4. High pH buffer: 100 mM Tris-HCl, pH 8.0, 1 M NaCl.
5. Storage buffer: 50 mM Tris-HCl, pH 8.0, 1 mM ethlenediaminetetraacetic acid (EDTA), disodium salt, 1 mM dithiothreitol (DTT), 100 mM NaCl.
6. Bacteriophage PBS2 uracil–DNA glycosylase inhibitor protein Ugi and Ugi E28L (fraction IV) were purified as previously described *(26,27)*. Prior to initiating the ligand-coupling procedure, Ugi and Ugi E28L were diafiltered into the coupling buffer using an Amicon stirred cell (YM3 membrane) driven by N$_2$ gas (55 psi).
7. CNBr-activated Sepharose 4B (Amersham Pharmacia Biotech) freeze-dried 4% agarose beads (45–165 µm) were stored at 4°C.

2.2. Determination of the Ung Binding Capacity of Ugi–Sepharose

1. Pyrex fiberglass (Corning) with 8-µm silver fibers was siliconized using Sigma-cote (Sigma).
2. Micro-chromatographic columns were prepared using APEX microcapillary pipet tips (capacity 0–200 µL) with stubby stalk (West Coast Scientific, MC-10BR).
3. 10 mg/mL Bovine serum albumin (BSA) essentially fatty acid free (Sigma, A 7030).
4. Formula 989 scintillation fluor (Beckman).
5. DAB buffer: 30 mM Tris-HCl, pH 7.4, 1 mM EDTA, 1 mM DTT, 5 % (w/v) glycerol.

6. SDS buffer: 50 mM Tris-HCl, pH 6.8, 1% sodium dodecyl sulfate (SDS), 143 mM β-mercaptoethanol, 10% (w/v) glycerol, 0.04% bromphenol blue (3′,3′,5′,5′-tetrabromophenolsulfonphtalein).
7. *Escherichia coli* [leucine-³H]Ung (fraction V) with a specific activity of 13.5 cpm/pmol was purified as previously described *(10,26)*.

2.3. Isolation of Pig Liver Mitochondria

1. Motor-driven tissue homogenizer.
2. Homogenization buffer: 50 mM Tris-HCl, pH 8.0, 0.25 M sucrose, 25 mM KCl, 1 mM EDTA, 5 mM β-mercaptoethanol, 5 mM MgCl₂.

2.4. Preparation of Mitochondrial Extract

1. Levigated alumina, type A5 (Sigma).
2. Ammonium sulfate, Ultra Pure enzyme grade (Life Technologies).
3. Porcelain mortar (21 cm diameter) chilled overnight at –20°C.
4. Extraction buffer: 50 mM Tris-HCl, pH 8.0, 5 mM β-mercaptoethanol, 1 mM EDTA, 1 M NaCl, 1 mM phenylmethylsulfonyl fluoride (PMSF).
5. TMEG buffer: 20 mM Tris-HCl, pH 8.0, 5 mM β-mercaptoethanol, 1 mM EDTA, 10% (w/v) glycerol.
6. TEDG buffer: 20 mM Tris-HCl, pH 8.0, 1 mM EDTA, 1 mM DTT, 10% (w/v) glycerol.

2.5. Purification of Mitochondrial Uracil–DNA Glycosylase Using Ugi–Sepharose Affinity Chromatography

1. Micro Bio-Spin column (Bio-Rad).
2. 10 mM Potassium phosphate buffer, pH 7.2.
3. 0.5 M NaCl.
4. 10 mM Potassium phosphate buffer, pH 7.2, 5 M MgCl₂.

3. Methods

3.1. Preparation of Ugi–Sepharose

This section describes the procedure for preparing the Ugi–Sepharose affinity resin used to isolate mitochondrial uracil–DNA glycosylase. An immobilization strategy was developed that involved linking Ugi to imidocarbonate groups located on the Sepharose beads. The coupling process was designed to preferentially react primary amino groups of Ugi with the matrix in order to immobilize the protein through an isourea linkage. This approach was selected because none of the lysine residues (K10, K14, K36, K66, K80, K82) of Ugi have been implicated as essential in formation of the stable Ung–Ugi complex *(12)*. Following the coupling reaction with Ugi, residual active groups on the matrix were hydrolyzed to block nonspecific absorption. In addition, small amounts of uncoupled ligand were removed from the Ugi–Sepharose by cycling the resin through a series of washes of alternating low, pH 2.5, and high, pH 8.0, buffer. A procedure is also described for determining the Ugi-coupling efficiency. The immobilization method described for coupling the ligand to CNBr-activated Sepharose 4B applies equally to Ugi and Ugi E28L. Use of Ugi E28L–Sepharose is advocated for applications that require recovery of Ung with high specific activity. Preparation of a mock-treated Sepharose affinity column is recommended in order to determine whether proteins recovered following column elution are contaminants resulting from nonspecific absorption.

1. CNBr-activated Sepharose 4B (6 g) was suspended in 40 mL of 1 mM HCl and allowed to swell for 5 min at room temperature.
2. The hydrated agarose beads were placed into a 30-mL sintered glass (10–15 M) filter funnel and washed with 1160 mL of 1 mM HCl under mild vacuum produced by an aspirator (NALGENE).
3. After removal of the 1-mM HCl wash solution, the beads were resuspended in 5 mL of coupling buffer, which was then removed by filtration.
4. Sepharose beads were transferred from the filter funnel using a spatula to a 50-mL conical centrifuge tube (Falcon) and resuspended in 7 mL of coupling buffer. After resuspension, the preparation was divided into two equal batches based on the settled gel volume. One batch was used in the ligand-coupling reaction with Ugi protein and the other batch was used in a mock reaction.
5. In the mock reaction, an equal volume of coupling buffer was added to the tube containing the Sepharose beads. In the Ugi-coupling reaction, an equal volume of coupling buffer containing Ugi or Ugi E28L (6–10 mg of protein/mL of gel) was added to prepare the affinity resin.
6. The contents of both tubes were gently mixed for 2 h at room temperature using a three-dimensional-rotator table.
7. Following the coupling reaction, the gel samples were transferred into sintered glass filter funnels and the coupling buffer was removed by filtration. The filtrate was retained for further analysis to determine the coupling efficiency (*see* **Note 1**).
8. Each gel sample was washed with 5 gel volumes of coupling buffer at room temperature and the unbound Ugi was removed by filtration. As noted above under **Subheading 3.1., step 7**, the filtrate was saved for analysis of unbound Ugi.
9. Each gel preparation was resuspended in 35 mL of blocking buffer and incubated without agitation for 16 h at 4°C to inactivate any remaining imidocarbonate groups in the gel.
10. After decanting the blocking buffer from the gel samples, fresh blocking buffer (35 mL) was added and the settled beads were incubated for 1 h at 4°C. The blocking buffer was then removed by filtration. Both the decanted and filtered solutions were saved to detect unbound Ugi.
11. Both mock-treated and Ugi-reacted Sepharose beads were washed with three alternating cycles of low and high pH buffers to remove any uncoupled ligand.

Each wash cycle involved filtering 5 gel volumes of each low-pH and high-pH buffer through the gel matrix.

12. The final gel preparation was washed with 5 gel volumes of storage buffer and the resin was stored as a 25% (v/v) slurry at 4°C (*see* **Note 2**).

3.2. Determination of Ung-Binding Capacity of Ugi–Sepharose

This section describes a procedure for quantitatively determining the Ung-binding capacity of the Ugi–Sepharose resin produced in **Subheading 3.1.** Each preparation of affinity resin should be characterized to determine its binding capacity, prior to preparative use in isolating mitochondrial uracil–DNA glycosylase. *Escherichia coli* Ung was used as a standard to assess the binding capacity of the resin because (1) the Ung–Ugi interaction is well characterized *(10,12,21,25–27)*; (2) the Ung protein can be overproduced and isolated in large quantities, and then radiolabeled using [³H]leucine *(25,26)*. Although use of [³H]Ung is preferred for quantitative assessment of the binding capacity, an alternative approach can be adapted using nonradioactively labeled Ung in conjunction with the SDS–polyacrylamide gel electrophoresis (PAGE) method described as follows.

1. Micro-chromatographic columns were prepared by plugging the orifice of gel-loading micropipet tips with siliconized glass wool (*see* **Note 3**). After adding 100 μL of 10 m*M* potassium phosphate buffer, pH 8.0, containing 100 μg/mL BSA to the tip, a 10-min incubation was carried out at 25°C to block nonspecific binding sites. The buffer was then removed by brief (approx 30 s) centrifugation in a clinical centrifuge.

2. Binding reactions (200 μL) were prepared by mixing 100 μL of Ugi–Sepharose (25% [v/v] slurry) with an equal volume of DAB buffer containing various amounts of *E. coli* [³H]Ung (*see* **Fig. 1**). The suspension was gently mixed for 10 min at 25°C on a three-dimensional rotator table, and then applied to the micro-chromatographic column described under **Subheading 3.2., step 1**.

3. The binding buffer was removed from the Ugi–Sepharose resin by centrifugation as described under **Subheading 3.2., step 1**, and designated the flow-through fraction. The resin was then washed twice with 200 μL of DAB buffer. After centrifugation, the micropipet tips were sealed using a hot hemostat and 100 μL of SDS buffer was added to elute the [³H]Ung. The sealed micropipet was incubated at 95°C for 10 min, the tip was then cut with a razor blade, and the liquid contents removed by centrifugation. This elution process was repeated twice.

4. Flow-through, wash, and elution fractions were analyzed by 12.5% SDS-PAGE *(25,28)* and the [³H]Ung band was visualized following Coomassie Blue staining (*see* **Fig. 1A**). To quantitatively assess the amount of [³H]Ung that had bound to the Ugi–Sepharose, 25μL of the eluted fractions were added to 500 μL of distilled water and vigorously mixed with 5 mL of Formula 989 scintillation fluor (Beckman). The amount of ³H-radioactivity in each fraction was then determined by liquid scintillation spectrometry (*see* **Fig. 1B**).

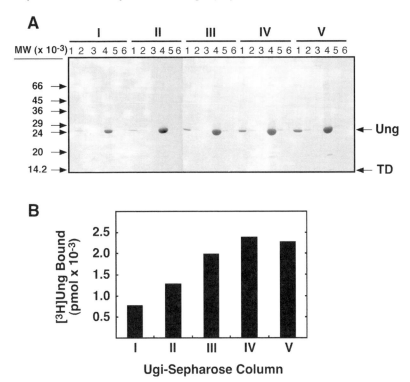

Fig. 1. Determination of Ugi–Sepharose binding capacity. (**A**) Five samples (100 μL) of Ugi–Sepharose (25% [v/v] slurry) equilibrated in DAB buffer were incubated with 975 (I), 1950 (II), 2930 (III), 3900 (IV), and 4870 pmol (V) of *E. coli* [³H]Ung for 10 min at 25°C. Following the Ung-binding reaction (see **Subheading 3.2.** for details), Ugi–Sepharose micro-columns were prepared, washed twice with 200 μL of DAB buffer, and eluted with three 100-μL additions of SDS buffer. Samples (25 μL) of the flow-through (lane 1) and wash fractions (lanes 2 and 3), along with samples (50 μL) of the elution fractions (lanes 4–6) were resolved by 12.5% SDS–polyacrylamide gel electrophoresis; protein bands were detected by Coomassie Blue (G250) staining. The location of molecular-weight standards BSA, ovalbumin, glyceraldehyde-3-phosphate dehydrogenase, carbonic anhydrase, trypsinogen, trypsin inhibitor, and α-lactalbumin are indicated by arrows from top to bottom, respectively. The location of Ung and the tracking dye (TD) are also indicated by arrows. (**B**) Flow-through (I), wash (II), and elution fractions (III–V) (25 μL) were analyzed for ³H-radioactivity using a liquid scintillation spectrometer. The amount of [³H]Ung bound to each Ugi–Sepharose column was determined by summing the radioactivity detected in the three elution fractions.

5. The binding specificity of the Ugi–Sepharose preparation should be examined before use in purification of uracil–DNA glycosylase from a complex protein mixture. To test the specificity of binding, various amounts of purified Ung were

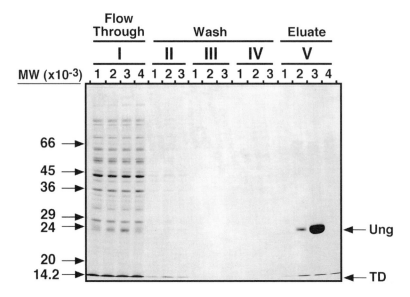

Fig. 2. Purification of Ung from a crude extract of *E. coli* using Ugi–Sepharose chromatography. Four samples (100 µL) of *E. coli* crude extract were mixed with either 0 µg (1), 5 µg (2), or 50 µg of purified *E. coli* Ung (3, 4) and incubated with 100 µL of Ugi-Sepharose (1–3) or mock-treated Sepharose (4) for 10 min at 25°C. Four micro-columns were prepared, the resin was washed three times with DAB buffer and the bound protein was eluted with SDS buffer as described in **Subheading 3.2.** Samples (50 µL) of the flow-through (I), wash (II–IV), and elution fractions (V) were analyzed by 12.5% SDS-PAGE as described in **Fig. 1.** The location of molecular standards and tracking dye (TD) are indicated by arrows.

added to an *E. coli* crude extract and subjected to Ugi–Sepharose chromatography, as described in **Subheading 3.2.** Binding specificity was evaluated using 12.5% SDS-PAGE to analyze the flow-through, wash, and eluted fractions (*see* **Fig. 2**). In addition, a similar analysis was conducted with the mock-treated Sepharose to determine whether nonspecific proteins were retained because of interaction with the Sepharose matrix (**Fig. 2**, lanes I4 and V4). Under the binding, washing, and elution conditions described, Ugi–Sepharose demonstrated highly specific Ung binding with minimal nonspecific protein absorbance.

3.3. Isolation of Pig Liver Mitochondria

This protocol describes a method for isolating large quantities of liver mitochondria and has been successfully used with calf, pig, and rat tissue sources. Briefly, the isolation procedure involves (1) use of a meat grinder to masticate liver tissue, (2) cell lysis with a tissue homogenizer, (3) removal of unbroken cells and nuclei by low-speed centrifugation, and (4) isolation of the mitochondrial fraction by differential centrifugation. All operations should be carried out at 0–4°C, unless otherwise indicated, and the procedure should be conducted as rapidly as possible during a single day. When the final mitochondrial pellet is acquired and frozen in liquid nitrogen, storage at –80°C for several months does not appear to adversely affect subsequent isolation of mitochondrial uracil–DNA glycosylase.

1. A fresh pig liver was obtained (Clark Meat Laboratory, Oregon State University) and the gallbladder removed. The tissue was lacerated several times, placed in a plastic bag that was surrounded by ice and contained 500 mL of 0.25 M sucrose, and transported to the laboratory.
2. Liver tissue was minced into approx $2 \times 2 \times 2$ cm slices, briefly placed in ice-cold 0.25 M sucrose, and then pulverized using a hand-operated meat grinder. Approximately 1200 g (wet weight) of ground liver paste was collected into 2.5 L of 0.25 M sucrose at 4°C.
3. The liver tissue was passed two or three times through a custom-built motor-driven homogenizer apparatus with a conical (10-cm diameter) Teflon pestle (designed by Dan Ziegler, University of Texas). The resulting homogenate was monitored by microscopic examination for cellular disruption (*see* **Note 4**). Following cell lysis, 200 mL of 1 M potassium phosphate buffer, pH 7.5, was added and the total volume was adjusted to 4 L with 0.25 M sucrose.
4. The cellular homogenate was centrifuged at 800g for 10 min in a GSA rotor (Sorvall) at 4°C. The supernatant fraction was pressed through two layers of cheesecloth and then centrifugation was performed at 8500g for 30 min in a SA-600 rotor (Sorvall) at 4°C.
5. The mitochondrial pellet was resuspended in 800 mL of homogenization buffer using a Dounce homogenizer (pestle B).
6. Following resuspension, the mitochondrial preparation was again centrifuged to remove whole cells and nuclei at 800g for 10 min, as described in **step 4**. The supernatant fraction was obtained and centrifuged at 8500g for 30 min as described in **step 4**.
7. Mitochondrial pellets were resuspended in an equal volume of homogenization buffer (800 mL) and centrifuged at 8500g for 30 min.
8. **Step 7** was repeated three times; however, the resuspension volume was sequentially reduced by 50% during the third and fourth cycles.
9. The final mitochondrial pellets were scraped from the centrifuge tubes into 50-mL polyethylene tubes, frozen in liquid nitrogen, and stored at –80°C.

3.4. Preparation of Mitochondrial Extract

The following procedure details the method used to obtain a soluble protein fraction from mitochondria that can be successfully used in Ugi–Sepharose chromatography to isolate uracil–DNA glycosylase. Frozen mitochondria are partially thawed and lysed by grinding in the presence of alumina power. The alumina is then removed to obtain a crude extract of soluble mitochondrial proteins. Following two ammonium sulfate precipitation steps (0–35% and 35–65%), the precipitated proteins from the second step are resuspended in TEDG buffer, dialyzed, and prepared for affinity chromatography. Inclusion of the ammonium sulfate fractionation procedure is critical for optimal Ugi–Sepharose chromatography results.

1. Frozen pig or rat liver mitochondrial pellets (approx 200 g) were partially thawed, placed into a chilled (–20°C) porcelain mortar, and rapidly fractured into ≤1-cm³ nuggets using the pestle. After gradual addition of 125 g of levigated alumina with grinding, the mitochondria were forcefully pulverized into a smooth, creamy paste. Ice-cold extraction buffer (500 mL) was slowly added during the grinding process, which was continued for approx 5 min following the final buffer addition.
2. The alumina was removed from the extract by centrifugation at 450g in a SA-600 rotor for 10 min at 4°C.
3. The supernatant fraction was decanted from the white pellet and centrifuged at 100,000g in a 50.2 Ti rotor (Beckman) for 1 h at 4°C. The golden-brown supernatant was isolated by siphon using a tube attached to a 60-mL syringe; care was taken to avoid the floating lipid layer.
4. The supernatant was subjected to dialysis at 4°C accompanied by three consecutive changes (2 L each) of TMEG buffer containing 150 mM NaCl (*see* **Note 5**). The resulting crude mitochondrial extract was designated fraction I.
5. Two ammonium sulfate precipitation steps were used to fractionate mitochondrial proteins. The first step involved precipitation of proteins by 0–35% (saturation) ammonium sulfate fractionation. The crude mitochondrial extract was incubated on ice with gentle stirring, and powdered ammonium sulfate was gradually added (19.4 g/100 mL) over a 30-min period. Following the ammonium sulfate addition, stirring was continued for 15 min, after which the precipitate was collected by centrifugation at 10,000g in a SA-600 rotor for 15 min at 4°C. The supernatant fraction was isolated with care to avoid the floating coagulated material and subjected to 35–65% ammonium sulfate fractionation (18.4 g/100 mL) following the procedure described above in this step. Precipitated protein was collected by centrifugation as described, and the pellets were resuspended in 30 mL of TEDG buffer and dialyzed extensively against the same buffer (*see* **Note 5**). The resulting sample constituted fraction II.

3.5. Purification of Mitochondrial Uracil–DNA Glycosylase Using Ugi–Sepharose Chromatography

The following chromatographic procedure utilizes Ugi–Sepharose to purify mitochondrial uracil–DNA glycosylase (UNG1) from the mitochondrial extract (fraction II) prepared in **Subheading 3.4.** Column chromatography was conducted using a four-step procedure that involved (1) binding of UNG1 to the affinity matrix, (2) washing unbound mitochondrial proteins from the resin, (3) removing nonspecifically bound proteins from the column, and (4) recovering UNG1 in the eluted fraction. The efficiency of UNG1 binding was improved by cycling the mitochondrial extract through the micro-column prior to the first wash step. Following ligand binding, the majority of the mitochondrial proteins were eliminated from the affinity column by washing the resin with five sequential aliquots of 10 mM potassium phosphate buffer, pH 7.2. Proteins that bound weakly or nonspecifically to the Ugi–Sepharose are removed from the column by a wash solution containing 0.5 M NaCl. Finally, UNG1 protein was eluted from the column with buffer containing 10 mM potassium phosphate, pH 7.2, and 5 M MgCl$_2$. This purification scheme was applied to both rat and pig liver mitochondrial extracts (*see* **Fig. 3**).

1. Rat and pig liver mitochondrial uracil–DNA glycosylase (fraction II) were assayed for enzyme activity as previously described *(14)*. Based on the activity of the preparations, samples determined to contain approx 4 µg of uracil–DNA glycosylase (*see* **Note 7**) were chosen for further purification.
2. Ugi–Sepharose affinity chromatography was performed using a Micro Bio-Spin (Bio-Rad) column containing 0.1 mL (column volume) of resin equilibrated with DAB buffer. The sample was cycled through the column for 2 h in a closed loop using a peristaltic pump (30 mL/h).
3. After ligand binding, the column was washed with 5 mL of 10 mM potassium phosphate buffer, pH 7.2, in 1-mL aliquots, followed by three applications (100 µL) of 0.5 M NaCl and one application (100 µL) of 10 mM potassium phosphate buffer, pH 7.2.
4. Bound mitochondrial uracil–DNA glycosylase was eluted with 100 µL of 10 mM potassium phosphate buffer, pH 7.2, containing 5 M MgCl$_2$ (*see* **Note 8**).

4. Notes

1. The coupling efficiency of Ugi to the Sepharose 4B resin was determined based on the amount of Ugi added to the coupling reaction minus the amount of unbound Ugi detected in the postcoupling filtrates. Ugi concentration was measured by absorbance spectroscopy using a molar extinction coefficient of 1.2×10^4 L/mol cm at a wavelength of 280 nm *(10)*.
2. A coupling efficiency of approx 200 nmol of Ugi bound per milliliter of packed Ugi–Sepharose can be obtained, based on the amount of unbound Ugi recovered following the ligand immobilization reaction. This value should not be used to calculate the binding capacity of the resin because not all of the immobilized Ugi protein retains the ability to bind the enzyme. Thus, the binding capacity of Ugi–Sepharose must be separately determined as described in **Subheading 3.2.**
3. In order to reduce nonspecific protein binding, glass wool was siliconized with a 5% (v/v) solution of dichlorodimethyl silane in heptane. The fiberglass stock was soaked in Sigmacote solution for approx 10 min at room temperature, thoroughly rinsed with distilled water, and cured overnight in a drying oven before use.

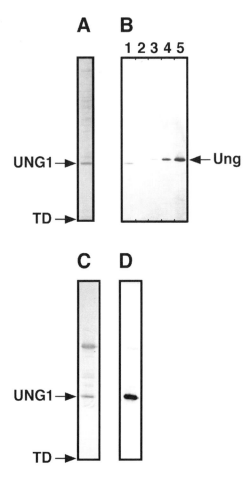

Fig. 3. Purification of mitochondrial uracil–DNA glycosylase using Ugi–Sepharose chromatography. Rat and pig liver mitochondrial uracil–DNA glycosylase (fraction II) were purified as described in **Subheading 3.5.**, and samples (2 µL) of the eluted fraction (**step 4**) from the rat (**A** and **B**, lane 1) and pig (**C** and **D**) uracil–DNA glycosylase preparations were resolved by 12.5% SDS-PAGE. In addition, samples containing 0 ng, 5 ng, 30 ng, and 150 ng of purified *E. coli* Ung (fraction V) were similarly processed (**B**, lanes 2–5, respectively). Following electrophoresis, protein bands were visualized using ICN Rapid Ag-Stain (**A** and **C**) or detected by Western blot analysis (**B** and **D**) with anti-Ung antibody (*see* **Note 6**). The location of *E. coli* Ung, UNG1, and the tracking dye (TD) are indicated by arrows.

4. A standard light microscope (100× magnification) was used to assess the integrity of the cell preparation during the tissue homogenization process. The homogenate was processed until approx 70% of the liver cells appeared to be disrupted. The

degree of cell breakage was estimated by side-by-side inspection of the cells before and after each homogenization step.

5. Conductivity measurements (μS/cm) were determined for the dialysis buffer and mitochondrial protein sample (before and after dialysis) using a Radiometer CDM-80 conductivity meter. A sample (100 μL) of each solution was diluted 10-fold with distilled water and its conductivity determined at 25°C. Dialysis was considered to be complete when the conductivity of the initial dialysis buffer approximately equaled that of the protein sample.

6. Rabbit polyclonal antibodies were raised against purified *E. coli* uracil–DNA glycosylase (fraction V). Antisera were diluted 1:5000 prior to use in Western blot analysis. The membrane was developed in 10 mL of 100 mM NaHCO$_3$, pH 9.8, using alkaline phosphatase-conjugated secondary antibody (goat anti-rabbit IgG) at 1:2000 dilution, and 100 μL each of 30 mg/mL nitroblue tetrazolium in 70% dimethylformamide and 15 mg/mL of 5-bromo-4-chloro-3-indolyl phosphate in 100 % dimethylformamide *(29)*.

7. A specific activity of 2×10^6 units/mg was used to calculate the amount (μg) of mitochondrial uracil–DNA glycosylase in the preparation. Activity was determined in a standard reaction mixture (100 μL) containing 70 mM HEPES-KOH, pH 7.5, 1 mM DTT, 1 mM EDTA, 25 mM NaCl, 10 mmol of calf thymus DNA containing [^{3}H]uracil (approx 250 cpm/pmol of uracil), and 0.002–0.2 units of enzyme. Incubation was conducted for 30 min at 37°C and terminated on ice with the addition of 250 μL of ammonium formate, pH 4.2. Free [^{3}H]uracil was resolved from nonhydrolyzed [uracil-^{3}H]DNA by application of 300 μL of the reaction mixture to a Bio-Rad 1-X8 (formate form) column (0.2 cm $\times$ 2 cm) equilibrated in 10 mM ammonium formate, pH 4.2. The column was washed with 1.7 mL of equilibration buffer to elute [^{3}H]uracil. Fractions (2×1 mL) were collected, combined with 5 mL of Formula 989, and ^{3}H radioactivity was measured using a liquid scintillation spectrometer. One unit of uracil–DNA glycosylase activity is defined as the amount of enzyme that catalyzes the release of 1 nmol of uracil per hour under standard reaction conditions.

8. Ugi E28L–Sepharose is recommended for applications that require recovery of uracil–DNA glycosylase with high specific activity. Optimal recovery of enzyme was observed using 100 mM triethylamine, pH 11.7, as the elution agent.

References

1. Ostrove, S. (1990) Affinity chromatography: general methods, in *Methods in Enzymology* (Deutscher, M. P., ed.), Academic, San Diego, CA, vol. 182, pp. 357–371.
2. Pharmacia Fine Chemicals (1979) *Affinity Chromatography: Principles and Methods*, Ljungfoeretagen AB, Obrebro, Sweden.
3. Prasad, R., Singhal, R. K., Srivastava, D. K., Molina, J. T., Tomkinson, A. E., and Wilson, S. H. (1996) Specific interaction of DNA polymerase beta and DNA ligase I in a multiprotein base excision repair complex from bovine testis. *J. Biol. Chem.* **271**, 16,000–16,007.

4. Singhal, R. K., Prasad, R., and Wilson, S. W. (1995) DNA polymerase beta conducts the gap-filling step in uracil-initiated base excision repair in a bovine testis nuclear extract. *J. Biol. Chem.* **270,** 949–957.
5. Hall, M. C. and Matson, S. W. (1999) The *Escherichia coli* MutL protein physically interacts with MutH and stimulates the MutH-associated endonuclease activity. *J. Biol. Chem.* **3,** 1306–1312.
6. Mosbaugh, D. W. and Bennett, S. E. (1994) Uracil–excision DNA repair. *Prog. Nucleic Acid Res. Mol. Biol.* **48,** 315–370.
7. Duncan, B. K. and Weiss, B. (1982) Specific mutator effects of *ung* (uracil-DNA glycosylase) mutations in *Escherichia coli. J. Bacteriol.* **151,** 750–755.
8. Radany, E. H., Dornfeld, K. J., Sanderson, R. J., Savage, M. K., Majumdar, A., Seidman, M. M., et al. (2000) Increased spontaneous mutation frequency in human cells expressing the phage PBS2-encoded inhibitor of uracil-DNA glycosylase. *Mutat. Res.* **461,** 41–58.
9. Lindahl, T., Ljungquist, S., Siegert, W., Nyberg, B., and Sperens, B. (1977) DNA N-glycosidases. Properties of a uracil–DNA glycosidase from *Escherichia coli. J. Biol. Chem.* **252,** 3286–3294.
10. Bennett, S. E., Jensen, O. N., Barofsky, D. F., and Mosbaugh, D. W. (1994) UV-Catalyzed cross-linking of *Escherichia coli* uracil–DNA glycosylase to DNA. *J. Biol. Chem.* **269,** 21,870–21,879.
11. Mol, C. D., Arvai, A. S., Sanderson, R. J., Slupphaug, G., Kavli, B., Krokan, H. E., et al. (1995) Crystal structure of human uracil–DNA glycosylase in complex with a protein inhibitor: protein mimicry of DNA. *Cell* **82,** 701–708.
12. Putnam, C. D., Shroyer, M. J., Lundquist, A. J., Mol, C. D., Arvai, A. S., Mosbaugh, D. W., et al. (1999) Protein mimicry of DNA from crystal structures of the uracil-DNA glycosylase inhibitor protein and its complex with *Escherichia coli* uracil–DNA glycosylase. *J. Mol. Biol.* **287,** 331–346.
13. Upton, C., Stuart, D. T., and McFadden, G. (1993) Identification of a poxvirus gene encoding a uracil DNA glycosylase. *Proc. Natl. Acad. Sci. USA* **90,** 4518–4522.
14. Domena, J. D. and Mosbaugh, D. W. (1985) Purification of nuclear and mito-chondrial uracil–DNA glycosylase from rat liver. Identification of two distinct subcellular forms. *Biochemistry* **24,** 7320–7328.
15. Domena, J. D., Timmer, R. T., Dicharry, S. A., and Mosbaugh, D. W. (1988) Purification and properties of mitochondrial uracil–DNA glycosylase from rat liver. *Biochemistry* **27,** 6742–6751.
16. Gupta, P. K. and Sirover, M. A. (1981) Stimulation of the nuclear uracil DNA glycosylase in proliferating human fibroblasts. *Cancer Res.* **41,** 3133–3136.
17. Nilsen, H., Otterlei, M., Haug, T., Solum, K., Nagelhus, T. A., Skorpen, F., et al. (1997) Nuclear and mitochondrial uracil–DNA glycosylases are generated by alternative splicing and transcription from different positions in the *UNG* gene. *Nucleic Acids Res.* **25,** 750–755.
18. Nilsen, H., Steinsbekk, K. S., Otterlei, M., Slupphaug, G., Aas, P. A., and Krokan, H. E. (2000) Analysis of uracil–DNA glycosylases from the murine Ung gene reveals differential expression in tissues and in embryonic development and a

subcellular sorting pattern that differs from the human homologues. *Nucleic Acids Res.* **28**, 2277–2285.

19. Caradonna, S., Ladner, R., Hansbury, M., Kosciuk, M., Lynch, F., and Muller, S. (1996) Affinity purification and comparative analysis of two distinct human uracil–DNA glycosylases. *Exp. Cell. Res.* **222**, 345–359.

20. Bharati, S., Krokan, H., Kristiansen, L., Otterlei, M., and Slupphaug, G. (1998) Human mitochondrial uracil–DNA glycosylase preform (UNG1) is processed to two forms one of which is resistant to inhibition by AP sites. *Nucleic Acid Res.* **26**, 4953–4959.

21. Bennett, S. E. and Mosbaugh, D. W. (1992) Characterization of the *Escherichia coli* uracil–DNA glycosylase/inhibitor protein complex. *J. Biol. Chem.* **267**, 22,512–22,521.

22. Wang, Z. and Mosbaugh, D. W. (1989) Uracil–DNA glycosylase inhibitor gene of bacteriophage PBS2 encodes a binding protein specific for uracil–DNA glycosylase. *J. Biol. Chem.* **264**, 1163–1171.

23. Karran, P., Cone, R., and Friedberg, E. C. (1981) Specificity of the bacteriophage PBS2 induced inhibitor of uracil–DNA glycosylase. *Biochemistry* **20**, 6092–6096.

24. Cone, R., Bonura, T., and Friedberg, E. C. (1980) Inhibitor of uracil–DNA glycosylase induced by bacteriophage PBS2. Purification and preliminary characterization. *J. Biol. Chem.* **255**, 10,354–10,358.

25. Bennett, S. E., Schimerlik, M. I., and Mosbaugh, D. W. (1993) Kinetics of the uracil–DNA glycosylase/inhibitor protein association. *J. Biol. Chem.* **268**, 26,879–26,885.

26. Lundquist, A. J., Beger, R. D., Bennett, S. E., Bolton, P. H., and Mosbaugh, D. W. (1997) Site-directed mutagenesis and characterization of uracil-DNA glycosylase inhibitor protein. *J. Biol. Chem.* **272**, 21,408–21,419.

27. Sanderson, R. J. and Mosbaugh, D. W. (1996) Identification of specific carboxyl groups on uracil–DNA glycosylase inhibitor protein that are required for activity. *J. Biol. Chem.* **271**, 29,170–29,181.

28. Laemmli, U. K. (1970) Cleavage of structural proteins during the assembly of the heads of bacteriophage T4. *Nature* **227**, 680–685.

29. Shroyer, M. J. N. (1999) *Escherichia coli* uracil–DNA glycosylase: DNA binding, catalysis, and mechanism of action, M.S. thesis, Oregon State University.

15

Characterization of Specialized mtDNA Glycosylases

Simon G. Nyaga and Vilhelm A. Bohr

1. Introduction

Mitochondria are cytoplasmic organelles that generate cellular energy in the form of ATP by the process of oxidative phosphorylation. Mitochondria contain multiple copies of a 16.5-kb circular genome. Mitochondrial DNA (mtDNA) encodes a subset of 13 structural, 22 transfer (tRNA), and 2 ribosomal (rRNA) genes. These few polypeptides encoded by mtDNA are essential components of oxidative phosphorylation. The other proteins involved in the metabolism and processing of the mtDNA, including mtDNA regulatory factors, are encoded in the nuclear genome. Transcription of these genes takes place in the nucleus and translation in the cytoplasm is then followed by selective transport to the mitochondria.

Unlike the nuclear genome, the mitochondrial genome is not protected by histones and lacks introns. This lack of histone protection and closeness to the electron transport system renders the mitochondrial genome susceptible to damage by reactive oxygen species (ROS) produced as byproducts of oxidative phosphorylation. Reactive oxygen species include superoxide anion ($^{\bullet}O_2^-$), hydrogen peroxide (H_2O_2), and hydroxyl radicals ($^{\bullet}OH$). Other notable sources of ROS include lipid peroxidation, radiation, redox cycling compounds, various drugs such as AZT, metabolism of phagocytic leukocytes, and heat *(1)*.

Early work probing the levels of ROS-induced lesions revealed that mtDNA contained high steady-state levels of oxidative damage *(2)*. However, recent data have revealed that full-length mitochondrial genomes of young animals do not carry particularly high levels of oxidative damage *(3,4)*. This discrepancy in the quantitated levels of oxidative damage indicates the unreliability of the earlier methods of detecting oxidative damage in DNA. Thus, extra care must be exercised in quantitating mtDNA damage. This is especially important when

From: *Methods in Molecular Biology, vol. 197: Mitochondrial DNA: Methods and Protocols*
Edited by: W. C. Copeland © Humana Press Inc., Totowa, NJ

Nuclear DNA Repair

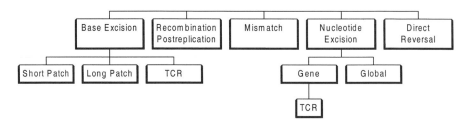

Mitochondrial DNA Repair

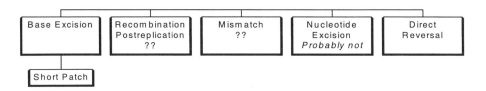

Fig. 1. Mammalian nuclear and mitochondrial DNA repair pathways.

purifying mtDNA, during which oxidative DNA base lesions are invariably introduced into the DNA during the extraction procedure.

Oxidative damage to mtDNA results in various DNA alterations. These alterations include abasic sites, sugar damage, single-strand breaks, and double-strand breaks. Some common base modifications produced by oxidative damage to mtDNA include 8-hydroxyguanine (8-oxoG) and thymine glycol. These modified bases have been detected in human cells *(5,6)*. Unrepaired modified bases can mispair and thereby cause mutations. Mutations in the mtDNA are thought to cause various rare human diseases and may potentially contribute to the pathogenesis of several degenerative diseases associated with aging and cancer *(7)*. Some of the mtDNA mutations are carcinogenic and various cancers have been shown to harbor mitochondrial mutations that are consistent with the ones caused by ROS *(8,9)*. These and many other observations raise the question as to whether mitochondria have the capacity to repair their DNA and, if so, what are the prevailing mechanisms.

The mtDNA repair pathways reported to date are far fewer than those involving nuclear DNA *(see* **Fig. 1**). Studies performed utilizing purified enzymes or whole-cell extracts have revealed two major DNA repair pathways: nucleotide excision repair (NER) and base excision repair (BER) (for a review, see **ref.** *10*). The mechanism of NER is well characterized and it involves multiple proteins and factors *(11)*. In the process of NER, the damaged base is removed along

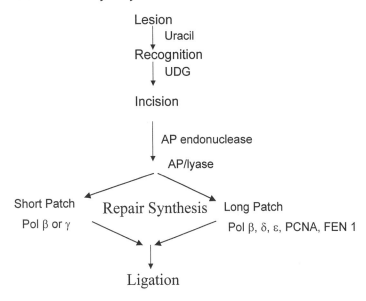

Fig. 2. Major BER steps.

with a stretch of normal bases. Early experiments addressing the question of the ability of mtDNA to repair cyclobutane pyrimidine dimers suggested that mitochondria did not have the ability to repair such lesions *(12)*. These results led to the generization that mitochondria were devoid of DNA repair. Although various DNA repair pathways have been reported in the mitochondria, there is no tangible evidence for NER in mitochondria to date.

Base excision repair, on the other hand, involves far fewer proteins/factors than NER and, in this case, only the damaged base is removed. Several studies have shown that BER pathway exists in mitochondria (for a review, see **refs. 13** and **14**). The major steps involved in BER are outlined in **Fig. 2**. BER is initiated by a class of enzymes known as DNA glycosylases that incise the *N*-glycosyl bond between the damaged base and the sugar, creating an apurinic/apyrimidinic (AP) site (reviewed in **ref. 15**). Two classes of glycosylases have been characterized: pure glycosylases that lack an associated AP lyase activity and glycosylases with an associated AP lyase activity (reviewed in **ref. 16**). AP endonucleases or glycosylases with an associated AP lyase activity incise the DNA backbone 3′ or 5′ to the AP site, leaving a 5′ sugar–phosphate moiety. This sugar–phosphate moiety is removed by deoxyribophosphodiesterase activity if present. The resulting one-nucleotide gap is filled by the action of a DNA polymerase and the nick is sealed by DNA ligase. The mechanisms of nuclear BER have been the subject of intense study and, so far, numerous enzymes that participate in this process have been characterized (*see* **Fig. 1**). Interestingly, many more nuclear BER enzymes are still being discovered. Conversely, the mechanisms governing mitochondrial BER are much less well understood. Characterization of the enzymes that participate in mitochondrial BER has been rather slow and, therefore, the mechanisms of mitochondrial BER have remained obscure. These characterizations and mechanistic studies have been hindered by several factors: (1) limited number of mitochondria in most cells with an exception of skeletal muscle tissues and other tissues requiring high expenses of energy, (2) low abundance of mitochondrial enzymes, (3) lack of standardized characterization methods, and (4) whereas, there are in vitro assays for nuclear DNA repair studies, no such assays existed for mtDNA repair until recently. These limitations and others not mentioned in this chapter have confined mtDNA repair studies to the use of mitochondrial extracts rather than to purified proteins.

Recently, some progress has been made in the area of characterization of specialized mitochondrial DNA glycosylases despite the hinderances outlined earlier. A limited understanding of the DNA repair mechanisms and the glycosylases present in the mitochondria has been achieved by taking advantage of various biochemical techniques and unique protein features. These techniques include affinity chromatography, histochemical studies, mitochondrial localization signal(s) (MLS), in vitro repair assays, the use of crude mitochondrial extracts, and inference from the lesions repaired by specific enzymes. Sequence analysis of potential genes encoding nuclear glycosylases with MLS have been used to discover more mtDNA glycosylases. Differential splicing has also contributed to the pool of mtDNA glycosylases. Mitochondrial uracil–DNA glycosylase (UDG) is one such glycosylase that is a splice variant of nuclear UDG *(17)*. Whereas the presence of MLS may be taken to imply that a glycosylase will end up in the mitochondria, there is a likelihood that most molecules may not reach their destination for various reasons such as degradation, unfavorable protein interactions, faulty transport process, or even unfavorable modifications along the way. The ability of MLS to direct a protein effectively to the mitochondria can be tested by tagging it to a reporter gene such as green fluorescent protein (GFP).

One of the first mtDNA glycosylase to be detected in mitochondrial extracts was mitochondrial UDG *(18)*. Mitochondrial UDG was later purified from human cells by affinity chromatography *(19)*. The years that followed saw the characterization of other mammalian mtDNA glycosylases. The 30-kDa mitochondrial UDG is encoded by the same gene that encodes the 36-kDa nuclear UDG *(20)*. The mouse and human homologs of yeast Ogg1 that preferentially incises 8-oxoG opposite cytosine from DNA have been shown to contain MLS *(21)*. This observation suggests that 8-oxoG glycosylase may be present in the mitochondria. Recently, we have observed that Ogg1 is the sole

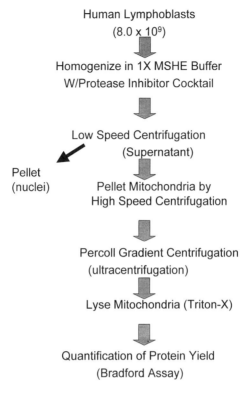

Fig. 3. Mitochondrial extract preparation.

glycosylase responsible for the repair of 8-oxoG:C in mouse mitochondrial extracts *(21a)*. Several other glycosylases, including hNTH1 and hMYH, glycosylases that are specific for the incision of thymine glycol and adenine opposite 8-oxoG in DNA, respectively, have been shown to contain the N-terminal MLS *(22)*. Furthermore, the MLS sequences of hNth1 and hMYH have been shown to direct GFP to the mitochondria *(22)*. Our laboratory has data showing efficient incision of thymine glycol by mitochondrial extracts from rats and humans *(23)* [Nyaga and Bohr, unpublished results].

The ability to prepare mitochondrial extracts free of nuclear contamination and the development of an in vitro mitochondrial repair assay in this laboratory has been central in the effort to characterize mitochondrial DNA glycosylases *(13,23)*. A schematic of mitochondrial extract preparation is shown in **Fig. 3**. These two techniques in conjunction with affinity chromatography were used to partially purify a 25- to 30-kDa mitochondrial oxidative damage endonuclease from rat liver, named mtODE *(13)*. This enzyme is not an endonuclease as the name suggests, rather it is a glycosylase with an associated AP lyase activity. Following the same techniques used to purify mtODE, a putative glycosylase/AP lyase enzyme specific for incision of thymine glycol was partially purified from rat liver *(23)*. This glycosylase/AP lyase has an apparent molecular weight of 37 kDa and was named mitochondrial thymine glycosylase (mtTGendo). The two enzymes, mtODE and mtTGendo, are yet to be purified to physical homogeneity, the genes encoding them cloned, and the recombinant enzymes further characterized.

Oxidative damage as recognized by formamidopyrimidine DNA glycosylase (Fpg), the bacterial enzyme that incises 8-oxoG and oxidized purines, was shown to be repaired in rat cells and in Chinese hamster ovary cells *(24,25)*. The original gene-specific repair assay (GSR) *(26)* has been modified in order to detect oxidative damage *(24)*. This assay uses damage–specific enzymes to recognize and cleave the entire or parts of mitochondrial genome at sites of lesions without isolating mitochondrial DNA *(27)*. The same assay can also be applied to a known gene encoded by the nuclear genome with an aim of detecting oxidative lesions for comparison purposes. Fpg has been used to detect Fpg-sensitive sites in the mitochondrial genome in comparison to the nuclear dihydrofolate reductase gene *(28)*. The repair kinetics of acridine orange plus light-induced oxidative damage followed the same kinetics both on mitochondrial genome and in nuclear dihydrofolate reductase gene. The efficient repair of Fpg-sensitive sites suggested the existence of a repair pathway for oxidative damage and/or Fpg homolog in mitochondria. The GSR of a number of DNA damaging agents have been used to demonstrate repair of strand breaks and alkali-sensitive sites in rodents and human mitochondria *(24,29)*. By using a combination of alkali and endonuclease III or Fpg, it was demonstrated that AP sites or strand breaks are efficiently repaired in the mitochondria *(25,28,29)*. These results were corroborated by the partial purification of an AP endonuclease from mitochondria *(30)*. Polymerase chain reaction (PCR) has been used to analyze damage induction and removal in nuclear and mtDNA by hydrogen peroxide *(31)*. The PCR assay examines a multitude of polymerase-blocking lesions. This study revealed that mitochondrial DNA damage repair after brief exposure to hydrogen peroxide was as efficient as in nuclear DNA. The incorporation assay was used to demonstrate that plasmid DNA treated with hydrogen peroxide was efficiently repaired by mitochondrial extracts *(32)*. Another oxidative lesion that is repaired from mitochondrial DNA is 4-nitroquinoline *(32)*, a lesion repaired by NER. Mitochondria have also been shown to contain a MutT homolog for hydrolyzing 8oxodGTP to 8-oxodGMP *(33)*.

Contrary to the earlier misconception that mitochondria lacked DNA repair activities, it is now quite clear that many lesions are efficiently repaired in mtDNA. However, much work is needed to characterize and elucidate the mechanisms of mtDNA repair. Some methods that have been used successfully in the characterization of mtDNA repair pathways and mechanisms are described in detail in this chapter.

2. Materials

All protease inhibitors were purchased from (Boehringer) unless indicated.

1. Mannitol, ACS reagent (Sigma).
2. Sucrose crystals (J.T. Baker Inc., Phillipsburg, NJ).
3. 1 *M* HEPES, pH 7.4.
4. 0.5 *M* Ethylenediaminetetraacetic acid (EDTA), pH 8.0, (Quality Biologicals Inc., Gaithersburg, MD).
5. Ethylene glycol-bis [β-aminoethyl ether] *N*,*N*,*N'*,*N'*-tetraacetic acid (EGTA) (Sigma).
6. Spermidine (Sigma).
7. Spermine (Sigma).
8. 1 mg/mL Pepstatin (Boehringer Mannheim).
9. 1 mg/mL Chymostatin A.
10. 2 mg/mL Leupeptin.
11. Benzamide–HCl, hydrate (Sigma).
12. 100 m*M* Phenylmethylsulfonyl fluoride (PMSF) (Boehringer Mannheim).
13. 10 m*M* E-64.
14. 1.0 *M* Dithiothreitol (DTT), (ICN Biomedicals, Aurora, OH).
15. 100% Glycerol (Gibco-BRL).
16. 10% SDS (sodium dodecyl sulfate) (Quality Biologicals Inc., Gaithersburg, MD).
17. T4 Polynucleotide kinase (New England Biolabs, Beverly, MA).
18. T4 Polynucleotide kinase buffer (New England Biolabs, Beverly, MA).
19. [^{32}P-γ]ATP, 3000 Ci/mmol (Amersham).
20. *Hae*III (Sigma).
21. 10X *Hae*III buffer (Sigma).
22. *Hpa*II (Sigma).
23. 10X *Hpa*II buffer (Sigma).
24. 5 mg/mL Proteinase K (Gibco-BRL).
25. 20 mg/mL Glycogen (Roche).
26. Ultrapure urea (Gibco-BRL).
27. Phenol:chloroform:isoamyl alcohol (25:24:1) (Sigma).
28. Chloroform:isoamyl alcohol (24:1) (Sigma).
29. Formamide (Sigma).
30. 100% Ethanol (The Warner-Graham Co., Cockeysville, MD).
31. 10X Tris-borate EDTA (TBE) (Quality Biologicals Inc., Gaithersburg, MD).
32. 40% Polyacrylamide solution, acrylamide:*N*,*N'*-methylene-bis-acrylamide (19:1) (Bio-Rad).
33. 10% Ammonium persulfate (APS) (Sigma).

34. *N,N,N',N'*-tetramethylenediamine (TEMED) (Gibco-BRL).
35. 100 mM MgCl$_2$ (Sigma).
36. [α-^{32}P]dCTP (3000 Ci/mMol) (Amersham).
37. 100 mM dNTPs (dATP, dTTP, dGTP, dCTP) (Amersham Pharmacia).
38. 500 mM Phosphocreatine (Sigma).
39. 2.5 mg/mL Phosphocreatine kinase (Sigma).
40. 2 mg/mL BSA (bovine serum albumin) (BRL).
41. 40 mM ATP (Pharmacia Biotech).
42. 0.5 mg/mL Bestatin.
43. Tris-EDTA buffer, pH 8.0, (Quality Biologicals Inc., Gaithersburg, MD).
44. Percoll (Amersham Pharmacia, Biotech, Uppsala, Sweden).
45. 11 M Ammonium acetate (Sigma).
46. Lamin B antibodies (Calbiochem).
47. 1X MSHE buffer: 0.21 M mannitol, 0.07 M sucrose, 10 mM HEPES, pH 7.4, 1 mM EDTA, 1 mM EGTA, 0.15 mM spermine, 0.75 mM spermidine containing protease inhibitors (1 µg/mL aprotinin, 1 µg/mL pepstatin, 1 µg/mL chymostatin, 2 µM bestatin, 2 µg/mL leupeptin, 2 µg/mL benzamide-HCl, 1 µM PMSF, and 1 µM E-64). DTT should be added just before use to a final concentration of 1 mM.
48. Mitochondrial lysis buffer: 20 mM HEPES, pH 7.4, 5 mM DTT, 1 mM EDTA, 5% glycerol and protease inhibitors with final concentrations given in **item 47**.

3. Methods

3.1. In Vitro mtDNA Repair Assays

In this section, we will describe in detail some of the methods that have been used to characterize mtDNA glycosylases. The mtDNA in vitro incision assay has been instrumental in elucidating the mechanisms of DNA repair in mitochondria. The assay utilizes mitochondrial extracts prepared as in **Subheading 2.** and lesion-containing oligonucleotides.

3.1.1. Mitochondrial Extract Preparation from Human Lymphoblasts

Mitochondrial extracts are prepared by a modified version of Croteau et al. *(13)* (*see* **Fig. 3**). All of the procedures are performed at 4°C except where indicated.

1. The cells are pelleted by centrifugation at 500g for 5 min.
2. Resuspend cell pellet in an equal volume of 1X MSHE buffer.
3. The cells are lysed by 35 strokes of a Dounce homogenizer.
4. Dilute the homogenate threefold to fourfold or more with 1X MSHE buffer containing protease inhibitors and DTT.
5. Centrifuge the homogenate at 500g for 7 min at 4°C. This step pellets the nuclei leaving the mitochondria in the supernatant.

6. Transfer the supernatant containing mitochondria carefully into a fresh tube and save the nuclear pellet at –80°C (*see* **Note 1**).
7. Centrifuge the supernatant at 10,000g for 7 min to pellet the mitochondria.
8. Wash the mitochondrial pellet twice with 1X MSHE buffer at 10,000g for 7 min each. The mitochondrial pellet is approximately 2 mL and roughly beige in color.
9. Resuspend the mitochondrial pellet in 1 mL of 1X MSHE buffer and layer onto 2X Percoll gradient mixture (1 part Percoll and 1 part 2X MSHE) (*see* **Note 2**).
10. Seal the tubes (Polyallomer Quick seal tubes) with heat and spin at 51,000g (27,000 rpm) at 4°C for 1 h in the Beckman ultracentrifuge (fixed angle 60.1 Ti rotor) (*see* **Note 3**).
11. Remove the mitochondria layer carefully and dilute threefold or higher with 1X MSHE buffer by spinning at 1400g (3000 rpm) (tabletop centrifuge), 4°C for 10 min (*see* **Note 4**).
12. Pour off supernatant and dry the pellet.
13. The mitochondrial pellet can be stored in –80°C or can be lysed immediately in mitochondrial lysis buffer (20 mM HEPES, pH 7.4, 5 mM DTT, 1 mM EDTA, 5% glycerol, and protease inhibitors with concentrations as in **Subheading 2., item 47**).

3.1.2. Mitochondrial Lysis

1. The mitochondrial pellet is resuspended in a small volume of mitochondrial lysis buffer at 4°C.
2. Dropwise add Triton X-100 to a final concentration of 0.05%.
3. Add KCl to a final concentration of 300 mM and stir slowly at 4°C for 1 h.
4. Quantify the total protein yield. Several methods exist for quantifying total protein concentration, including Bradford assay, Lowry assay, Biuret assay, and BCA assay. We will now describe the Bradford assay because we use it more often.

3.1.3. Total Protein Quantification by Bradford Assay

This assay is based on the calorimetric reaction between Coomassie Brilliant Blue G-250 and the protein in a sample. The complex is formed in about 2 min and is stable for 1 h. The method can be used to measure protein concentrations ranging from 1 to 140 μg/mL. Known protein standards such as BSA are used to generate a curve of protein concentration versus ultraviolet (UV) absorbance units at 595 nm. The concentration of the unknown (total mitochondrial protein) is obtained by extrapolation from the standard curve.

1. Prepare a BSA stock of 200 μg/mL in 1X MSHE buffer without protease inhibitors.
2. Dilute the mitochondrial proteins to 1:5, 1:10, 1:50, or as appropriate in 1X MSHE buffer without protease inhibitors. We routinely perform these reactions in 40-μL total reaction volumes.

3. Aliquot 0, 25, 50, 100, and 200 μg/mL of the BSA dilutions from the 200-μg/mL stock.
4. Add 1X MSHE buffer without protease inhibitors or DTT to a final volume of 40 μL.
5. Add 1.0 mL of Bio-Rad dye concentrate (1:5 dilution in water) to each of the tubes.
6. Prepare a blank with buffer only.
7. Read absorbance at 595 nm for the standards and the protein sample.
8. Plot absorbance (nm) versus protein concentration for the standards and extrapolate the protein concentration of the mitochondrial extracts. This procedure yields mitochondrial protein free of nuclear and cytosolic protein contamination with concentrations in the range of 2–10 mg/mL (*see* **Note 5**).

3.1.4. Oligonucleotide Labeling

1. Oligonucleotides of a desired length containing single lesions of interest at specific sites are designed and synthesized commercially (Midland Reagent Inc.).
2. The oligonucleotides (100 ng) are labeled on the 5'-end using [^{32}P-γ]ATP (Amersham) and T4 polynucleotide kinase (New England Biolabs, Beverly, MA) at 37°C for 1 h.
3. A typical labeling reaction contains the following reagents in a 20-μL reaction volume: single-stranded oligonucleotide (100 ng), T4 polynucleotide kinase (10 units), 1X T4 polynucleotide kinase buffer, 30 μCi [^{32}P-γ]ATP, and water to bring the reaction to a 20 μL total volume.
4. The reaction is incubated at 37°C for 1 h.
5. Complementary oligonucleotide is added in excess, usually threefold to fourfold.
6. KCl or NaCl is added to a final concentration of 100 mM. The reaction is incubated at 85–90°C for 5 min and then slowly cooled to room temperature.
7. The reaction is stored in –20°C.

3.1.5. Mitochondrial In Vitro Incision Assay

The diagramatic illustration of the in vitro repair assay is as shown in **Fig. 4**.

1. An appropriate amount (0.2–0.5 ng) of duplex 5'-end-labeled oligonucleotide containing a single lesion of interest is incubated with mitochondrial extracts at 37°C in the presence of incision buffer containing 20 mM HEPES, pH 7.4, 5 mM EDTA, 5 mM DTT, 5% glycerol, and 100 μg/mL BSA).
2. In the case of kinetics experiments, the extracts and the oligonucleotides are incubated, and aliquots are removed at specified time-points.
3. Stop the reactions by the addition of 0.2 μg/μL Proteinase K and 0.4% SDS and heating at 55°C for 15 min followed by phenol:chloroform:isoamyl alcohol (25:24:1) extraction (*see* **Subheading 3.1.6.** and **Note 6**).

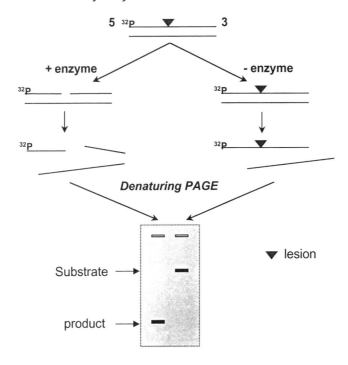

Fig. 4. A schematic of mitochondrial incision assay.

4. If necessary, precipitate the DNA overnight by adding 1/5 volume of 11 *M* ammonium acetate, 1 μg glycogen, and 2.5 volumes of cold 70% ethanol (*see* **Note 7**).
5. Incubate the DNA overnight in –20°C.
6. Spin down at 16,000*g* (14,000 rpm) (Eppendorf centrifuge model 5415C) for 10 min.
7. Discard the supernatant and wash the DNA with 100% ethanol by spinning down at 16,000*g* (14,000 rpm) (Eppendorf centrifuge model 5415C) for 10 min.
8. Dry the DNA by speed vacuum centrifugation for 20–30 min.
9. Dissolve the DNA in formamide loading dye.
10. Separate the samples by electrophoresis through 20% polyacrylamide gel containing 7 *M* urea at a constant 15 W for 1 h, 10 min.
11. Remove the gel, cover with Saran Wrap, and expose for 1 h in a PhosphorImager cassette at room temperature.
12. Quantify the bands using ImageQuant software (PhosphorImager, Molecular Dynamics, Sunnyvale, CA) to determine percent incision.

3.1.6. Protein Extraction

1. Add an equal volume of phenol:chloroform:isoamyl alcohol (25:24:1) to the reaction mixture and vortex.

2. Spin-down in microfuge for 4 min at 16,000*g* (14,000 rpm).
3. Remove 40 μL of the aqueous layer (top layer).
4. To the aqueous layer, add an equal volume (40 μL) of chloroform:isoamyl alcohol (24:1) and vortex.
5. Spin-down in an Eppendorf centrifuge at 16,000*g* (14,000 rpm) for 4 min.
6. Remove 35 μL of the aqueous layer and add an equal volume of 1X formamide loading dye.

3.1.7. Gel Electrophoresis

The reaction products are resolved by electrophoresis through 20% polyacrylamide gel containing 7 *M* urea, 89 m*M* Tris-borate, pH 8.0, and 2 m*M* EDTA.

3.1.8. Preparation of 20% Polyacrylamide Stock Solution (1 L) Containing 7 M urea and 10X TBE

1. Pour 500 mL of 40% polyacrylamide solution, acrylamide:*N*, *N'*-methylene-bis-acrylamide (19:1), respectively, into a 1-L beaker.
2. Add 100 mL of 10X Tris-borate-EDTA (TBE).
3. Weigh 420.42 g of urea and add to the beaker.
4. Stir until the urea dissolves completely.
5. Add water to bring the volume up to 1 L.
6. Mix well and filter through 0.2-μm millipore filters. Use dark bottles or cover the containers with alluminium foil because acrylamide is light sensitive. Store at room temperature.

3.1.9. Gel Preparation of 20% Denaturing Polyacrylamide Gel

1. Clean glass plates and spacers thoroughly with soapy water, clear distilled water, and 70% ethanol.
2. Dry the glass plates and assemble them as follows: put two clean spacers of appropriate thickness between a clean pair of glass plates.
3. Attach two clean clumps to the sides of the glass plates and tighten the setup, making sure that the glass plates and the spacers are level (use a clean flat surface).
4. Transfer the setup to a stable base and tighten the setup.
5. Attach an appropriate precleaned comb equipped with an appropriate number of teeth on the open surface of the glass plate setup.
6. Pour 40 mL of 20% polyacrylamide solution prepared as described in **Subheading 3.1.8.** to a clean beaker.
7. To this solution, add 300 μL of 10% ammonium persulfate (APS) and 30 μL TEMED.
8. Mix well and pour the solution quickly into the assembled glass plate setup before polymerization begins (*see* **Note 8**).
9. When the gel polymerizes, remove the glass plates from the base and assemble onto the gel running apparatus equipped with a bottom buffer chamber, an upper buffer chamber, and electrodes for connections to an electric power supply.

3.1.10. Electrophoresis

1. Add buffer to both chambers (1X TBE) and prerun the gel for about 30 min at constant 15 W.
2. Remove the upper chamber and flush the wells with a syringe and needle to get rid of the unpolymerized material and loosened urea.
3. Load the samples into the wells, replace the upper chamber buffer, and separate by electrophoresis for 1 h, 15 min at a constant 15 W.
4. Remove one glass plate carefully to avoid breakage of the gel and replace with a semirigid paper (Whatman).
5. Mark the orientation on the gel by cutting one corner and cover the other surface of the gel with Saran Wrap.
6. Expose in PhosphorImager cassette for 1 h or as desired.
7. Transfer the band images to a PhosphorImager and quantify the bands.
8. Analyze the data graphically or by any other desired means.

These in vitro assays have been used to detect mitochondrial DNA glycosylases specific for various lesions, including 8-oxoG, thymine glycol, and uracil. We have shown that mitochondria prepared from human lymphoblasts harbors multiple pathways for initiating the repair of a number of oxidative lesions (Nyaga and Bohr, unpublished results). The in vitro incision assay can be modified in order to study repair synthesis after the initial *N*-glycosyl bond incision.

3.2. Repair Synthesis Assay

3.2.1. Duplex DNA Formation

1. Add 225 ng of unlabeled oligonucleotide containing a single lesion such as uracil to an Eppendorf tube.
2. Add an equal amount of complementary oligonucleotide. Add 100 m*M* NaCl or KCl to the reaction tube in a 50-µL reaction volume.
3. Heat the reaction to 85–90°C for 5 min and slow cool to room temperature. Store in –20°C until ready to use.

3.2.2. Repair Synthesis Reaction

1. Standard reaction conditions contain the following: 40 m*M* HEPES, pH 7.8, 75 m*M* KCl, 1 m*M* DTT, 5 m*M* MgCl$_2$, 0.1 m*M* EDTA, 0.2 mg/mL BSA, 4 µCi [α-^{32}P]dCTP, 2 µ*M* dCTP, 20 µ*M* dGTP, dTTP and dATP, 25 m*M* phosphocreatine, 50 µg/mL phosphocreatine kinase, 50 µg human mitochondrial protein extracts, and 2.5 pmol duplex uracil containing oligonucleotide or 200 ng of a single uracil containing M13 plasmid DNA and 1 µg of a 53-mer single-stranded competitor oligonucleotide *(34,35)*.
2. The oligonucleotides and the plasmid DNAs are designed in such a way that after incision of the lesion, a recognition sequence for *Hpa*II (Sigma) is established (*see* **Fig. 5**).

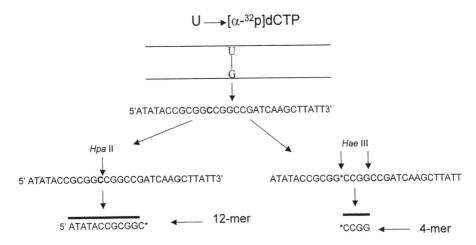

Fig. 5. Analysis of uracil repair patch size by restriction digestion.

3. In addition, the oligonucleotides are synthesized in such a way that after incision of uracil and incorporation of radiolabeled dCTP during short-patch BER, a restriction site for *Hae*III (Sigma) is established (*see* **Fig. 5**).
4. The reactions are incubated at 32°C and aliquots are taken out at desired time-points.
5. The reactions are stopped by phenol : chloroform : isoamyl alcohol (25 : 24 : 1) extraction followed by chloroform : isoamyl alcohol extraction (24 : 1).
6. Add an equal volume of formamide loading dye.
7. The samples are separated at a constant 15 W for 1 h, 10 min on a denaturing 20% polyacrylamide gel containing 7 *M* urea.
8. The gels are exposed on PhosphorImager cassettes for at least 1 h and the images scanned and quantified.

3.3. Determination of Patch Sizes for Uracil Repair

3.3.1. Patch Size for Uracil Repair in Oligonucleotides

1. Repair synthesis reactions are performed as described in **Subheading 3.2.2.**
2. The reactions are stopped by addition of 0.2 μg/μL Proteinase K and 0.4% SDS and incubated at 55°C for 15 min.
3. The proteins are then extracted as described above for repair synthesis.
4. The DNA is precipitated for 1 h in –20°C in the presence of 5 *M* NaCl to a final concentration of 300 m*M* and 3 volumes of ice-cold 100% ethanol.
5. The samples are washed twice with 100% ethanol, dried for 20 min by vacuum centrifugation, and then dissolved in TE buffer, pH 7.8.
6. About 2 pmol of the repaired oligonucleotide is digested with 30 units of either *Hpa*II or *Hae*III for 1 h at 37°C. The total reaction volume is usually 20 μL and contains 1/10 of 10X restriction enzyme buffer supplied with the enzyme.

7. The reaction is terminated by heating for 2–5 min at 80–90°C and adding an equal volume of formamide loading dye.

8. The products are separated by electrophoresis through 20% polyacrylamide gel containing 7 M urea.

3.3.2. Determination of Patch Size for Uracil Repair in Plasmid DNA

1. Repair synthesis reaction is performed as described in **Subheadings 3.2.1.** and **3.2.2.** except that the substrate is either a control or single lesion containing M13 plasmid DNA *(34,35)*.

2. The reactions are performed for shorter time than with oligonucleotide assay so as to control for nonspecific degradation of the plasmid DNA by endonuclease G *(36,37)*.

3. The reaction is quenched with 20 mM EDTA.

4. The samples are treated with 0.1 µg/µL Rnase for 10 min at 37°C followed by addition of 0.2 µg/µL proteinase K and 0.4% SDS.

5. The reactions are heated to 55°C for 15 min, and 300 mM NaCl is added.

6. The samples are extracted with an equal volume of phenol:chloroform:isoamyl alcohol (25:24:1) followed by a chloroform:isoamyl alcohol (24:1) extraction.

7. DNA is precipitated as described in **Subheading 3.3.1., steps 4** and **5**, washed with 80% ethanol, dried by vacuum centrifugation, and dissolved in TE buffer, pH 7.8.

8. About 100 ng of the repaired plasmid is digested with 20 units of *Hind*III and 20 units of *Hpa*II at 37°C for 1 h in a 20-µL reaction containing 1/10 of 10X restriction enzyme buffer supplied with the enzyme.

9. The samples are heated for 2–5 min at 80–90°C and the products separated by electrophoresis in TBE through 20% polyacrylamide gel containing 7 M urea.

10. The band images are scanned, quantified, and analyzed.

4. Notes

1. Extreme care must be exercised in separating supernatant containing mitochondria after pelleting the nuclei because this step can easily result in nuclear protein contamination. A precautionary technique would be to avoid pipetting the last 1–2 mL of the supernatant.

2. Percoll must be diluted using 2X MSHE buffer otherwise the gradient will not form properly, thereby making it impossible to recover the mitochondria band.

3. Cool the top of the polyallomer Quick seal tubes containing mitochondria with ice cubes to avoid warming the mitochondria. Failure to cool the tubes might result in heat inactivation of the mitochondrial proteins.

4. Mitochondria must be diluted with 1X MSHE buffer after Percoll gradient purification, otherwise sedimentation will not take place during subsequent centrifugation to pellet mitochondria.

5. Mitochondrial extracts are easily contaminated by the abundant nuclear or cytosolic proteins. Western blot analysis using antibodies against proteins found exclusively in the nucleus or cytoplasm is commonly used to control for this

kind of contamination. We have used lamin B antibodies effectively to check for nuclear contamination in mitochondrial preparations. Some groups have used lactate dehydrogenase as a cytosolic marker and one of the mitochondrial enzymes such as cytochrome oxidase c to check for mitochondrial activity and PCNA (proliferating nuclear antigen) *(13,38)* as a nuclear fraction marker.

6. Pipetting of the aqueous layer during the extraction process should be performed with extreme caution to avoid picking up the organic layer that would result in loss of samples.

7. DNA precipitation and protein extraction may be necessary in order to obtain sharp bands and for efficient cutting by restriction endonucleases. Precipitation may result in loss of DNA; therefore, care must be exercised so as to minimize losses through inefficient precipitation and during the extraction process. A longer incubation time is better.

8. Polyacrylamide solution must be homogeneous for effective separation of substrate and products.

References

1. Wagner, J. R., Hu, C. C., and Ames, B. N. (1992) Endogenous oxidative damage of deoxycytidine in DNA. *Proc. Natl. Acad. Sci. USA* **89(8),** 3380–3384.

2. Richter, C., Park, J. W., and Ames, B. N. (1988) Normal oxidative damage to mitochondrial and nuclear DNA is extensive. *Proc. Natl. Acad. Sci. USA* **85(17),** 6465–6467.

3. Anson, R. M., Senturker, S., Dizdaroglu, M., and Bohr, V. A. (1999) Measurement of oxidatively induced base lesions in liver from Wistar rats of different ages. *Free Radical Biol. Med.* **27(3–4),** 456–462.

4. Suter, M. and Richter, C. (1999) Fragmented mitochondrial DNA is the predominant carrier of oxidized DNA bases. *Biochemistry* **38(1),** 459–464.

5. Ames, B. N. (1989) Endogenous DNA damage as related to cancer and aging. *Mutat. Res.* **214(1),** 41–46.

6. Dizdaroglu, M., Olinski, R., Doroshow, J. H., and Akman, S. A. (1993) Modification of DNA bases in chromatin of intact target human cells by activated human polymorphonuclear leukocytes. *Cancer Res.* **53(6),** 1269–1272.

7. Wallace, D. C. (1999) Mitochondrial diseases in man and mouse. *Science* **283(5407),** 1482–1488.

8. Fliss, M. S., Usadel, H., Caballero, O. L., Wu, L., Buta, M. R., Eleff, S. M., et al. (2000) Facile detection of mitochondrial DNA mutations in tumors and bodily fluids. *Science* **287(5460),** 2017–2019.

9. Polyak, K., Li, Y., Zhu, H., Lengauer, C., Willson, J. K., Markowitz, S. D., et al. (1998) Somatic mutations of the mitochondrial genome in human colorectal tumours. *Nature Genet.* **20(3),** 291–293.

10. Friedberg, E. C., Walker, G. C., and Siede, W. (1995) *DNA Repair and Mutagenesis,* American Society for Microbiology, Washington, DC.

11. Wood, R. D. (1997) Nucleotide excision repair in mammalian cells. *J. Biol. Chem.* **272(38),** 23,465–23,468.

12. Clayton, D. A., Doda, J. N., and Friedberg, E. C. (1974) The absence of a pyrimidine dimer repair mechanism in mammalian mitochondria. *Proc. Natl. Acad. Sci. USA* **71(7)**, 2777–2781.

13. Croteau, D. L. and Bohr, V. A. (1997) Repair of oxidative damage to nuclear and mitochondrial DNA in mammalian cells. *J. Biol. Chem.* **272(41)**, 25,409–25,412.

14. Bogenhagen, D. F. (1999) Repair of mtDNA in vertebrates. *Am. J. Hum. Genet.* **64(5)**, 1276–1281.

15. Krokan, H. E., Standal, R., and Slupphaug, G. (1997) DNA glycosylases in the base excision repair of DNA. *Biochem. J.* **325,** 1–16.

16. Sun, B., Latham, K. A., Dodson, M. L., and Lloyd, R. S. (1995) Studies on the catalytic mechanism of five DNA glycosylases. Probing for enzyme–DNA imino intermediates. *J. Biol. Chem.* **270(33)**, 19,501–19,508.

17. Slupphaug, G., Markussen, F. H., Olsen, L. C., Aasland, R., Aarsaether, N., Bakke, O., Krokan, H. E., et al. (1993) Nuclear and mitochondrial forms of human uracil–DNA glycosylase are encoded by the same gene. *Nucleic Acids Res.* **21(11)**, 2579–2584.

18. Anderson, C. T. and Friedberg, E. C. (1980) The presence of nuclear and mitochondrial uracil–DNA glycosylase in extracts of human KB cells. *Nucleic Acids Res.* **8(4)**, 875–888.

19. Caradonna, S., Ladner, R., Hansbury, M., Kosciuk, M., Lynch, F., and Muller, S. (1996) Affinity purification and comparative analysis of two distinct human uracil–DNA glycosylases. *Exp. Cell. Res.* **222(2)**, 345–359.

20. Nilsen, H., Otterlei, M., Haug, T., Solum, K., Nagelhus, T. A., Skorpen, F., et al. (1997) Nuclear and mitochondrial uracil–DNA glycosylases are generated by alternative splicing and transcription from different positions in the UNG gene. *Nucleic Acids Res.* **25(4)**, 750–755.

21. Nishioka, K., Ohtsubo, T., Oda, H., Fujiwara, T., Kang, D., Sugimachi, K., et al. (1999) Expression and differential intracellular localization of two major forms of human 8-oxoguanine DNA glycosylase encoded by alternatively spliced OGG1 mRNAs. *Mol. Biol. Cell* **10(5)**, 1637–1652.

21a. Souza-Pinto, N., Eide, L., Hogue, B. A., et al. Repair of 8-oxodeoxyguanosine lesions in mitochondrial DNA depends on the ogg1 gene and 8-oxoguanine accumulates in the mitochondrial DNA of ogg1 defective mice. *Cancer Res., Adv. In Brief,* **61,** 5378–5381.

22. Takao, M., Aburatani, H., Kobayashi, K., and Yasui, A. (1998) Mitochondrial targeting of human DNA glycosylases for repair of oxidative DNA damage. *Nucleic Acids Res.* **26(12)**, 2917–2922.

23. Stierum, R. H., Croteau, D. L., and Bohr, V. A. (1999) Purification and characterization of a mitochondrial thymine glycol endonuclease from rat liver. *J. Biol. Chem.* **274(11)**, 7128–7136.

24. Driggers, W. J., Grishko, V. I., LeDoux, S. P., and Wilson, G. L. (1996) Defective repair of oxidative damage in the mitochondrial DNA of a xeroderma pigmentosum group A cell line. *Cancer Res.* **56(6)**, 1262–1266.

25. Taffe, B. G., Larminat, F., Laval, J., Croteau, D. L., Anson, R. M., and Bohr, V. A. (1996) Gene-specific nuclear and mitochondrial repair of formamidopyrimidine DNA glycosylase-sensitive sites in Chinese hamster ovary cells. *Mutat. Res.* **364(3)**, 183–192.

26. Bohr, V. A., Smith, C. A., Okumoto, D. S., and Hanawalt, P. C. (1985) DNA repair in an active gene: removal of pyrimidine dimers from the DHFR gene of CHO cells is much more efficient than in the genome overall. *Cell* **40(2)**, 359–369.

27. Anson, R. M. and Bohr, V. A. (1999) Gene-specific and mitochondrial repair of oxidative DNA damage, in *DNA Repair Protocols* (Henderson, D. S., ed.), Humana, Totowa, NJ, Vol. 113, pp. 257–279.

28. Driggers, W. J., LeDoux, S. P., and Wilson, G. L. (1993) Repair of oxidative damage within the mitochondrial DNA of RINr 38 cells. *J. Biol. Chem.* **268(29)**, 22,042–22,045.

29. Pettepher, C. C., LeDoux, S. P., Bohr, V. A., and Wilson, G. L. (1991) Repair of alkali-labile sites within the mitochondrial DNA of RINr 38 cells after exposure to the nitrosourea streptozotocin. *J. Biol. Chem.* **266(5)**, 3113–3117.

30. Tomkinson, A. E., Bonk, R. T., and Linn, S. (1988) Mitochondrial endonuclease activities specific for apurinic/apyrimidinic sites in DNA from mouse cells. *J. Biol. Chem.* **263(25)**, 12,532–12,537.

31. Yakes, F. M. and Van Houten, B. (1997) Mitochondrial DNA damage is more extensive and persists longer than nuclear DNA damage in human cells following oxidative stress. *Proc. Natl. Acad. Sci. USA* **94(2)**, 514–519.

32. Ryoji, M., Katayama, H., Fusamae, H., Matsuda, A., Sakai, F., and Utano, H. (1996) Repair of DNA damage in a mitochondrial lysate of *Xenopus laevis* oocytes. *Nucleic Acids Res.* **24(20)**, 4057–4062.

33. Kang, D., Nishida, J., Iyama, A., Nakabeppu, Y., Furuichi, M., Fujiwara, T., et al. (1995) Intracellular localization of 8-oxo–dGTPase in human cells, with special reference to the role of the enzyme in mitochondria. *J. Biol. Chem.* **270(24)**, 14,659–14,665.

34. Dianov, G., Price, A., and Lindahl, T. (1992) Generation of single-nucleotide repair patches following excision of uracil residues from DNA. *Mol. Cell. Biol.* **12(4)**, 1605–1612.

35. Stierum, R. H., Dianov, G. L., and Bohr, V. A. (1999) Single-nucleotide patch base excision repair of uracil in DNA by mitochondrial protein extracts. *Nucleic Acids Res.* **27(18)**, 3712–3719.

36. Cummings, O. W., King, T. C., Holden, J. A., and Low, R. L. (1987) Purification and characterization of the potent endonuclease in extracts of bovine heart mitochondria. *J. Biol. Chem.* **262(5)**, 2005–2015.

37. Gerschenson, M., Houmiel, K. L., and Low, R. L. (1995) Endonuclease G from mammalian nuclei is identical to the major endonuclease of mitochondria. *Nucleic Acids Res.* **23(1)**, 88–97.

38. Chen, D., Lan, J., Pei, W., and Chen, J. (2000) Detection of DNA base-excision repair activity for oxidative lesions in adult rat brain mitochondria. *J. Neurosci. Res.* **61(2)**, 225–236.

16

Purification, Separation, and Identification of the Human mtDNA Polymerase With and Without Its Accessory Subunit

Matthew J. Longley and William C. Copeland

1. Introduction

The human mitochondrial genome encodes a variety of genes required for oxidative phosphorylation, and loss of these essential gene functions induces a multitude of severe metabolic disorders *(1–4)*. Mitochondrial DNA (mtDNA) is replicated by the nuclear encoded DNA polymerase γ *(5)*, and pol γ is the only replicative eukaryotic DNA polymerase to display sensitivity to a wide range of antiviral nucleotide analogs *(6–12)*. Consequently, long-term treatment of patients with antiviral nucleotide analogs such as zidovudine (AZT), zalcitabine (ddC), and didanosine (ddI) induces a characteristic collection of mitochondrial dysfunctions that mimics mitochondrial genetic diseases *(13–16)*. To aid investigations into the properties and subunit composition of this important enzyme, we present a simple biochemical assay to distinguish two forms of pol γ and a purification scheme to resolve them.

Animal cell DNA polymerase γ was first purified from *Drosophila melanogaster* by Kaguni and colleagues and was shown to consist of 125-kDa and 35-kDa subunits *(17)*. Pol γ from *Xenopus laevis* contains two subunits of 140 kDa and 50 kDa *(18)*, and an initial report on human HeLa cell pol γ identified 140-kDa and 54-kDa polypeptides in the most purified fraction *(19)*. Originally cloned from *Saccharomyces cerevisiae (20)*, the coding sequences for the larger, catalytic subunit have been isolated from human, mouse, chicken, *X. laevis*, *D. melanogaster*, *Schizosaccharomyces pombe*, and *Pychia pastoris (21–24)*. The predicted sizes for these proteins range from 115 kDa for

From: *Methods in Molecular Biology, vol. 197: Mitochondrial DNA: Methods and Protocols*
Edited by: W. C. Copeland © Humana Press Inc., Totowa, NJ

S. pombe to 143 kDa for *S. cerevisiae*, and all the genes contain conserved sequence motifs for polymerase and 3′→5′ exonuclease functions. The 3′→5′ exonuclease activity has been shown to enhance the high-fidelity DNA synthesis exhibited in vitro by purified γ-polymerases *(25–28)*. Alteration of the conserved exonuclease motif in yeast results in a mitochondrial mutator phenotype *(29)*, and expression of exonuclease-deficient pol γ fusion proteins in cultured human cells also results in the accumulation of point mutations in mitochondrial DNA *(30)*. Further, loss of the exonuclease function in transgenic mice results in the rapid accumulation of point mutations and deletions in cardiac mtDNA, and the mutagenesis is accompanied by cardiomyopathy *(31)*.

Human DNA polymerase γ is composed of two subunits: a 140-kDa polypeptide containing the catalytic activities *(22,32)* and a 55-kDa accessory subunit required for highly processive DNA synthesis *(33)*. Previously, we purified both the native and recombinant forms of the human catalytic subunit *(32)*. Purification to homogeneity required five consecutive chromatographic steps and yielded approx 15% of the total soluble polymerase γ activity (*see* **Fig. 1**). We have also cloned the full-length cDNA for the accessory subunit, overexpressed the protein in *E. coli* and purified the protein to homogeneity *(33)* (**Fig. 2,** lane 2). To isolate the native form of the human accessory subunit, we re-examined the purification of native DNA polymerase γ from HeLa mitochondrial lysates *(32)*. Although the 55-kDa band cochromatographed with the 140-kDa catalytic subunit over the first two columns, the 55-kDa protein eluted from single-stranded DNA–cellulose at a significantly higher salt concentration than monomeric p140. Assay of DNA polymerase activity across this profile revealed an additional, cleanly resolved peak of polymerase activity coincident with the 55-kDa subunit. Immunoblot analysis confirmed that the single-stranded DNA–cellulose column cleanly resolved two forms of native DNA polymerase γ: a low salt form eluting at approx 0.22 *M* NaCl containing only the catalytic subunit and a high salt form eluting at approx 0.38 *M* NaCl containing both the catalytic subunit and the accessory subunit (*see* **Figs. 2** and **3**).

The ability to detect and to resolve the two forms is entirely dependent on the selection of chromatographic resins, DNA substrates, and in vitro assay conditions. For example, among all of the chromatographic resins we have tested, gradient elution from single-stranded (ss) DNA–cellulose is the only procedure that cleanly resolves the two native forms of pol γ. Utilization of other resins results in the monomeric and heterodimeric forms remaining mixed or only being partially resolved. Also, the two forms exhibit marked differences in activity under any single set of assay conditions. For example, p55 raises the salt optimum and stimulates p140 activity on all substrates tested. Although both forms of the enzyme are active on poly(rA)•oligo(dT) at 75 m*M*

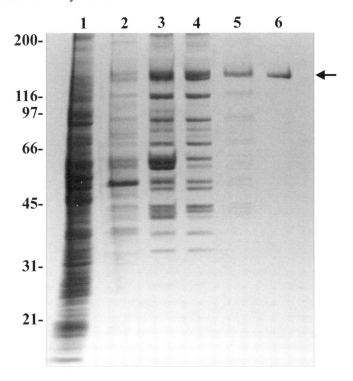

Fig. 1. Purification profile of recombinant human pol γ. Proteins were resolved by 4–20% sodium dodecyl sulfate–polyacrylamide gel electrophoresis (SDS-PAGE) and stained with Coomassie Blue. Lane 1: cleared lysate (2000 U p140); lane 2: phosphocellulose (4000 U p140); lane 3: phenyl–Sepharose (40,000 U p140); lane 4: single-stranded DNA–cellulose (40,000 U p140); lane 5: hydroxylapatite (40,000 U p140); lane 6: MonoQ (40,000 U p140). Positions of molecular-weight standards (kDa) are indicated. The arrow indicates the position of the pol γ catalytic subunit.

NaCl, isolated p140 is virtually inactive on activated DNA at physiological concentrations of salt. Accordingly, assay of pol γ activity with this substrate would not reveal the presence of free p140, and the two forms would remain mixed unless resolved on ssDNA–cellulose. The choice of assay conditions is therefore critical, because biochemical analysis of mixed preparations would yield composite values, depending on the relative abundance and inhibition of the two forms.

To distinguish it from the other eukaryotic DNA polymerases, pol γ has been characterized as a processive, salt-tolerant, dideoxynucleotide-sensitive, aphidicolin-resistant, Family A type DNA polymerase that can utilize a wide variety of DNA substrates including poly(rA)•oligo(dT) *(6)*. Traditionally, DNA polymerase γ has also been viewed as sensitive to the sulfhydryl-blocking

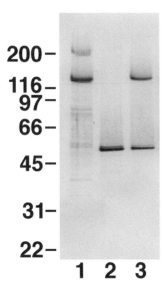

Fig. 2. DNA polymerase γ subunit composition. The two forms of human pol γ were resolved by SDS-PAGE and stained with silver. Lane 1: free catalytic subunit (150 ng p140); lane 2: recombinant accessory subunit (40 ng p55); lane 3: two-subunit polymerase γ complex (approx 120 ng recombinant p140•p55) from DNA–cellulose (Fig. 3, fraction 20).

agent *N*-ethylmaleimide (NEM), and this sensitivity is one of the few inhibitory characteristics that set pol γ apart from the NEM-resistant DNA polymerase β *(6,9)*. NEM irreversibly and covalently modifies sulfhydryl groups in proteins, and inhibition by NEM indicates solvent accessible cysteine residue(s), which are critical for activity. Therefore, we were not surprised to observe extreme sensitivity of free p140 to NEM *(32)*. Native heterodimeric HeLa pol γ exhibited the classical inhibitory profile (sensitive to dideoxynucleotides and resistant to aphidicolin), but the complex was unexpectedly resistant to NEM *(33)*. We sought to exploit the ability to reconstitute the pol γ complex from individual components to investigate the cause of this NEM resistance. Both the native and recombinant p140•p55 complexes display nearly complete resistance to NEM up to 1 m*M* (*see* **Fig. 4**) and still retain >50% activity with 10 m*M* NEM (data not shown). In marked contrast, the single catalytic subunit was inhibited to 50% with less than 0.1 m*M* NEM and >90% inhibited at 0.5 m*M* NEM. Thus, the p55 accessory subunit protects the catalytic subunit from NEM inhibition by over 100-fold. This inhibition was independent of ionic strength, and the order of addition experiments demonstrated that DNA binding was not required for this protection. We believe that previous reports of NEM

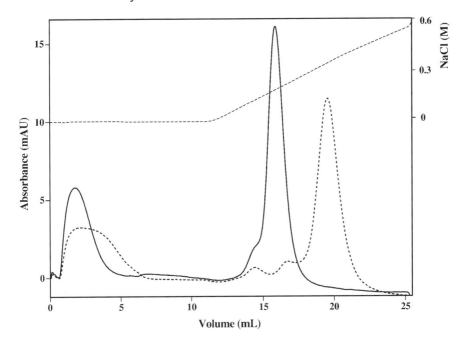

Fig. 3. Single-stranded DNA–cellulose chromatography of human DNA polymerase γ polypeptides. Purified recombinant p140 and p55 were resolved by single-stranded DNA–cellulose chromatography as described in **Subheading 3.1.** The absorbance of the column effluent was monitored at 280 nm. Protein samples were p140 (35 μg, solid line) or an equal molar mixture or p140 and p55 (42 μg, dotted line).

sensitivity for DNA polymerase γ are the expected conclusions for preparations containing excess catalytic subunit *(17,19,34).*

We present here a protocol for purification of DNA polymerase γ from HeLa mitochondrial lysates that resolves the free catalytic subunit and the heterodimeric complex. We also describe a simple enzymatic assay that utilizes differential inhibition by NEM to distinguish the two native forms.

2. Materials

2.1. Purification of the Single Catalytic Subunit from Baculoviral Infected Insect Cells

1. Buffer A: 0.05 *M* potassium phosphate buffer, pH 7.5, 10% glycerol, 1 m*M* EDTA, 1 m*M* of 2-mercaptoethanol, 0.1 m*M* phenylmethylsulfonyl fluoride (PMSF), and 1 μg/mL leupeptin.
2. Lysis buffer: Buffer A also containing 0.5% Nonidet P-40 (NP-40) and 0.1 *M* NaCl.

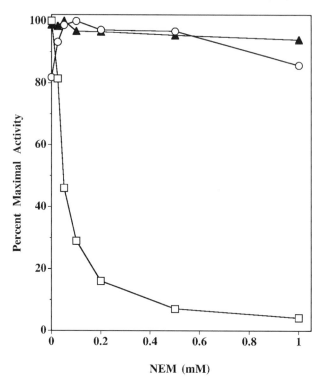

Fig. 4. The accessory subunit protects DNA polymerase γ from inhibition by *N*-ethylmaleimide. DNA polymerase activity of 4 ng p140 (open squares), p140•p55 (closed triangles), or the native HeLa pol γ complex (open circles) was assayed as described in **Subheading 3.4.** Reactions were performed in the absence of sulfhydryl-reducing agents and contained the indicated quantities of *N*-ethylmaleimide. All values are an average of two or more determinations.

3. Buffer B: 0.3 *M* potassium phosphate buffer, pH 7.5, 10% glycerol, 1 m*M* EDTA, 1 m*M* of 2-mercaptoethanol, 0.1 m*M* PMSF, 1 µg/mL leupeptin, and 0.05 *M* NaCl.
4. Buffer C: 0.025 *M* Tris-HCl, pH 7.5, 10% glycerol, 0.1 m*M* EDTA, 1 m*M* of 2-mercaptoethanol.
5. Buffer D: 0.05 *M* Tris-HCl, pH 7.5, 10% glycerol, 1 m*M* of 2-mercaptoethanol, 0.01% NP-40, and 0.1 *M* KCl.
6. Buffer E: 0.05 *M* Tris-HCl, pH 7.5, 10% glycerol, 1 m*M* EDTA, 1 m*M* of 2-mercaptoethanol, and 0.01% NP-40.
7. Phosphocellulose P-11 resin (Whatman), washed and equilibrated in buffer A also containing 0.05 *M* NaCl.
8. Phenyl–Sepharose HP (Pharmacia) placed in a column (5.1 cm × 0.79 cm²), washed and equilibrated in buffer B.

9. Single-stranded DNA cellulose (Sigma) placed in a column (1 mL), washed and equilibrated in buffer C also containing 0.05 M NaCl.
10. Ceramic hydroxylapatite column (CHT-II, Bio-Rad, 1 mL) equilibrated in buffer D.
11. HR 5/5 Mono Q (Pharmacia) fast-protein liquid chromatography (FPLC) column.
12. FPLC chromatography system placed in a cold box or cold room.

2.2. Additional Solutions for Purification of the Two Subunit Native DNA Polymerase γ from Tissue Culture Cells

1. Dulbecco's minimum essential medium (DMEM) media supplemented with high glucose, sodium pyruvate, and pyridoxine HCl, supplied from LifeTechnologies (cat. no. 11995 or equivalent).
2. Mitochondrial lysis buffer: 0.02 M Tris-HCl, pH 8.0, 10% glycerol, 0.1% of 2-mercaptoethanol, 0.3 M NaCl, 0.5% Triton X-100, 0.1 mM PMSF, and 1 µg/mL leupeptin.

2.3. DNA Polymerase Assay

1. Poly(rA)•oligo(dT)$_{12-18}$ was purchased from Pharmacia (cat. no. 27-7878-01) and dissolved in distilled H$_2$O (dH$_2$O) to a concentration of 2.5 mg/mL.
2. Radioisotopes ^{32}P- α-dTTP (3000 Ci/mmol, 10 mCi/mL) were from Amersham. Unlabeled dTTP was purchased from Pharmacia as a 10 mM stock.
3. 2X Assay mix: 50 mM HEPES-KOH, pH 8.0, 150 mM NaCl, 5 mM of 2-mercapto-ethanol, 200 µg/mL acetylated bovine serum albumin (BSA), 1.0 mM MnCl$_2$, 50 µM of ^{32}P-α-dTTP (2000 cpm/pmol), 100 µg/mL poly(rA)$_{290}$•oligo(dT)$_{12-18}$. (*See* **Notes 1** and **2**.)
4. Enzyme dilution buffer: 50 mM Tris-HCl, pH 7.5, 10% glycerol, 1 mM EDTA, 1 mM of 2-mercaptoethanol, 0.01% NP-40, and 100 µg/mL acetylated BSA.
5. Stop solution: 20 g NaOH, 44.6 g NaPP$_i$, 0.5 g BSA dissolved in 1 L of deionized water and supplemented with 100 mg of sonicated calf thymus DNA as a carrier.
6. DNA precipitation buffer: 20% (w/v) trichloroacetic acid (TCA), made by diluting a 100% solution. 100% (w/v) TCA is made by dissolving 500 g of crystalline TCA (Fisher) into 227 mL of dH$_2$O.
7. Filter manifold (Hoefer FH 225 Filtration manifold, 25 mm, Pharmacia cat. no 80-6024-14) connected to a vacuum source.
8. Whatman GF/C filters, 24-mm circles.
9. Assay tubes: 12 × 75 mm disposable borosilicate glass culture tubes (Kimble).
10. Wash solutions: 1 N HCl, 95% ethanol.

2.4. N-Ethylmaleimide Inhibition Assay

1. 10 mM N-Ethylmaleimide stock dissolved in water.
2. Enzyme dilution buffer from **Subheading 2.3.** without 2-mercaptoethanol or dithiothreitol (DTT).
3. Assay buffer from **Subheading 2.3.** without 2-mercaptoethanol or DTT.

3. Methods

3.1. Purification of Recombinant DNA Polymerase γ

The following purification protocol is used to isolate the untagged mature catalytic subunit of pol γ (p140) from *Sf*9 insect tissue culture cells that have been infected for 48 h with a recombinant baculovirus expressing the human p140 cDNA *(32)*. All chromatographic steps were performed at 4°C.

1. A thawed or fresh *Sf*9 insect cell pellet derived from 20–30 T_{175} flasks infected with baculovirus expressing the catalytic subunit of human pol γ is resuspended in 3 volumes of lysis buffer.
2. Lysed cells were centrifuged at 20,000*g* for 10 min, and the supernatant (Fraction I) was diluted to 40 mL with buffer A also containing 0.05 *M* NaCl prior to the addition of 20 mL phosphocellulose P-11 equilibrated in this buffer.
3. Following 45 min of end-over-end mixing, the resin is collected by vacuum filtration, washed three times with 50 mL of buffer A also containing 0.05 *M* NaCl, and placed in a glass column (4.1 cm × 4.9 cm²). The column is developed with a 200-mL linear gradient of buffer A also containing 0.05 *M* NaCl to buffer B.
4. Fractions are assayed for reverse transcriptase activity (*see* **Subheading 3.3.**). Fractions containing activity (Fraction II) are adjusted to 0.3 *M* potassium phosphate, pH 7.5, and applied to a phenyl–Sepharose column equilibrated in buffer B.
5. The phenyl–Sepharose column is washed with 10 mL of buffer B and developed with a 4.0-mL linear gradient from buffer B to buffer A containing 0.05 *M* NaCl.
6. Fractions are assayed for reverse transcriptase activity (*see* **Subheading 3.3.**) and active fractions (Fraction III) are applied directly to a 1-mL column of single-stranded DNA–cellulose equilibrated in buffer C also containing 0.05 *M* NaCl. The column is washed with 9 mL of equilibration buffer and developed with a 15-mL linear gradient of NaCl (0.05–0.6 *M*) in buffer C. Free catalytic subunit elutes at approx 0.22 *M* NaCl (Fraction IV).
7. Fraction IV is applied directly to a 1-mL column of ceramic hydroxylapatite equilibrated in buffer D. The column is washed with 4 mL of buffer D and developed with a 20-mL linear gradient of buffer D containing 0–0.36 *M* potassium phosphate, pH 7.5.
8. DNA pol γ activity elutes at approx 0.27 *M* potassium phosphate and Fraction V is diluted with buffer E until the conductivity was equivalent to buffer E also containing 0.1 *M* KCl. The enzyme was applied to a 1.0-mL MonoQ FPLC column equilibrated in buffer E containing 0.1 *M* KCl. The column is washed with 3 mL of equilibration buffer and eluted with a 20-mL linear gradient of KCl (0.1 – 0.5 *M*) in buffer E.
9. Pol γ activity elutes at approx 0.22 *M* KCl, and Fraction VI was frozen in small aliquots with liquid nitrogen and stored at –80°C.

3.2. Native DNA polymerase γ purification from HeLa cells

The native forms of mitochondrial DNA polymerase are purified from tissue culture cells by the same method used to purify the recombinant single catalytic subunit, with the following additions and exceptions.

1. HeLa S_3 cells are grown at 37°C in 15- to 30-L batches in spinner cultures to approx 5×10^5 cells/mL in DMEM supplemented with 10% calf serum prior to harvest by centrifugation.
2. Cells are lysed by hypotonic Dounce homogenization, and mitochondria are isolated by the two-step discontinuous sucrose gradient method *(35)*.
3. Mitochondria are lysed in 3 volumes of mitochondrial lysis buffer. Lysates are centrifuged at 20,000g for 10 min. If necessary, the supernatant (Fraction I) may be stored at –80°C after freezing in liquid nitrogen.
4. Fresh or thawed lysates derived from 30 to 100 L of HeLa cells (150–500 mg protein) are pooled and adjusted by dilution with buffer A until the conductivity was equivalent to buffer A containing 0.05 *M* NaCl. All subsequent purification steps were the same as those for the recombinant protein, except 50 mL of phosphocellulose P-11 was utilized and the volumes of the three washes and the gradient were doubled for this first column.
5. Two peaks of DNA polymerase γ activity may elute from the single-stranded DNA–cellulose column. The form eluting at approx 0.22 *M* NaCl is the free catalytic subunit, which is purified by hydroxylapatite and MonoQ column chromatography as stated above in **Subheading 3.1.** The form eluting at approx 0.38 *M* NaCl is the two-subunit complex, which is reasonably pure but may be further purified by MonoQ chromatography as stated above in **Subheading 3.1.** (*See* **Notes 3–6.**)

3.3. Enzymatic Assay of DNA Polymerase γ

1. Set up the 12 × 75-mm borosilicate tubes in a test tube rack embedded in ice.
2. Calculate the amount of H_2O needed and add H_2O first, then 25 µL of 2X mix; add enzyme last. (Example: 20 µL dH_2O + 25 µL 2X mix + 5 µL enzyme, usually diluted 1 : 100 in enzyme dilution buffer.)
3. The final volume of each reaction is 50 µL. Mix gently and incubate the tubes at 37°C for 10 min.
4. Add 1.0 mL of Stop Solution, mix, and add 1.0 mL of 20% TCA. Mix again and hold tubes on ice for 5 min to precipitate DNA.
5. Prepare the filtering apparatus by placing GF/C filters onto screens and setting metal chimneys onto each filter.
6. Apply a vacuum and add HCl (0.1 *N*) to each chimney to wet and settle the filter onto the screen.
7. Pour assay tube's contents onto filter, generously rinse tube three times with HCl.
8. Wash filter 5 to 10 times with HCl, then once with ethanol.

9. Remove filters onto a drying board and dry under a heat lamp for 5 min.

10. Count the dried filters in a scintillation counter with an appropriate scintillation cocktail.

11. To convert radioactivity to units of polymerase activity, count 5 μL of the 2X assay mix by spotting on a dry GF/C filter and counting in the same scintillation cocktail. Because 5 μL of 2X mix represents 250 pmol of labeled TTP, the specific activity of the mix is the "cpm"/250 pmol. Convert the radioactivity of each sample to units by dividing the counts per minute for each samples by the specific activity of the mix. One unit of polymerase is defined by as the amount of enzyme to incorporate 1 pmol of labeled TTP into acid insoluble DNA in 1 h at 37°C.

3.4. NEM Inhibition Assay

The two forms of pol γ have distinctly different responses to *N*-ethylmaleimide. The isolated catalytic subunit of pol γ is almost completely inhibited by 1.0 m*M* NEM, whereas the two-subunit pol γ retains more than 90% of its activity at this concentration (*see* **Fig. 4**).

1. The reverse transcriptase assay in **Subheading 3.3.** is repeated, but reactions also contain increasing concentrations of NEM, ranging from 0 to 1 m*M*. The enzyme dilution buffer and the assay buffer are made without 2-mercaptoethanol or DTT. (*See* **Note 7.**)

2. Pure isolated catalytic subunit will exhibit an inhibition curve similar to the open squares in **Fig. 4**.

3. If both subunits are present in the correct stoichiometry, then an inhibition curve similar to the closed triangles or open circles in **Fig. 4** will result.

4. Preparations of pol γ containing a mixture of the two forms exhibit an intermediate level of inhibition, inversely proportional to the relative abundance of the accessory subunit.

4. Notes

1. DNA pol γ can utilize a wide variety of DNA substrates for detection of activity, such as activated DNA or primed M13, as well as many homopolymeric substrates like poly(dA)•oligo(dT) and poly(dC)•oligo(dG). However, many of the nuclear DNA polymerases can also utilize these substrates quite efficiently. We describe here the use of poly(rA)•oligo(dT) in a reverse transcriptase assay, because pol γ is known to possess reverse transcriptase activity. Thus, this assay is much more specific for pol γ.

2. The reverse transcriptase assay described here uses 75 m*M* NaCl, because both forms of pol γ activity can be detected at this salt concentration. Once the subunit composition of a given preparation of pol γ is identified, the optimal salt concentration can be used. The single catalytic subunit prefers 50 m*M* NaCl, whereas the two-subunit complex prefers 75–150 m*M* NaCl on this substrate.

3. It is imperative to use a shallow salt gradient to separate the two forms of pol γ by ssDNA chromatography. Steep gradients will not resolve the two forms.

4. Typical yields from 100 L of HeLa cells are approx 4 μg of total purified polymerase represented in both forms. This represents roughly 10% of the DNA polymerase activity in the original mitochondrial lysate.

5. The native pol γ purification can also be used with most nonhuman tissue culture cells such as CHO or COS cells. Noncultured human tissue sources, such as placenta, may be used, but mechanical homogenization will be necessary. Also, carefully maintaining cold temperatures and the use of a wider range of protease inhibitors will be necessary to reduce proteolysis and/or fragmentation of the polymerase complex.

6. The purification should be performed without stopping to maximize the quality and activity of the enzyme preparation. If this is not possible, then harvest of tissue culture cells, banding of mitochondria, and preparation of mitochondrial lysates should be done in 1 d. Mitochondrial lysates can be frozen with liquid nitrogen, stored at –80°C, and used at a later date. Purification from mitochondrial lysates should be carried out as quickly as possible. If pausing is necessary, we recommend stopping after the single-stranded DNA–cellulose column, so the enzyme is held at higher ionic strength and in a substantially pure form. The final two columns should be finished on the next day.

7. Sulfhydryl-reducing agents such as 2-mercaptoethanol and dithiothreitol quantitatively react with NEM, thereby shifting the inhibition curve. Thus, 2-mercaptoethanol or dithiothreitol should not be included in the reaction mixture or enzyme dilution buffers.

References

1. Anderson, S., Bankier, A. T., Barrell, B. G., de Bruijn, M. H., Coulson, A. R., Drouin, J., et al. (1981) Sequence and organization of the human mitochondrial genome. *Nature* **290,** 457–465.

2. Wallace, D. C. (1992) Diseases of the mitochondrial DNA. *Annu. Rev. Biochem.* **61,** 1175–1212.

3. Luft, R. (1994) The development of mitochondrial medicine. *Proc. Natl. Acad. Sci. USA* **91,** 8731–8738.

4. Wallace, D. C. (1999) Mitochondrial diseases in man and mouse. *Science* **283,** 1482–1488.

5. Clayton, D. A. (1982) Replication of animal mitochondrial DNA. *Cell* **28,** 693–705.

6. Kornberg, A. and Baker, T. A. (1992) *DNA Replication, 2nd Ed., W.H. Freeman, New York.*

7. Fry, M. and Loeb, L. A. (1986) *Animal Cell DNA Polymerases*, CRC, Boca Raton, FL.

8. Lewis, W., Meyer, R. R., Simpson, J. F., Colacino, J. M., and Perrino, F. W. (1994) Mammalian DNA polymerases alpha, beta, gamma, delta, and epsilon incorporate

fialuridine (FIAU) monophosphate into DNA and are inhibited competitively by FIAU Triphosphate. *Biochemistry* **33,** 14,620–14,624.

9. Wang, T. S.-F. (1991) Eukaryotic DNA polymerases. *Annu. Rev. Biochem.* **60,** 513–552.

10. Cherrington, J. M., Allen, S. J., McKee, B. H., and Chen, M. S. (1994) Kinetic analysis of the interaction between the diphosphate of (*S*)-1-(3-hydroxy-2-phosphonylmethoxypropyl)cytosine, ddCTP, AZTTP, and FIAUTP with human DNA polymerases beta and gamma. *Biochem. Pharmacol.* **48,** 1986–1988.

11. Faraj, A., Fowler, D. A., Bridges, E. G., and Sommadossi, J. P. (1994) Effects of 2′,3′-dideoxynucleosides on proliferation and differentiation of human pluripotent progenitors in liquid culture and their effects on mitochondrial DNA synthesis. *Antimicrob. Agents Chemother.* **38,** 924–930.

12. Eriksson, S., Xu, B., and Clayton, D. A. (1995) Efficient incorporation of anti-HIV deoxynucleotides by recombinant yeast mitochondrial DNA polymerase. *J. Biol. Chem.* **270,** 18,929–18,934.

13. Dalakas, M. C., Illa, I., Pezeshkpour, G. H., Laukaitis, J. P., Cohen, B., and Griffin, J. L. (1990) Mitochondrial myopathy caused by long-term zidovudine therapy [see comments]. *N. Engl. J. Med.* **322,** 1098–1105.

14. Arnaudo, E., Dalakas, M., Shanske, S., Moraes, C. T., DiMauro, S., and Schon, E. A. (1991) Depletion of muscle mitochondrial DNA in AIDS patients with zidovudine-induced myopathy. *Lancet* **337,** 508–510.

15. Peters, B. S., Winer, J., Landon, D. N., Stotter, A., and Pinching, A. J. (1993) Mitochondrial myopathy associated with chronic zidovudine therapy in AIDS. *Q. J. Med.* **86,** 5–15.

16. Lewis, W. and Dalakas, M. C. (1995) Mitochondrial toxicity of antiviral drugs. *Nat. Med.* **1,** 417–422.

17. Wernette, C. M. and Kaguni, L. S. (1986) A mitochondrial DNA polymerase from embryos of *Drosophila melanogaster*. Purification, subunit structure, and partial characterization. *J. Biol. Chem.* **261,** 14,764–14,770.

18. Insdorf, N. F. and Bogenhagen, D. F. (1989) DNA polymerase gamma from *Xenopus laevis*. I. The identification of a high molecular weight catalytic subunit by a novel DNA polymerase photolabeling procedure. *J. Biol. Chem.* **264,** 21,491–21,497.

19. Gray, H. and Wong, T. W. (1992) Purification and identification of subunit structure of the human mitochondrial DNA polymerase. *J. Biol. Chem.* **267,** 5835–5841.

20. Foury, F. (1989) Cloning and sequencing of the nuclear gene MIP1 encoding the catalytic subunit of the yeast mitochondrial DNA polymerase. *J. Biol. Chem.* **264,** 20,552–20,560.

21. Ropp, P. A. and Copeland, W. C. (1995) Characterization of a new DNA polymerase from *Schizosaccharomyces pombe*: a probable homologue of the *Saccharomyces cerevisiae* DNA polymerase gamma. *Gene* **165,** 103–107.

22. Ropp, P. A. and Copeland, W. C. (1996) Cloning and characterization of the human mitochondrial DNA polymerase, DNA polymerase gamma. *Genomics* **36,** 449–458.

23. Ye, F., Carrodeguas, J. A., and Bogenhagen, D. F. (1996) The gamma subfamily of DNA polymerases: cloning of a developmentally regulated cDNA encoding *Xenopus laevis* mitochondrial DNA polymerase gamma. *Nucleic Acids Res.* **24,** 1481–1488.

24. Lewis, D. L., Farr, C. L., Wang, Y., Lagina, A. T. R., and Kaguni, L. S. (1996) Catalytic subunit of mitochondrial DNA polymerase from *Drosophila* embryos. Cloning, bacterial overexpression, and biochemical characterization. *J. Biol. Chem.* **271,** 23,389–23,394.

25. Kunkel, T. A. and Soni, A. (1988) Exonucleolytic proofreading enhances the fidelity of DNA synthesis by chick embryo DNA polymerase-gamma. *J. Biol. Chem.* **263,** 4450–4459.

26. Kunkel, T. A. and Mosbaugh, D. W. (1989) Exonucleolytic proofreading by a mammalian DNA polymerase. *Biochemistry* **28,** 988–995.

27. Insdorf, N. F. and Bogenhagen, D. F. (1989) DNA polymerase gamma from *Xenopus laevis*. II. A 3′→5′ exonuclease is tightly associated with the DNA polymerase activity. *J. Biol. Chem.* **264,** 21,498–21,503.

28. Olson, M. W. and Kaguni, L. S. (1992) 3′→5′ exonuclease in *Drosophila* mitochondrial DNA polymerase. Substrate specificity and functional coordination of nucleotide polymerization and mispair hydrolysis. *J. Biol. Chem.* **267,** 23,136–23,142.

29. Foury, F. and Vanderstraeten, S. (1992) Yeast mitochondrial DNA mutators with deficient proofreading exonucleolytic activity. *EMBO J.* **11,** 2717–2726.

30. Spelbrink, J. N., Toivonen, J. M., Hakkaart, G. A., Kurkela, J. M., Cooper, H. M., Lehtinen, S. K., et al. (2000) In vivo functional analysis of the human mitochondrial DNA polymerase POLG expressed in cultured human cells [in process citation]. *J. Biol. Chem.* **275,** 24,818–24,828.

31. Zhang, D., Mott, J. L., Chang, S. W., Denniger, G., Feng, Z., and Zassenhaus, H. P. (2000) Construction of transgenic mice with tissue-specific acceleration of mitochondrial DNA mutagenesis. *Genomics* **69,** 151–161.

32. Longley, M. J., Ropp, P. A., Lim, S. E., and Copeland, W. C. (1998) Characterization of the native and recombinant catalytic subunit of human DNA polymerase gamma: identification of residues critical for exonuclease activity and dideoxynucleotide sensitivity. *Biochemistry* **37,** 10,529–10,539.

33. Lim, S. E., Longley, M. J., and Copeland, W. C. (1999) The mitochondrial p55 accessory subunit of human DNA polymerase gamma enhances DNA binding, promotes processive DNA synthesis, and confers *N*-ethylmaleimide resistance. *J. Biol. Chem.* **274,** 38,197–38,203.

34. Mosbaugh, D. W. (1988) Purification and characterization of porcine liver DNA polymerase gamma: utilization of dUTP and dTTP during in vitro DNA synthesis. *Nucleic Acids Res.* **16,** 5645–5659.

35. Bogenhagen, D. and Clayton, D. A. (1974) The number of mitochondrial deoxyribonucleic acid genomes in mouse L and human HeLa cells. Quantitative isolation of mitochondrial deoxyribonucleic acid. *J. Biol. Chem.* **249,** 7991–7995.

17

Assay of mtDNA Polymerase γ from Human Tissues

Robert K. Naviaux

1. Introduction

Mitochondrial DNA (mtDNA) replication is a complex process involving over 20 proteins organized along the inner mitochondrial membrane as a multienzyme complex called the mtDNA replisome, or replication factory *(1–3)*. **Figure 1** illustrates some of the protein components that participate in mitochondrial DNA replication. A principal component of the mtDNA replisome is the mtDNA polymerase γ. This enzyme is found as an αβ heterodimer and as an α monomer associated with at least four other unidentified cellular proteins *(4)*. Both α- and β-subunits have been cloned *(5–7)*. The α-subunit is catalytic and contains both the polymerase and the 3′ to 5′ proofreading exonuclease activities. It is 140,000 Daltons in size.

The substrate virtuosity displayed by DNA polymerase γ is unique among eukaryotic polymerases. The mitochondrial enzyme is an effective reverse transcriptase (RNA-directed DNA polymerase) when measured on homopolymeric RNA templates. It is also an effective DNA polymerase when directed by single-stranded homopolymeric DNA, single-stranded heteropolymeric DNA, or activated double-stranded DNA templates *(8)*. The experimental choice of template radically influences both the measured kinetic performance of the enzyme and its response to specific inhibitors, such as azidothymidine (AZT) *(9)*. The most sensitive assays developed to date have exploited the RNA-directed DNA polymerase activity of the enzyme. These assays are typically about 10–50 times more sensitive than conventional assays, which are performed with activated double-stranded DNA templates *(9)*.

Mitochondrial disorders are among the most recently recognized groups of diseases known to medicine. They affect both children and adults and are

From: *Methods in Molecular Biology, vol. 197: Mitochondrial DNA: Methods and Protocols*
Edited by: W. C. Copeland © Humana Press Inc., Totowa, NJ

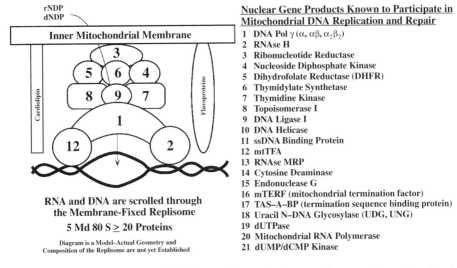

Fig. 1. The mitochondrial DNA replisome. The actual geometry and subunit composition of the mtDNA replisome is not yet known. The proteins shown are those that have been shown to participate in mtDNA replication or repair.

particularly difficult to diagnose because of a broad spectrum of phenotypic presentations *(10,11)*. Muscle biopsy is often obtained in the course of the diagnostic evaluation of patients with suspected mitochondrial disease. A number of inherited forms of mtDNA depletion have been described *(12–14)*. Depletion of mtDNA may also result from AZT *(15)* or fialuridine *(16)* toxicity. Deficiency in any 1 of the 20 or so components of the mtDNA replisome could theoretically lead to mtDNA depletion. In addition, Alpers syndrome is associated with mtDNA depletion and mtDNA polymerase γ deficiency *(17)*. This assay has been previously described *(18)*.

2. Materials
2.1. Stock Solutions

Only the highest-quality enzyme or spectral-grade reagents should be used. Water should be the purist available double-distilled, or distilled and reverse osmosis purified via a Milli-Q (Millipore, or similar) reagent water purification system.

1. 0.25 *M* Sucrose.
2. Protease XXVII (Nagarse Protease, Fluka cat. no. 82518).
3. Glycerol lysis buffer: 25 m*M* Tris-HCl, pH 8.0, 50 m*M* KCl, 2% Triton X-100, 50% glycerol.

4. 5X Buffer stock: 0.25 M Tris-HCl, pH 8.0, 25 mM MgCl$_2$, 0.5% Triton X-100, 50 mM dithiothreitol (DTT), and 25 mM KCl (*see* **Note 1**).
5. 5X Template stock: 50 µg/mL poly(rA) (Pharmacia cat. no. 27-4110-02, mean length = 500 nucleotides (nt), molecular weight [MW] = 1.85×10^5, 54 nM, yielding a final [1X] nucleotide concentration of 27 µM), 25 µg/mL oligo(dT$_{12-18}$) (Pharmacia cat. no. 27-7858-02, MW = 4905, 1.0 µM, yielding a final [1X] nucleotide concentration of 15 µM, and a primer to template molar ratio of 18.5 : 1), and 25 µM dTTP (*see* **Note 1**).
6. 50 mM MnCl$_2$ (Alfa Aesar, cat. no. 10804, Ward Hill, MA).
7. ^{32}P-α-dTTP (3000 Ci/mmol, 10 µCi/µL, NEN BLUE/NEG/005H or similar).
8. 2X SSC: 0.3 M NaCl, 0.03 Na^3 citrate, pH 7.0.
9. 95% Ethanol.
10. RNAse-free (diethylpyrocarbonate-treated) Mili-Q water.
11. 10/9X Reaction buffer (0.45 mL, enough for 45 reactions) is prepared fresh, on the day of the assay: 100 µL of 5X buffer stock, 100 µL 5X template stock, 5 µL of 50 mM MnCl$_2$, 25 µL of ^{32}P-α-dTTP (3000 Ci/mmol, 10 µCi/µL, 0.167 µM final radionuclide concentration), and 220 µL of diethylpyrocarbonate (DEPC)-treated water. (*See* **Note 2**.)
    a. When 9 µL of 10/9X reaction buffer is combined with 1 µL of mitochondria, the final (1X) reaction conditions are 50 mM Tris-HCl, pH 8.0, 100 mM NaCl, 2.5 mM KCl, 5 mM MgCl$_2$ and 0.5 mM MnCl$_2$, 10 mM DTT, 0.1% Triton X-100, 2.5% glycerol, 5 µM dTTP, 10 µg/mL poly(rA), 5 µg/mL oligo(dT)$_{12-18}$, 500 µCi/mL ^{32}P-α-dTTP (3000 Ci/mmol).

2.2. Other Materials

1. Biospec Tissue Tearor, Tissue Homgenizer (Fisher cat. no. 15-338-55, or similar).
2. GeNunc Module reaction trays (48-well, 4 × 12 tray, Nunc cat. no. 2-32549).
3. DE81 anion-exchange paper (Whatman, cat. no. 3658-915).
4. BioMax MS^{TM} film with matched intensifying screen (Kodak).
5. Ecolume scintillation fluid (ICN).

3. Methods

3.1. Isolation of Mitochondria (see Note 3)

A variety of methods are available for the isolation of mitochondria from solid tissues, blood, and cells in culture *(19–21)*. Mitochondria from different tissue sources have different purification properties and must be treated differently to achieve optimum yields and quality. In particular, mitochondria from cells in culture and platelets, brain, liver, and skeletal muscle all require distinct purification protocols. We have focused on skeletal muscle because it is commonly obtained in the course of a diagnostic evaluation for mitochondrial disease. Skeletal muscle mitochondria were isolated from 250–800 mg of fresh (unfrozen) human anterior quadriceps (vastis intermedius) skeletal muscle by a modification of the method of Hatefi et al. *(22)*. All procedures are performed on ice, with prechilled buffers and instruments. A minimum of 250 mg of good quality skeletal muscle, free of fascia and fat, is required. This is processed as follows:

1. The muscle is weighed and minced into 2- 3-mm cubes with fine forceps, scissors, and scalpel.
2. Save 25–50 mg of muscle at –80°C for DNA purification and mtDNA quantitation.
3. Resuspend up to 500 mg of muscle in 1 mL of 0.25 M cold sucrose in a 6-mL sterile Falcon tube. Additional muscle is processed in parallel 1-mL samples.
4. Homogenize with a Biospec Tissue Tearor (Fisher cat. no. 15-338-55) or similar electric homogenizer at a setting of 5 for 6 s.
5. Transfer the muscle homogenate to a chilled Falcon 2059 tube, using a 1-mL pipet as a filter to excluded any tissue that was not homogenized.
6. Add 1 mL of 0.25 M sucrose and repeat **step 4**.
7. Collect all homogenates from the biopsy into a single Falcon 2059 tube.
8. Add 1/40 volume of 2 M Tris-HCl, pH 7.6, (at 20°C) to bring the homogenate to 50 mM Tris-HCl, pH 8.0, (at 4°C).
9. Add 1 mg/mL Nagarse Protease (Fluka cat. no. 82518).
10. Incubate on ice for 20 min, mixing every 5 min.
11. Transfer the homogenate to a prechilled Potter-Elvehjem Teflon-glass, small-volume homogenizer attached to an electric hand drill.
12. Homogenize with two passes of the Teflon pestle.
13. Add an equal volume of cold 0.25 M sucrose and repeat the homogenization with an additional two passes of the Teflon pestle.
14. Pool the homogenates in a 40-mL polycarbonate tube and collect residual homogenate by rinsing the Potter-Elvehjem homogenizer with 10–15 mL of 0.25 M sucrose.
15. Centrifuge in a precooled SS-34 rotor or equivalent, at 600g for 15 min to sediment nuclei and unlysed cells
16. Remove the overlaying lipid layer, which appears as a white to tan pellicle overlying the supernatant, by aspiration and carefully transfer the mitochondria-containing supernatant to a fresh, chilled, polycarbonate tube, leaving the pellet undisturbed.
17. Centrifuge the post-600g supernatant at 10,000g for 15 min.
18. Carefully remove the supernatant and discard. Resuspend the mitochondria-containing pellet in 1 mL of 0.25 M sucrose with 50 mM Tris-HCl, pH 8.0, (at 4°C)
19. Bring the mitochondrial suspension up to a total volume of 40 mL with the same solution to wash.
20. Centrifuge at 17,000g for 15 min.
21. Discard the supernatant and carefully resuspend the mitochondria-containing pellet in a volume of 0.25 M sucrose equal to the original mass of the biopsy.

For example, mitochondria from an 800-mg biopsy would be resuspended in 800 μL (0.8 mL) of 0.25 *M* sucrose to achieve a final protein concentration of approximately 1–5 mg/mL.

22. Divide the sample into four labeled aliquots and freeze at –80°C until further use.
23. A large-scale (50–100 g of muscle) preparation of beef heart (this can often be obtained from a local slaughterhouse) mitochondria or human muscle mitochondria is prepared and frozen at 2 mg/mL in multiple 20-μL aliquots for use as a standard, internal control, to be run with each DNA polymerase γ assay performed on different days.

3.2. Protein and Citrate Synthase Determinations (see Note 5)

Mitochondrial protein is determined from thawed samples by a small-scale modification of the Lowry method (Bio-Rad kit cat. no. 500-0112). This requires about 20 μg of mitochondrial protein. Mitochondrial yield is quantified by measurement of the mitochondrial matrix protein citrate synthase (CS) by the method of Sheperd and Garland *(23)*, which requires about 5 μg of mitochondrial protein. One unit of CS is defined as 1 pmol/min of acetyl CoA consumed.

3.3. DNA Polymerase Assays

1. Thaw 5–10 μg of isolated skeletal muscle mitochondria, lyse, and stabilize in an equal volume of glycerol lysis buffer.
2. Mix the lysed mitochondria a 10- to 20-μL micropipet, bring to a concentration of at most 1 mg/mL with additional glycerol lysis buffer, and place on ice.
3. Prepare four serial two-fold dilutions of lysed skeletal muscle mitochondria in glycerol lysis buffer, to yield final protein concentrations of 1.0, 0.5, 0.25, and 0.125 μg/μL.
4. Pipet 1 μL of each dilution into the bottom of a 25-μL GeNunc Module (48-well, 4 × 12 tray, Nunc cat. no. 2-32549) reaction trays at room temperature.
5. Pipet an equal volume of glycerol lysis buffer into four separate wells that will be used as the quadruplicate blank.
6. Thaw a tube of large-scale control mitochondria (from beef heart or human skeletal muscle) containing 10–20 μL of mitochondrial protein, lyse this in glycerol lysis buffer, and prepare as in **steps 1** and **2** as an internal control.
7. Place the loaded reaction trays behind a radiation shield for all subsequent steps of the reaction.
8. Pipet 9 μL of 10/9X reaction buffer into to each well and mix the full 10-μL reaction by rapid micropipetting six times.
9. Seal the GeNunc reaction trays with an adhesive top (supplied by the manufacturer) and float in a 37°C water bath for 30 min.
10. Carefully touch the bottoms of the GeNunc trays to a paper towel to remove any water and place the trays behind a shield. Do not tap the trays, as this will contaminate the tape covers with radioactivity.

11. Carefully remove the tape covers from the GeNunc trays and stop the reactions by pipetting 5 µL of each reaction onto 0.5-in. squares drawn and numbered with a pencil (conveniently done during the 30-min incubation period) on DE81 anion-exchange paper (Whatman, cat. no. 3658 915). (*See* **Note 4**.)
    a. Take care not to indent the DE81 paper with the tip of the micropipet, as this will produce spurious spots ("zits") on the autoradiograph.
12. Wash the DE81 paper three times for 5 min each in 250 mL of 2X SSC with gentle agitation.
13. Wash the DE81 paper once in 250 mL of 95–100% ethanol with gentle agitation, and air-dry for 30 min.
14. Wrap the dried DE81 paper in Saran Wrap and tape this to a thin piece of cardboard backing that has been marked at the corners with fluorescence spots of paint for orientation (the cardboard that comes in each case of x-ray film works nicely for this purpose)
15. Place the cardboard and DE81 paper in a film cassette, overlay with BioMax MS™ film (or equivalent) and a matched intensifying screen, and expose for 1 h at room temperature. (Exposure at –80°C is not necessary.)
16. During the film exposure, spot 1 µL of 10/9X reaction buffer in quadruplicate on precut 0.5-cm squares of DE81 paper, to be used directly, without washing, as an internal control for radioisotopic specific activity and decay.
17. After the film is exposed and developed, cut the DE81 paper into 0.5-cm squares along the pencil markings.
18. Quantify the high-molecular-weight polymerase reaction products captured to the DE81 paper squares by scintillation counting in 5 mL of Ecolume (ICN).
19. Subtract the mean background (determined from quadruplicate blanks) from each reaction and calculate the total dTMP incorporation into high-molecular-weight products (radioactive and nonradioactive)

3.3.1. Activity Calculations (see Note 6)

1. Calculate the blank by taking the mean of the quadruplicate blanks and subtract this from the scintillation cpm obtained from each sample.
2. Calculate a regression curve from the four serial dilutions (cpm minus the blank) for each sample, with cpm on the y axis, and micrograms of mitochondrial protein on the x axis.
3. The slope of the regression will be in cpm/µg. **Table 1** gives some real experimental data.
4. Using the data in **Table 1**, a least squares regression line can be calculated and is found to be as follows:
    a. $y = 116,904x + 40,117$. The regression coefficient $R^2 = 0.9974$.
    b. Therefore, the slope of the line is 116,904 cpm/µg, and the y intercept is 40,117 cpm. This is illustrated in **Fig. 2**.
5. Assuming the ^{32}P-dTTP had a specific activity of 3000 Ci/mmol (3000 pCi/fmol) is used on its reference date and does not require correction for decay, and the counting efficiency of ^{32}P is 100% in scintillation fluid, the specific activity of

Table 1
Sample Data

Protein (μg)	Sample cpm	Quadruplicate blank cpm	Sample – blank[a]
1	182,843	4,446	178,725
0.5	118,437	2,397	114,319
0.25	78,943	5,189	74,755
0.125	59,893	4,721	55,705

[a]Mean blank = 4188

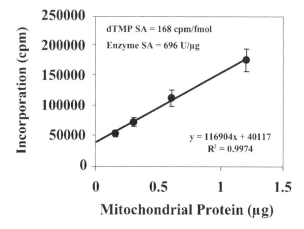

Fig. 2. Sample data regression analysis: the intercept problem. The data from Table 1 are analyzed by least squares linear regression (SA: specific activity).

dTMP nucleotides incorporated into the DNA polymerase γ reaction products can be calculated:

a. The ratio of total nucleotides to radioactive (countable) nucleotides is Total dTTP concentration ÷ radionuclide concentration = (5 μM + 0.167 μM) ÷ 0.167 μM = 30.9

b. Reaction product specific activity = 3000 pCi/fmol × 2.2 cpm/pCi ÷ 30.9 Total dTTP/^{32}P-dTTP = 214 cpm/fmol.

c. If the assay is not performed on the reference date for ^{32}P-dTTP, then the specific activity must be adjusted. For example, the half-life of ^{32}P is 14 d and the exponential decay constant (k) is 0.048472; therefore, the fractional decay on d 5 past the reference date is $e^{(-0.048472 \times 5\text{ d})} = 0.78$. 0.78×214 cpm/fmol = 168 cpm/fmol

6. You can use the four spots of the 10/9X reaction mix performed in **Subheading 3.3., step 16** to calculate the labeling error and counting efficiency. This is the fractional difference between the *expected* cpm based on decay and the *actual*

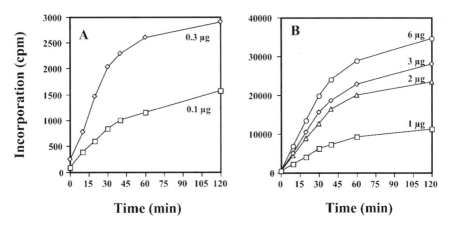

Fig. 3. DNA polymerase γ assay reaction progress curves. Enzyme activity was measured at 10-min intervals for 2 h, using mitochondrial protein loads of 0.1–6 μg. The assay was linear for at least 30 min at all protein concentrations.

measurements on the day of the assay. If everything is perfect, then 1 μL of 10/9X reaction mix should contain 556 μCi/mL (10/9 × 500 μCi/mL) or 0.556 μCi/μL on the reference date for ^{32}P-dTTP. This is 1,223,200 disintegrations per min (dpm). If your counting efficiency is less than 100% or if any errors were made during the preparation of the assay reagents for the day, this step will allow you to detect the error and to compensate for it.

7. Specific activities are expressed in units/μg mitochondrial protein. One unit of DNA polymerase γ activity is defined as 1 fmol of total nucleotides incorporated in 30 min at 37°C, using poly(rA) as a template. Because the slope of the regression line in **Fig. 2** was 116,904 cpm/μg, if the assay were performed on d 5 past the reference date, yielding a dTMP SA of 168 cpm/fmol, then the activity of this sample is 116,904 cpm/μg ÷ 168 cpm/fmol = 696 U/μg.

8. Moloney murine leukemia virus reverse transcriptase (Gibco-BRL) can be used as a high-specific-activity polymerase control to test the limits of linearity of this assay, with respect to enzyme activity. The specific activity of MoMuLV-RT was 350,000 BRL units/mg (1 BRL unit = 1 nmol/10 min). This is equivalent to 1.05×10^9 fmol/μg incorporated in a 30-min reaction.

3.3.2. Limits of Linearity, Sensitivity, and Reproducibility

1. Reaction progress curves showed that this assay is linear for 30 min (*see* **Fig. 3**).
2. Mitochondrial protein dose-response curves showed that his assay is linear for total mitochondrial protein loads of up to 2 μg (*see* **Fig. 4**).
3. Activity dose-response curves using MoMuLV-RT showed the assay was linear in the range of 30–30,000 U/μg protein.
4. Skeletal muscle mitochondrial samples with DNA polymerase γ activities of as little as 30 U/μg are distinguishable from background. Most muscle samples fall

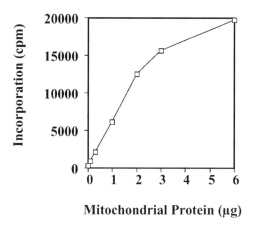

Mitochondrial Protein (μg)

Fig. 4. Protein dose-response curve. The mtDNA polymerase γ activities measured 30 min in Fig. 3 were replotted according to protein concentration. The assay was linear from less than 0.1–2 μg.

into the range of 400–800 U/μg. Samples with zero activity have been seen in Alpers syndrome *(17)*, whereas samples with as much as 2500 U/μg have been seen in disorders with secondary mitochondrial proliferation. Each laboratory must determine its own normal and abnormal ranges in appropriate control and patient populations when validating this assay for clinical use (*see* **Note 5**).

5. In our hands, the coefficient of variation (CV=sd/m) for this assay, using four-point serial dilution, and least squares regression, was 6.4% within days. The CV between days was 12%. A typical autoradiograph of DNA polymerase γ reaction products captured on DE81 paper is shown in **Fig. 5**. The results of five separate regression analyses of a single sample of skeletal muscle mitochondria are shown with a mean of 670 and a CV of 6.4%.

3.3.3. The Intercept Problem

1. Inspection of **Fig. 2** shows that despite subtraction of the assay blank shown in **Table 1**, the *y* intercept is often nonzero. The calculation of a regression line eliminates this problem as a source of error in the assay.

2. Serious inaccuracies are produced if quadruplicate measurements of a single dilution of mitochondria are used instead of serial dilutions, because the *y* intercept cannot be quantified without serial dilutions. For example, if we look at the data for 0.125 μg of mitochondrial protein shown in **Table 1**, 55,705 cpm were obtained after the subtraction of the blank. The specific activity of dTMP incorporated was 168 cpm/fmol. Quadruplicate measures of 0.125 μg protein would yield the following (incorrect) activity:

   a. 55,705 cpm ÷ 0.125 μg protein ÷ 168 cpm/fmol = 2,652 U/μg.

Autoradiography	Activity (U/µg)	Mean (sd)	SEM (%)
Dilution Series 0.125 0.25 0.5 1x	**Best Fit Regression (R²)**		

	Activity (U/µg)	Mean (sd)	SEM (%)
	Blank		
	653 (0.995)		
	747 (0.997)	670	19.2
		(43)	(2.9%)
	656 (0.992)	(6.4%)	
	617 (0.994)		
	672 (0.965)		

Fig. 5. Sample autoradiograph and assay reproducibility. The DNA polymerase γ activity in a single skeletal muscle mitochondrial sample was analyzed in five replicates of a four-point serial dilution.

b. The true activity, as shown in **Fig. 2**, was 696 U/µg. The use of quadruplicate determinations of a single dilution of mitochondria would have overestimated the actual value by over 380%.
c. As larger loads of mitochondrial protein are analyzed, the intercept of 40,117 cpm in **Fig. 2** becomes a smaller fraction of the true activity, so in the limit, if it were possible to use very high amounts of protein, the intercept would represent a negligible amount of the total and the observed quadruplicate result would approach the true activity asymptotically. Of course, in practice this is not possible, because the protein dose-response falls off from linear at loads of more than 2 µg of mitochondrial protein (*see* **Fig. 4**).
3. Application of this assay to standard Michaelis–Menten kinetic analysis shows a well-behaved catalytic system, with a hyperbolic catalytic response to increasing amounts of added substrate (total dTTP) that is linearized in a standard Hanes–Woolf *(24)* plot of the data (*see* **Fig. 6**). In the example shown, the mitochondrial DNA Polymerase γ displayed a K_m for dTTP of 1.48 µ*M*.

4. Notes

1. The 5X buffer and template stocks could be stored for up to 6 mo at –20°C. $MnCl_2$ could not be added until the day of the assay because of precipitation that occurred in the concentrated stocks.
2. We chose a concentration of nonradioactive dTTP of 5 µ*M* for our standard reaction conditions. This is greater than three times the observed K_m of 1.48 µ*M*, so the reaction velocity is > 75% of its V_{max}, but the nonradioactive substrate concentration is still small enough so as not to reduce the detection sensitivity in terms of actual cpm captured to DE81 paper, by competition with ^{32}P-dTTP.

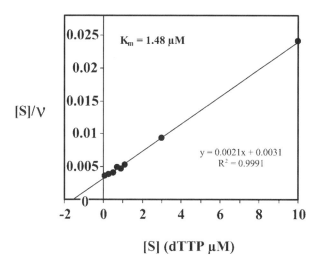

[S]/v

$K_m = 1.48 \, \mu M$

$y = 0.0021x + 0.0031$
$R^2 = 0.9991$

[S] (dTTP μM)

Fig. 6. Michaelis–Menton kinetics of DNA polymerase γ. The described assay conformed to conventional Michaelis–Menton kinetics as demonstrated by a good linear fit to Hanes–Woolf ([S]/v vs [S]) transfomation *(24)* under increasing concentrations of dTTP, from 0.1 to 10 μ*M*.

Under these conditions, nearly 1 in every 30 nucleotides incorporated is radioactive. By doubling the nonradioactive dTTP to 10 μ*M*, the assay sensitivity would be cut in half (because now 1 in 60 nucleotides is radioactive) while producing a theoretical increase in rate of just 10% (from 77% to 87% of the V_{max}). Therefore, dTTP of 5 μ*M* represents a favorable kinetic compromise between sensitivity, and maximum enzyme activity.

3. The isolation of mitochondria before the application of the assay for DNA polymerase γ eliminates a number of analytical obstacles that would otherwise complicate assays performed in whole-cell lysates. Cytoplasmic RNAses are largely removed, along with contaminating nuclear polymerases, other RNAs (tRNAs, mRNAs, and rRNAs), nucleotides, and inhitory small molecules. Nucleotides and cellular RNAs can compete with the defined pol γ reaction components, making accurate activity determinations impossible.

4. The capture of high-molecular-weight polynucleotide reaction products on anion-exchange paper (DE81) in this assay helped to avoid the tedious steps of trichloroacetic acid (TCA) precipitation, collection, and washing of individual spun-glass filters that are standard operating procedure in most other polymerase assays *(9)*. The simplicity of DE81 paper contributed to improved replicate precision and dramatically reduced the time and labor required to assay four-point dilutions and quadruplicate patient samples.

5. In practice, the polymerase activities from isolated mitochondria are expressed both as raw values and normalized for mitochondrial content by dividing by

the citrate synthase activity. Citrate synthase is a mitochondrial matrix protein that is used extensively as a marker of mitochondrial content of mitochondrial preparations.

6. As of this writing, 14 different DNA polymerases, named α, β, γ, δ, ε, and so on have been described. To date, only DNA polymerase γ has been shown to be present in mitochondria. However, if whole-cell lysates are used, there is a practical difficulty in designing reaction conditions that distinguish between mitochondrial γ and nuclear β, because β can also use poly(rA) as a substrate *(25)*. When mitochondria are isolated prior to assay, this biochemical difficulty is eliminated.

References

1. Reddy, G. P. V. and Pardee, A. B. (1980) Multienzyme complex for metabolic channeling in mammalian DNA replication. *Proc. Natl. Acad. Sci. USA* **77,** 3312–3316.
2. Shadel, G. S. and Clayton, D. A. (1997) Mitochondrial DNA maintenance in vertebrates. *Annu. Rev. Biochem.* **66,** 409–435.
3. Lemon, D. P. and Grossman, A. D. (1998) Localization of bacterial DNA polymerase: evidence for a factory model of replication. *Science* **282,** 1516–1519.
4. Longley, M. J., Ropp, P. A., Lim, S. E., and Copeland, W. C. (1998) Characterization of the native and recombinant catalytic subunit of human DNA polymerase γ: identification of residues critical for exonuclease activity and dideoxynucleotide sensitivity. *Biochemistry* **37,** 10,529–10,539.
5. Ropp, P. A. and Copeland, W. C. (1996) Cloning and characterization of the human mitochondrial DNA polymerase, DNA polymerase γ. *Genomics* **36,** 449–458.
6. Lecrenier, N., Van Der Bruggen, P., and Foury, F. (1997) Mitochondrial DNA polymerases from yeast to man: a new family of polymerases. *Gene* **185,** 147–152.
7. Wang, Y., Farr, C. L., and Kaguni, L. S. (1997) Accessory subunit of mitochondrial DNA polymerase from *Drosophila* embryos: cloning, molecular analysis, and association in the native enzyme. *J. Biol. Chem.* **272,** 13,640–13,646.
8. Fry, M. and Loeb, L. A. (1986) *Animal Cell DNA Polymerases*, CRC, Boca Raton, FL, pp. 91–101.
9. Lewis, W., Meyer, R. R., Simpson, J. F., Colacino, J. M., and Perrino, F. W. (1994) Mammalian DNA polymerases α, β, γ, δ, and ε incorporate fialuridine (FIAU) monophosphate into DNA and are inhibited competitively by FIAU triphosphate. *Biochemistry* **33,** 14,620–14,624.
10. De Vivo, D. C. (1993) The expanding clinical spectrum of mitochondrial diseases. *Brain Dev.* **15,** 1–22.
11. Kerr, D. S. (1997) Protean manifestations of mitochondrial diseases: a minireview. *J. Pediatr. Hematol./Oncol.* **19,** 279–286.
12. Moraes, C. T., Shanske, S., Tritschler, H.-J., Aprille, J. R., Andreetta, F., Bonilla, E., et al. (1991) mtDNA depletion with variable tissue expression: a novel genetic abnormality in mitochondrial diseases. *Am. J. Hum. Genet.* **48,** 492–501.

13. Ricci, E., Moraes, C. T., Servidei, S., Tonali, P., Bonilla, E., and DiMauro, S. (1992) Disorders associated with depletion of mitochondrial DNA. *Brain Pathol.* **2,** 141–147.

14. Poulton, J., Sewry, C., Potter, C. G., Bougeron, T., Chretien, D., Wijburg, F. A., et al. (1995) Variation in mitochondrial DNA levels in muscle from normal controls. Is depletion of mtDNA in patients with mitochondrial myopathy a distinct clinical syndrome? *J. Inherit. Metabol. Dis.* **18,** 4–20.

15. Dalakas, M. C., Illa, I., Pezeshkpour, G. H., Laukaitis, J. P., Cohen, B., and Griffin, J. L. (1990) Mitochondrial myopathy caused by long-term zidovudine therapy. *N. Engl. J. Med.* **322,** 1098–1105.

16. Mckenzie, R., Fried, M. W., Sallie, R., Conjeevaram, H., Di Bisceglie, A. M., Park, Y., et al. (1995) Hepatic failure and lactic acidosis due to fialuridine (FIAU), an investigational nucleoside analogue for chronic hepatitis B. *N. Engl. J. Med.* **333,** 1099–1105.

17. Naviaux, R. K., Nyhan, W. L., Barshop, B. A., Poulton, J., Markusic, D., Karpinski, N. C., et al. (1999) Mitochondrial DNA polymerase γ deficiency and mitochondrial DNA depletion in a child with Alpers syndrome. *Ann. Neurol.* **45,** 54–58.

18. Naviaux, R. K., Markusic, D., Barshop, B., Nyhan, W. L., and Haas, R. H. (1999) Sensitive assay for mitochondrial DNA polymerase γ. *Clin. Chem.* **45,** 1725–1733.

19. Attardi, G. M. and Chomyn, A., (eds.) (1995) *Mitochondrial Biogenesis and Genetics, Part A, Methods in Enzymology Vol. 260,* Academic, San Diego, CA.

20. Attardi, G. M. and Chomyn, A. (eds.) (1996) *Mitochondrial Biogenesis and Genetics, Part B, Methods in Enzymology Vol. 264,* Academic, San Diego, CA.

21. Taylor, R. W. and Turnbull, D. M. (1997) Laboratory diagnosis of mitochondrial disease, in *Organelle Diseases* (Applegarth, D. A., Dimmick, J. E., and Hall, J. G., eds.), Chapman & Hall, London, pp. 341–350.

22. Hatefi, Y., Jurtshuk, P., and Haavik, A. G. (1961) Studies on the electron transport system. 32. Respiratory control in beef heart mitochondria. *Arch. Biochem. Biophys.* **94,** 148–155.

23. Sheperd, D. and Garland, P. B. (1969) Citrate synthetase from rat liver. *Methods Enzymol.* **13,** 11–16.

24. Cornish-Bowden, A. (1995) *Fundamentals of Enzyme Kinetics,* rev. ed., Portland Press, London, pp. 30–37.

25. Copeland, W. C., Chen, M. S., and Wang, T. S. F. (1992) Human DNA polymerases α and β are able to incorporate anti-HIV deoxynucleotides into DNA. *J. Biol. Chem.* **267,** 21,459–21,464.

18

Purification Strategies for *Drosophila* mtDNA Replication Proteins in Native and Recombinant Form

DNA Polymerase γ

Carol L. Farr and Laurie S. Kaguni

1. Introduction

The identification of proteins involved in mitochondrial biogenesis remains an important challenge. In the past decade, it has become clear that a biochemical approach is limited by the investigator's ability to develop a highly specific assay that differentiates the mitochondrial form from what generally prove to be multiple, and much more abundant, nuclear counterparts. A case in point is the mitochondrial DNA polymerase, pol γ. Pol γ was the third DNA polymerase identified in animal cells *(1,2)* and remains the only DNA polymerase known to function in mitochondria of the now at least 12 cellular DNA-directed DNA polymerases *(3)*. Through the development of a differential assay using high salt, we were able to purify the *Drosophila* holoenzyme form to near-homogeneity for the first time from any source *(4)*. The use of partially purified mitochondria was also critical, because in this and subsequent purification schemes for other mitochondrial proteins, we have observed repeatedly that the use of gradient-purified mitochondria results in large losses in yield. Thus, the take-home lesson in the biochemical identification of new mitochondrial activity is to develop a differential assay and to optimize the yield of the mitochondria from which extraction is pursued. Regarding the purification protocols presented, we have found that the chromatographic matrices used are, in general, efficacious for proteins involved in nucleic acid metabolism. In addition, hydroxylapatite, double-stranded DNA cellulose, ATP agarose, and

From: *Methods in Molecular Biology, vol. 197: Mitochondrial DNA: Methods and Protocols*
Edited by: W. C. Copeland © Humana Press Inc., Totowa, NJ

protein affinity chromatography resins are also generally useful. The order of use can be varied to good purpose, but as a rule of thumb, phosphocellulose as a first step is effective to eliminate approx 90% of bulk mitochondrial protein while retaining nucleic acid binding proteins, and velocity sedimentation is efficacious as a final step to remove adventitious nuclease activities that are invariably of low molecular mass and to remove any small ligands introduced in affinity chromatography steps. The velocity sedimentation is, of course, also useful for linking activity and polypeptide profiles and for obtaining an *S* value of the native protein to help in the determination of subunit structure.

A great aid in the identification of new mitochondrial proteins has been both the exploitation of yeast genetics and the completion of several genome sequences. With the identification of new yeast proteins via genetics, we have the opportunity to search for homologs, and this was accomplished for example in the identification of animal forms of the catalytic core of pol γ *(5,6)*. With the genome database, we can now pursue a bioinformatics approach to search genomic sequences for putative mitochondrial proteins and then sort them by secondary searches for known motifs in both prokaryotic and nuclear proteins with similar functions. Here, both yeast and *Drosophila* provide powerful approaches to validate such data by virtue of complementation of genetic mutants, whereas cell culture systems from many species can be employed to confirm the mitochondrial localization of new proteins. Once cloned and overexpressed, the purification of the mitochondrial protein becomes more facile, and we have included here the streamlined schemes for purification of recombinant forms of *Drosophila* pol γ *(4,7*; this chapter) and mitochondrial single-stranded DNA-binding protein (mtSSB) *(8*; Chapter 19) that again can be used as general schemes for proteins involved in mitochondrial nucleic acid metabolism. The authors are pleased to address any queries by electronic mail and wish all success in the development of their purification strategies.

2. Materials

2.1. *Native* Drosophila melanogaster *DNA Polymerase* γ *Purification*

The ionic strength of buffers is determined using a Radiometer conductivity meter.

1. Phosphocellulose P-11 (Whatman), prepared according to the manufacturer's directions.
2. Octyl-Sepharose CL-4B (Amersham Pharmacia Biotech).
3. Single-stranded DNA cellulose prepared as per Alberts and Herrick *(9)*.
4. Blue Sepharose CL-6B (Amersham Pharmacia Biotech).
5. Phenylmethylsulfonyl fluoride (PMSF) (Sigma) prepared as a 0.2 *M* stock solution in isopropyl alcohol and stored at –20°C.

6. Sodium metabisulfite is prepared as a 1.0-*M* stock solution at pH 7.5 and stored at –20°C.
7. Leupeptin is prepared as a 1-mg/mL stock solution in 50 m*M* Tris-HCl, pH 7.5, and 2 m*M* EDTA, and stored at –20°C.
8. 1 *M* Sucrose, ultrapure.
9. 3 *M* Ammonium sulfate, ultrapure.
10. 1 *M* HEPES-OH, pH 8.0, stored at 4°C.
11. 1 *M* Tris-HCl, pH 7.5.
12. 0.5 *M* EDTA, pH 8.0.
13. 1 *M* Calcium chloride.
14. 2 *M* Potassium chloride.
15. 1 *M* Potassium phosphate, pH 7.6.
16. 5 *M* Sodium chloride.
17. 1 *M* Dithiothreitol (DTT). Store aliquots at –20°C.
18. 10% Triton X-100.
19. 20% Sodium cholate. Cholic acid is dissolved in hot ethanol, filtered through Norit A (J.T. Baker Chemical Co.) and recrystallized twice before titration to pH 7.4 with sodium hydroxide.
20. All potassium phosphate buffers are at pH 7.6 and contain 2 m*M* DTT, 2 m*M* EDTA, 1 m*M* PMSF, 10 m*M* sodium metabisulfite, leupeptin at 2 µg/mL, and 20% glycerol. Where indicated, buffers also contain 0.015% Triton X-100.
21. Collodion membranes (Schleicher & Schuell).
22. Polyallomer tubes (14 × 89 mm, Beckman).

2.2. Recombinant Drosophila *DNA Polymerase* γ *Purification*

1. Construction of recombinant transfer vectors and baculoviruses encoding the two subunits of *D. melanogaster* pol γ is described in **ref. 7**. Briefly, the complete coding sequences including the mitochondrial presequences were cloned into the baculovirus transfer vector pVL1392/1393 (PharMingen) and viruses were constructed using linearized wild-type baculovirus AcMNPV DNA (BaculoGold) (PharMingen). The recombinant baculoviruses used here, α, α$_{C-HIS}$ (α with a C-terminal hexahistidine tag inserted between Ser[1145] and the stop codon) and β may be obtained from the authors.
2. *Sf9 (Spodoptera frugiperda)* cells (PharMingen).
3. TC-100 insect cell culture medium and fetal bovine serum (Gibco-BRL).
4. Insect cell transfection buffer and Grace's medium (PharMingen).
5. Phosphate-buffered saline (PBS): 135 m*M* NaCl, 10 m*M* Na$_2$HPO$_4$, 2 m*M* KCl, 2 m*M* KH$_2$PO$_4$.
6. 1 *M* Imidazole.
7. 2-Mercaptoethanol.
8. 7 mL Dounce homogenizer.
9. Ni-NTA agarose (Qiagen).
10. Other materials are as in **Subheading 2.1.**

2.3. DNA Polymerase Assay

1. Pol γ, approx 0.1 unit/μL.
2. DNase-I activated calf thymus DNA, 2-mg/mL stock, prepared as described *(10)*.
3. 5X Polymerase buffer: 250 mM Tris-HCl, pH 8.5, 20 mM MgCl$_2$, 2 mg/mL BSA.
4. 1 M DTT (Sigma).
5. 2 M Potassium chloride.
6. Stock mix of deoxynucleoside triphosphates containing 1 mM each of dGTP, dATP, dCTP, and dTTP (Amersham Pharmacia Biotech).
7. [^{3}H]-dTTP (ICN Biochemicals).
8. 100% Trichloroacetic acid (TCA).
9. 0.1 mM sodium pyrophosphate (NaPP$_i$).
10. Wash solution: 1 M HCl, 0.1 mM sodium pyrophosphate (NaPP$_i$).
11. 95% Ethanol.
12. Glass fiber filter paper (Schleicher and Schuell)
13. 10 × 75-mm Disposable culture tubes (Fisher)

3. Methods

3.1. Purification of Native Drosophila melanogaster DNA Polymerase γ

3.1.1. Preparation of Partially Purified Mitochondria

1. *D. melanogaster* (Oregon R) embryos (average age, 9 h) are collected immediately before use and washed and dechorionated as described *(11)*. Processed embryos are suspended at a ratio of 4 mL/g, wet weight, in homogenization buffer containing 15 mM HEPES, pH 8.0, 5 mM KCl, 2 mM CaCl$_2$, 0.5 mM EDTA, 0.5 mM DTT, 0.28 M ultrapure sucrose, 1 mM PMSF, 10 mM sodium metabisulfite, 2 μg/mL leupeptin, and homogenized in 40-mL portions by three strokes of a stainless steel/Teflon homogenizer (a standard Dounce homogenizer may be substituted with care).
2. Filter the homogenate through a 75-μm Nitex screen into a centrifuge bottle.
3. Rehomogenize the retentate in the same buffer (1 mL/g, wet weight starting material), filter as in **step 2**, and combine with the original filtrate.
4. Centrifuge the combined filtrate at 1000g for 7 min at 3°C.
5. Aspirate the supernatant into a clean side-arm flask, transfer to a fresh centrifuge bottle, and repeat centrifugation as in **step 4**.
6. Repeat **step 5**.
7. Pellet mitochondria by centrifugation at 7400g for 10 min at 3°C. Aspirate supernatant and discard.
8. Resuspend the mitochondrial pellet at a ratio of 2 mL of homogenization buffer per gram of starting embryos, and centrifuge at 8000g for 15 min at 3°C.
9. Repeat **step 8**.

10. Resuspend the third pellet at a ratio of 0.5 mL/g, combine sample into one tube, and repeat centrifugation as in **step 8**.
11. Freeze the final mitochondrial pellet in liquid nitrogen and store at –80°C.

3.1.2. Preparation of the Mitochondrial Extract

Note: All operations are performed at 0–4°C.

1. Thaw frozen partially purified mitochondria from freshly harvested and dechorionated *Drosophila* embryos (200 g) on ice for at least 30 min.
2. Resuspend at a ratio of 0.5 mL/g of starting embryos in 25 mM HEPES, pH 8.0, 10% glycerol, 0.3 M NaCl, 1 mM EDTA, 1 mM DTT, 1 mM PMSF, 10 mM sodium metabisulfite, and 2 µg/mL leupeptin.
3. Add sodium cholate to a final concentration of 2% and incubate the suspension on ice for 30 min with gentle mixing by inversion at 5-min intervals.
4. Centrifuge the resulting extract at 96,000g for 30 min at 3°C.
5. Recover the supernatant fluid and add an equal volume of buffer containing 25 mM HEPES, pH 8.0, 2 mM EDTA, 80% glycerol.
6. Store the mitochondrial extract (fraction I) at –20°C.

3.1.3. Phosphocellulose Chromatography and Ammonium Sulfate Fractionation

1. Dilute fraction I to an ionic equivalent of 70–80 mM potassium phosphate and load onto a phosphocellulose column (6.5 × 4 cm; 6 mg of protein per packed milliliter of phosphocellulose) equilibrated with 80 mM potassium phosphate buffer at a flow rate of 100 mL/h.
2. Wash the column with 375 mL of 100 mM potassium phosphate buffer at a rate of 250 mL/h and then apply a 375-mL linear gradient from 150–350 mM potassium phosphate. The DNA polymerase activity elutes at approx 200 mM potassium phosphate.
3. Assay and pool active fractions (fraction II). Adjust with 80% sucrose to a final concentration of 10%.
4. Add 1.2 volumes of saturated ammonium sulfate, pH 7.5, to achieve 55% of saturation at 0°C, and incubate the suspension on ice for 2 h.
5. Collect the precipitate by centrifugation at 96,000g for 30 min at 3°C. Resuspend the pellet in 2.0 mL of 10 mM potassium phosphate buffer containing 45% glycerol, and store at –20°C (fraction IIb).

3.1.4. Single-Stranded DNA–Cellulose Chromatography

1. Dialyze fraction IIb against 10 mM potassium phosphate buffer in a collodion bag (molecular weight cutoff, 25,000) until an ionic equivalent of 60–70 mM KCl is reached.
2. Load dialyzed sample onto a single-stranded DNA–cellulose column (2.8 × 3.2 cm; 0.6 mg of protein per packed milliliter of DNA–cellulose)

equilibrated with 20 m*M* potassium phosphate buffer containing 10% sucrose at a flow rate of 10 mL/h.

3. At this step in the purification and for all subsequent steps, all laboratory ware (which is mostly polypropylene) is silanized as described by Maniatis et al. *(12)* to prevent enzyme loss by adsorption.

4. Wash the column with 40 mL of 20 m*M* potassium phosphate buffer containing 100 m*M* KCl at 30 mL/h followed by successive elution with buffers containing 250 m*M* KCl (80 mL at 40 mL/h), 600 m*M* KCl (60 mL at 30 mL/h), and 1 *M* KCl (40 mL at 60 mL/h). The enzyme elutes at approx 400 m*M* KCl.

5. Assay and pool the active fractions (fraction III).

3.1.5. Octyl–Sepharose Chromatography

1. To increase hydrophobic interactions, add solid ammonium sulfate (0.36 g/mL) to fraction III over 30 min and stir the suspension for an additional 20 min.

2. Load the suspension onto an Octyl–Sepharose column (0.6×2.5 cm) equilibrated with 20 m*M* potassium phosphate buffer at a flow rate of 1.4 mL/h.

3. Wash the column with 2.8 mL of equilibration buffer at 2.1 mL/h and then elute at the same flow rate with 2.5 mL of buffer containing 0.3% Triton X-100, 2.8 mL of buffer containing 1% Triton X-100, and 2.1 mL of buffer containing 2% Triton X-100. The enzyme eluted after application of the buffer containing 1% Triton X-100 (fraction IV). Although this step results in only a 1.2- to 1.8-fold increase in specific activity, it is required to both desalt and concentrate the enzyme fraction before further purification.

3.1.6. Blue Sepharose Chromatography

1. Apply fraction IV directly to a Blue Sepharose column (0.6×2.8 cm) equilibrated with 20 m*M* potassium phosphate buffer containing 0.015% Triton X-100.

2. Wash the column with 1.6 mL of buffer containing 50 m*M* KCl at 1.2 mL/h, followed by 2.4 mL each of buffers containing 100 m*M*, 600 m*M*, 1 *M*, and 2 *M* KCl at 2.4 mL/h. Enzyme activity elutes between 400 and 550 m*M* KCl (fraction V).

3.1.7. Glycerol Gradient Sedimentation

1. Layer fraction V onto two preformed 12–30% glycerol gradients containing 50 m*M* potassium phosphate, pH 7.6, 200 m*M* $(NH_4)_2SO_4$, 0.015% Triton X-100, 2 m*M* DTT, 2 m*M* EDTA, 1 m*M* PMSF, 10 m*M* sodium metabisulfite, 2 µg/mL leupeptin, prepared in polyallomer tubes for use in a Beckman SW 41 rotor.

2. Centrifuge at 170,000*g* for 60 h at 3°C, then fractionate by collecting eight-drop (200 µL) fractions.

3. Assay and pool active fractions, then add an equal volume of 25 m*M* HEPES, pH 8.0, 2 m*M* EDTA, 80% glycerol, 0.015% Triton X-100 and store the enzyme (fraction VI) at –20°C, –80°C, or under liquid nitrogen. Samples are analyzed by gel electrophoresis to evaluate purity and yield (*see* **Note 1**).

3.2. Purification of Recombinant Drosophila DNA Polymerase γ from the Cytoplasm of Sf9 Cells

Purification of recombinant pol γ holoenzyme is conducted as described in **Subheading 3.1.** with the following modifications.

All operations were all performed at 0–4°C.

3.2.1. Sf9 Cell Growth and Soluble Cytoplasmic Fraction Preparation

1. Grow *Sf9* cells (500 mL) in TC-100 insect cell culture medium containing 10% fetal bovine serum at 27°C to a cell density of 2×10^6, dilute to a cell density of 1×10^6 with TC-100 containing 10% fetal bovine serum, and then infect with recombinant α and β baculoviruses at a multiplicity of infection of 5. Harvest 48 h postinfection.
2. Pellet the cells at 400*g* for 5 min and wash with an equal volume of cold PBS, repeat the centrifugation and discard the supernatant.
3. Resuspend the cell pellet (approx 1×10^9 cells) in 10 mL of homogenization buffer (50 m*M* Tris-HCl, pH 7.5, 100 m*M* KCl, 280 m*M* ultrapure sucrose, 5 m*M* EDTA, 2 m*M* DTT, 10 m*M* sodium metabisulfite, 1 m*M* PMSF, and 2 μg/mL leupeptin).
4. Lyse cells by 20 strokes in a Dounce homogenizer.
5. Centrifuge the homogenate at 1000*g* for 7 min.
6. Resuspend the resulting pellet in 5 mL of homogenization buffer and rehomogenize and centrifuge as in **step 5**.
7. Centrifuge the combined supernatant fractions at 8000*g* for 15 min to pellet the mitochondria.
8. Remove the supernatant and centrifuge at 100,000*g* for 30 min to obtain the cytoplasmic soluble fraction (fraction I).

3.2.2. Phosphocellulose Chromatography and Ammonium Sulfate Fractionation

1. Adjust fraction I (70–90 mg protein) to an ionic equivalent of 80 m*M* potassium phosphate and load onto a phosphocellulose column (15 mL) equilibrated with 80 m*M* potassium phosphate buffer (80 m*M* potassium phosphate, pH 7.6, 20% glycerol, 2 m*M* EDTA, 2 m*M* DTT, 1 m*M* PMSF, 10 m*M* sodium metabisulfite, and 2 μg/mL leupeptin) at a flow rate of 12 mL/h.
2. Wash the column with 3 volumes of 100 m*M* potassium phosphate buffer at a flow rate of 30 mL/h.
3. Elute proteins with a 3-volume linear gradient from 150 to 350 m*M* potassium phosphate buffer at a flow rate of 30 mL/h.
4. Follow the gradient with a 2-volume high-salt wash of 600 m*M* potassium phosphate buffer.
5. Assay and pool active fractions (fraction II) and adjust to a final concentration of 10% sucrose.

6. Add solid ammonium sulfate (0.351 g/mL of fraction II) and incubate overnight on ice.
7. Collect the precipitate by centrifugation at 100,000g for 30 min at 3°C.
8. Resuspend the pellet in 2.0 mL of 10 mM potassium phosphate buffer containing 45% glycerol and store at –20°C (fraction IIb).

3.2.3. Single-Stranded DNA–Cellulose Chromatography

1. Dialyze fraction IIb against 10 mM potassium phosphate buffer in a collodion bag (molecular weight cutoff, 25,000 kDa) until an ionic equivalent of 85 mM KCl is reached.
2. Load fraction IIb onto a single-stranded DNA–cellulose column (1.8 mL) equilibrated with 20 mM potassium phosphate buffer at a flow rate of 1.3 mL/h.
3. Wash the column with 2 volumes of potassium phosphate buffer containing 100 mM KCl at 2.7 mL/h.
4. Follow by successive elution at 4 mL/h with potassium phosphate buffer containing 250 mM KCl (8 mL), 600 mM KCl (6 mL), and 1 M KCl (4 mL).
5. Assay and pool active fractions (fraction III).

3.2.4. Octyl–Sepharose Chromatography

1. To increase hydrophobic interactions, add solid ammonium sulfate to 0.2 g/mL.
2. Stir for 20 min on ice, then centrifuge the suspension for 10 min at 40,000g.
3. Load the supernatant onto an Octyl-Sepharose column (0.5 mL) equilibrated with 20 mM potassium phosphate buffer at a flow rate of 0.5 mL/h.
4. Wash the Octyl-Sepharose column with 4 volumes of equilibration buffer at 2 mL/h.
5. Elute successively at 2 mL/h with 4 volumes of 20 mM potassium phosphate buffer containing 9% glycerol and 0.3%, 1%, and 2% Triton X-100.
6. Assay and pool active fractions (fraction IV).

3.2.5. Glycerol Gradient Sedimentation

1. Layer fraction IV onto two preformed 12–30% glycerol gradients containing 50 mM potassium phosphate, pH 7.6, 200 mM $(NH_4)_2SO_4$, 0.015% Triton X-100, 2 mM DTT, 2 mM EDTA, 1 mM PMSF, 10 mM sodium metabisulfite, 2 μg/mL leupeptin, prepared in polyallomer tubes for use in a Beckman SW 41 rotor.
2. Centrifuge at 140,000g for 60 h at 3°C, then fractionate by collecting eight drop (200 μL) fractions.
3. Assay and pool active fractions, then add an equal volume of 25 mM HEPES, pH 8.0, 2 mM EDTA, 80% glycerol, 0.015% Triton X-100 and store the enzyme (fraction V) at –20°C, –80°C, or under liquid nitrogen.

3.3. Purification of His-Tagged Recombinant Drosophila *DNA Polymerase* γ *from the Cytoplasm of* Sf9 *Cells*

3.3.1. Ni-NTA Agarose Affinity Purification of Recombinant Pol γ

1. Prepare fraction I from 500 mL of cells coinfected with $\alpha_{C\text{-HIS}}$ and β baculoviruses, chromatograph on phosphocellulose, and precipitate with ammonium sulfate as described in **Subheading 3.2.**
2. Resuspend the pellet in 2.0 mL of 10 mM potassium phosphate buffer containing 45% glycerol and store at –20°C (fraction IIb).
3. Dialyze fraction IIb (3–5 mg protein) against 10 mM potassium phosphate buffer in a collodion bag (M_r cutoff, 25,000 kDa) until an ionic equivalent of 100 mM KCl is reached.
4. Mix dialyzed fraction IIb with 500 μL of precharged Ni-NTA agarose (Qiagen) equilibrated in a buffer containing 20 mM Tris-HCl, pH 7.5, 500 mM KCl, 8% glycerol, 5 mM of β-mercaptoethanol, 1 mM PMSF, 10 mM sodium metabisulfite, 2 μg/mL leupeptin, and 5 mM imidazole.
5. Incubate the suspension for 10 h on ice with gentle shaking.
6. Allow the beads to settle and then wash twice with 1 mL of equilibration buffer for 30 min with gentle shaking.
7. Washed beads are packed into a column (0.5 mL), and protein retained on the beads is eluted successively at 2.5 mL/h with equilibration buffer containing 25 mM imidazole (1.0 mL), 250 mM imidazole (1.0 mL) and 500 mM imidazole (1.0 mL).
8. Assay and pool active fractions (fraction III, approx 0.9 mL).
9. Load onto two 12–30% glycerol gradients as described in **Subheading 3.2.**
10. Assay and pool active fractions, then add an equal volume of 25 mM HEPES, pH 8.0, 2 mM EDTA, 80% glycerol, 0.015% Triton X-100, and store the enzyme (fraction IV) at –20°C, –80°C, or under liquid nitrogen.

3.4. DNA Polymerase Assay

This section describes an assay of DNA synthesis that utilizes a multiply-primed double-stranded DNA substrate and pol γ fractions. DNA polymerase activity can be measured at different salt concentrations to discriminate nuclear (low-salt stimulated) and mitochondrial (high-salt stimulated) DNA polymerase activities. The assay measures the incorporation of [³H]-labeled dTMP into the DNA substrate, such that 1 unit of activity is that amount that catalyzes the incorporation of 1 nmol of deoxyribonucleoside triphosphate into acid insoluble material in 60 min at 30°C.

1. Adjust the water bath to 30°C.
2. Dry down the radioactive substrate by lyophilizing to less than half the original volume.

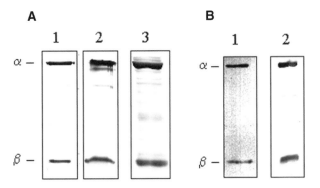

Fig. 1. SDS-PAGE gel electrophoresis of native and recombinant *Drosophila* pol γ. Near-homogeneous fractions of native and recombinant *Drosophila* pol γ were denatured and electrophoresed in 10% SDS–polyacrylamide gels and the proteins were stained with silver (**A**) or detected by immunoblotting (**B**). (**A**) Lane 1: native *Drosophila* pol γ (fraction VI, 10 units); lane 2: recombinant *Drosophila* pol γ (fraction V, 10 units); lane 3: recombinant C-terminal histidine-tagged *Drosophila* pol γ (fraction IV, 8 units). (**B**) Lane 1: recombinant *Drosophila* pol γ (fraction V, 1.5 units); lane 2: recombinant C-terminal histidine-tagged *Drosophila* pol γ (fraction IV, 2 units).

3. Prepare a master reaction mix using the stock solutions in **Subheading 2.3.** in a microcentrifuge tube on ice such that each reaction (0.05 mL) contains 1X polymerase buffer, 10 mM DTT, 200 mM KCl, 30 μM each of dGTP, dATP, dCTP, and [³H]dTTP (1000 cpm/pmol), and 250 μg/mL DNase I-activated calf thymus DNA. Vortex and spin briefly in the microcentrifuge.
4. Dispense the mix, 49 μL, to prechilled and numbered 10 × 75-mm tubes on ice.
5. Add the enzyme, approx 0.1 unit, to each tube avoiding bubbles and mix gently by flicking the tube three times.
6. Incubate the tubes for 30 min at 30°C, then transfer to ice.
7. Stop the reactions with 1 mL of 10% trichloroacetic acid, 0.1 mM NaPP$_i$, mix, and leave on ice ≥ 5 min.
8. Filter samples through glass fiber filters. Wash the tube twice with 1 M HCl, 0.1 mM NaPP$_i$, then wash the filtration funnel three times with 1 M HCl, 0.1 mM NaPP$_i$ and once with 95% ethanol.
9. Dry the filters under a heat lamp for approx 5 min, then count in scintillation fluid in a liquid scintillation counter.
10. Spot 1–2 μL of mix directly onto filters, dry, and count in scintillation fluid without filtration, to calculate the specific radioactivity of the mix.

4. Note

1. The purity and yield of the final preparations are determined by DNA polymerase assay, and sodium dodecyl sulfate–polyacrylamide gel electrophoresis (SDS-PAGE) followed by silver staining. **Figure 1** shows typical preparations. The

native and recombinant forms generally have a specific activity of 20,000–60,000 units/mg, depending on the preparation of DNase I-activated calf thymus DNA used for enzyme assay *(4,7)*.

References

1. Knopf, K. W., Yamada, M., and Weissbach, A. (1976) HeLa cell DNA polymerase gamma: further purification and properties of the enzyme. *Biochemistry* **15,** 4540–4548.
2. Bertazzoni, U., Scovassi, A. I., and Brun, G. M. (1977) Chick-embryo DNA polymerase gamma. Identity of polymerases purified from nuclei and mitochondria. *Eur. J. Biochem.*, **81,** 237–248.
3. Friedberg, E. C., Feaver, W. J., and Gerlach, V. L. (2000) The many faces of DNA polymerases: strategies for mutagenesis and for mutational avoidance. *Proc. Natl. Acad. Sci. USA* **97,** 5681–5683.
4. Wernette, C. M. and Kaguni, L. S. (1986) A mitochondrial DNA polymerase from embryos of *Drosophila melanogaster*. Purification, subunit structure, and partial characterization. *J. Biol. Chem.*, **261,** 14,764–14,770.
5. Ropp, P. A. and Copeland, W. C. (1996) Cloning and characterization of the human mitochondrial DNA polymerase, DNA polymerase γ. *Genomics* **36,** 449–458.
6. Ye, F., Carrodeguas, J. A., and Bogenhagen, D. F. (1996) The gamma subfamily of DNA polymerases: cloning of a developmentally regulated cDNA encoding *Xenopus laevis* mitochondrial DNA polymerase gamma. *Nucleic Acids Res.*, **24,** 1481–1488.
7. Wang, Y. and Kaguni, L. S. (1999) Baculovirus expression reconstitutes *Drosophila* mitochondrial DNA polymerase. *J. Biol. Chem.* **274,** 28,972–28,977.
8. Farr, C. L., Wang, Y., and Kaguni, L. S. (1999) Functional interactions of mitochondrial DNA polymerase and single- stranded DNA-binding protein. Template-primer DNA binding and initiation and elongation of DNA strand synthesis. *J. Biol. Chem.*, **274,** 14,779–14,785.
9. Alberts, B. M. and Herrick, G. (1971) DNA–cellulose chromatography. *Methods Enzymol.* **21,** 198–217.
10. Fansler, B. S. and Loeb, L. A. (1974) Sea urchin nuclear DNA polymerase. *Methods Enzymol.* **29,** 53–70.
11. Brakel, C. L. and Blumenthal, A. B. (1977) Multiple forms of *Drosophila* embryo DNA polymerase: evidence for proteolytic conversion. *Biochemistry* **16,** 3137–3143.
12. Maniatis, T., Fritsch, E. F., and Sambrook, J. (eds.) (1982) *Molecular Cloning: A Laboratory Manual.* Cold Spring Harbor Laboratory Press, Cold Spring Harbor, NY.

19

Purification Strategies for *Drosophila* mtDNA Replication Proteins in Native and Recombinant Form

Mitochondrial Single-Stranded DNA-Binding Protein

Carol L. Farr and Laurie S. Kaguni

1. Introduction

Single-stranded DNA-binding proteins (SSBs) serve critical roles in DNA replication, repair, and recombination *(1)*. Mitochondrial SSBs (mtSSBs) share similar physical and biochemical properties with *Escherichia coli* SSB *(2–7)*, with which they exhibit a high degree of amino acid sequence conservation *(7–9)*. Considering the roles served by *E. coli* SSB in bacterial replication in helix destabilization *(10)* and in enhancing DNA polymerase processivity *(11,12)* and fidelity *(13,14)*, we purified *Drosophila* mtSSB and studied its effects on in vitro DNA synthesis by pol γ, in an assay that mimics lagging DNA strand synthesis in mitochondrial replication *(6,15)*. Our biochemical data are consistent with an important role for mtSSB in initiation and elongation of DNA strands in mitochondrial DNA (mtDNA) replication, that has been documented genetically by the fact that a null mutation in the gene for the yeast homolog *(RIM1)* results in complete loss of mtDNA in vivo *(4)*. More recently, we have demonstrated that an insertion in the third intron of the *Drosophila* gene *(lopo)* results in developmental lethality, concomitant with the loss of mtDNA and respiratory capacity *(16)*.

Like *E. coli* SSB, mtSSBs are homotetramers of 13 to 16-kDa polypeptides. Not surprising given the conserved physical and biochemical properties of these *E. coli*-like SSBs, mtSSBs from yeast to man can be purified and characterized using similar schemes. In this chapter, we present purification schemes for

From: *Methods in Molecular Biology, vol. 197: Mitochondrial DNA: Methods and Protocols*
Edited by: W. C. Copeland © Humana Press Inc., Totowa, NJ

both native and recombinant forms of *Drosophila* mtSSB and a DNA synthesis stimulation assay that can be employed for mtSSB from any source.

2. Materials

2.1. Mitochondrial Single-Stranded DNA-Binding Protein Purification from Drosophila melanogaster Embryos

1. 1 *M* HEPES, pH 8.0, stored at 4°C.
2. 0.5 *M* Ethylenediaminetetraacetic acid (EDTA), pH 8.0.
3. 5 *M* Sodium chloride (NaCl).
4. 2 *M* Tris-HCl, pH 6.8, pH 7.5, and pH 8.8.
5. 1 *M* potassium phosphate, pH 7.6.
6. Ethylene glycol (Aldrich).
7. 0.2 *M* Phenylmethylsulfonyl fluoride (PMSF) in isopropanol. Store aliquots at –20°C.
8. 1 *M* sodium metabisulfite, prepared as a 1.0 *M* stock solution at pH 7.5 and stored at –20°C.
9. Leupeptin is prepared as a 1 mg/mL stock solution in 50 m*M* Tris-HCl, pH 7.5, 2 m*M* EDTA, and stored at –20°C.
10. 1 *M* Dithiothreitol (DTT). Store aliquots at –20°C.
11. 3 *M* Ammonium sulfate, ultrapure [(NH$_4$)$_2$SO$_4$].
12. 30% Polyacrylamide (29 : 1; acrylamide : bisacrylamide). Store at 4°C.
13. 100% Trichloroacetic acid (TCA). Store at 0–4°C.
14. 10% Sodium dodecyl sulfate (SDS).
15. 4X Resolving gel buffer: 1.5 *M* Tris-HCl, pH 8.8, 0.4% SDS.
16. 4X stacking gel buffer: 0.5 *M* Tris-HCl, pH 6.8, 0.4% SDS.
17. 5X SDS-PAGE running buffer: 0.125 *M* Tris base, 0.95 *M* glycine, 0.5% SDS.
18. 5X SDS-PAGE loading buffer: 50% glycerol, 2 *M* Tris base, 0.25 *M* DTT, 5% SDS, 0.1% bromophenol blue. Aliquots are stored at –20°C.
19. 10% Ammonium persulfate (APS).
20. 20% Sodium cholate, pH 7.5. Store at –20°C. Cholic acid (Sigma) was dissolved in hot ethanol, filtered through Norit A (J.T. Baker Chemical Co.), and recrystallized twice before titration to pH 7.4 with sodium hydroxide.
21. Mitochondrial extraction buffer (EBmitos): 25 m*M* HEPES, pH 8.0, 2 m*M* EDTA, 0.3 *M* NaCl.
22. Centricon-30 spin concentrators (Amicon/Millipore).
23. Single-stranded DNA cellulose prepared as described by Alberts and Herrick (*17*).
24. Polyallomer tubes (14 × 89 mm, Beckman).

2.2. Recombinant Drosophila mtSSB Purification

1. 3 *M* Sodium thiocyanate (NaSCN).
2. 1 *M* Spermidine trihydrochloride (SpCl$_3$, Sigma). Store at –20°C.
3. 10 mg/mL Lysozyme in 0.25 *M* NaCl, 5 m*M* DTT, 20 m*M* SpCl$_3$, 2 m*M* EDTA. Prepare immediately before use.

4. 5X Lysis solution: 1.25 *M* NaCl, 25 m*M* DTT, 100 m*M* SpCl$_3$, 10 m*M* EDTA, 1.5 mg/mL lysozyme.
5. 5% Tween-20.
6. Bacterial media (L broth): 1% tryptone (Difco), 0.5% yeast extract (Difco), 0.5% NaCl, pH 7.5.
7. 100 mg/mL Ampicillin. Filter-sterilize with a 0.2-µm syringe filter, aliquot, and store at –20°C.
8. 100 m*M* Isopropyl-β-D-thiogalactopyranoside (IPTG). Aliquot and store at –20°C.
9. Blue Sepharose CL-6B (Amersham Pharmacia Biotech).
10. *Escherichia coli* (*E. coli*) BL21 (λDE3) pLysS (Stratagene).
11. pET-11a encoding *Drosophila* mtSSB without the mitochondrial presequence, available from this lab.
12. Other materials as in **Subheading 2.1.**

2.3. DNA Polymerase Stimulation Assay

1. *Drosophila* pol γ, native or recombinant, at approx 0.1 unit/µL (*see* Chapter 18).
2. *Drosophila* mtSSB, native or recombinant, used at 0.4 mg/mL.
3. Singly-primed recombinant M13 DNA prepared by standard laboratory methods.
4. 5X Polymerase buffer: 250 m*M* Tris-HCl, pH 8.5, 20 m*M* MgCl$_2$, 2 mg/mL bovine serum albumin (BSA, Sigma).
5. 1 *M* DTT. Store aliquots at –20°C.
6. 2 *M* Potassium chloride (KCl).
7. Stock mix of deoxynucleoside triphosphates containing 1 m*M* each dGTP, dATP, dCTP, and dTTP (Amersham Pharmacia Biotech).
8. [^{3}H]-dTTP (ICN Biomedicals, Inc.).
9. 100% TCA.
10. Glass fiber filter paper (Schleicher and Schuell).
11. 10 × 75-mm Disposable culture tubes (Fisher).
12. 1 *M* HCl, 0.1 m*M* sodium pyrophosphate (NaPP$_i$,).

3. Methods
3.1. Purification of Mitochondrial Single-Stranded DNA-Binding Protein from Drosophila melanogaster Embryos
3.1.1. Mitochondrial Extraction

Perform all steps at 0–4°C.

1. Purification begins with the extraction of partially purified mitochondria from *Drosophila melanogaster* embryos (*see* Chapter 18, **Subheading 3.1.1.**).
2. Thaw crude mitochondrial pellets derived from 150 g of embryos on ice for at least 30 min.

3. Resuspend the pellets using 0.5 mL/g embryos EB-mitos containing 1 m*M* PMSF, 10 m*M* sodium metabisulfite, 2 µg/mL leupeptin, and 2 m*M* DTT. First, resuspend pellet completely in one-half of the total volume, and transfer to a plastic beaker on ice. Wash tubes with remaining one-half of the volume and combine.

4. Distribute the mitochondrial fraction to prechilled 45 Ti tubes (Beckman) and balance. Lyse mitochondria by adding sodium cholate to a final concentration of 2% in each tube. Mix gently and thoroughly by inversion.

5. Incubate 30 min on ice, mixing at 5-min intervals by inversion.

6. Spin lysed mitochondria in 45 Ti rotor, 140,000*g*, for 30 min at 4°C.

7. Collect soluble fraction by carefully pipetting the supernatant, avoiding loose pellet material and the upper lipid layer. Note recovered volume.

8. Filter recovered supernatant through eight layers of cheese cloth. Filter again through 0.45-µm nitrocellulose.

3.1.2. Single-Stranded DNA Cellulose Chromatography

All steps involving chromatography should be performed at 0–4°C in a timely manner to avoid degradation of the target protein. All chromatography buffers contain 1 m*M* PMSF, 10 m*M* sodium metabisulfite, 2 µg/mL leupeptin, and 2 m*M* DTT. The ionic strength of buffers is determined using a Radiometer conductivity meter.

1. Adjust the soluble mitochondrial extract (approx 350 mg protein) to a final concentration of 0.5 *M* NaCl and 8 m*M* EDTA.

2. Load adjusted mitochondrial extract onto 5 mL of single-stranded DNA (ssDNA) cellulose equilibrated with 10 column volumes (CV) of 30 m*M* Tris-HCl, pH 7.5, 10% glycerol, 500 m*M* NaCl, 2 m*M* EDTA at a flow rate of 1 CV per hour.

3. Wash the column at a flow rate of 2 CV per hour with 4 CV of 30 m*M* Tris-HCl, pH 7.5, 10% glycerol, 500 m*M* NaCl, 2 m*M* EDTA, collecting 1-CV fractions.

4. Elute the column in two steps at a flow rate of 2 CV per hour, collecting 1/4 CV fractions: 4 CV of 30 m*M* Tris-HCl, pH 7.5, 10% glycerol, 750 m*M* NaCl, 2 m*M* EDTA, and then 3 CV of 30 m*M* Tris-HCl, pH 7.5, 10% glycerol, 1.5 *M* NaCl, 2 m*M* EDTA, 50% ethylene glycol. mtSSB elutes in the 1.5 *M* NaCl, 50% ethylene glycol step.

3.1.3. SDS-PAGE Analysis of Column Fractions

Depending on the abundance of the target protein, column fractions may require concentration prior to sodium dodecyl sulfate–polyacrylamide gel electrophoresis (SDS-PAGE). Most mitochondrial proteins following the protocol in **Subheading 3.1.2.** will require concentration. We use TCA precipitation followed by pellet resuspension in 1X SDS loading buffer.

1. Pour 2-3, 17% SDS–polyacrylamide resolving gels (8 cm × 10 cm × 1 mm) with 4% stacking gels.

2. Precipitate 70-µL aliquots of column fractions by adding TCA to a final concentration of 12%. Incubate on ice at least 1.5 h. Spin 30 min in microcentrifuge,

then carefully remove supernatant. Resuspend precipitates in 1X SDS-PAGE loading buffer.

3. Boil samples 2 min at 100°C, then load immediately onto SDS gels.
4. Electrophorese the samples through the stacking gel at 10 mA per gel for 30 min or until samples reach the resolving gel, then increase current to 20 mA per gel for 1.5 h or until dye reaches the bottom of the gel.
5. Stain gels with silver nitrate to detect mtSSB and pool mtSSB-containing peak fractions.

3.1.4. Glycerol Gradient Sedimentation

Although the mtSSB fractions following ssDNA–cellulose chromatography are highly purified, glycerol gradient sedimentation is required to obtain the tetrameric form and to eliminate any carryover ssDNA. The elution step containing 1.5 M NaCl and 50% ethylene glycol partially disrupts the native homotetramer and dialysis does not reconstitute it completely.

All buffers contain 1 mM PMSF, 10 mM sodium metabisulfite, 2 µg/mL leupeptin, and 2 mM DTT. Steps are performed at 0–4°C.

1. Dialyze ssDNA cellulose pool against 300 volumes of buffer containing 8% glycerol, 50 mM Tris-HCl, pH 7.5, 200 mM $(NH_4)_2SO_4$, 2 mM EDTA using two changes of 1 h each.
2. Prepare three (10 mL) 12–30% glycerol gradients in 50 mM Tris-HCl, pH 7.5, 200 mM $(NH_4)_2SO_4$, 2 mM EDTA and chill on ice at least 1 h.
3. Layer 1–1.2 mL of dialyzed mtSSB fraction onto each gradient and spin in a SW41 rotor at 170,000g for 62 h at 4°C.
4. Fractionate gradients by collecting 225-µL aliquots. Analyze 30 µL per fraction by TCA precipitation followed by SDS-PAGE on 17% gels and staining with silver nitrate.
5. Pool peak fractions of tetrameric mtSSB avoiding fractions that sediment slowly, trailing toward the top of the gradient, that are likely to contain dimers and monomers. Tetrameric mtSSB typically sediments in fractions 25–31. Concentrate gradient pool using a Centricon-30 spin concentrator (pretreated with 5% Tween-20 to reduce nonspecific binding).
6. Aliquot concentrated mtSSB, freeze in liquid nitrogen, and store at –80°C. Alternatively, mtSSB may be stabilized by the addition of an equal volume of stabilization buffer (20 mM KPO$_4$, pH 7.6, 80% glycerol, 2 mM EDTA) and stored at –20°C.

3.2. Purification of Recombinant Drosophila mtSSB

3.2.1. Bacterial Cell Growth and Protein Overproduction

1. Inoculate 1 L of L broth containing 100 µg/mL ampicillin with *E. coli* BL21 (λDE3) pLysS containing pET11a–mtSSB (complete cDNA without the mito-chondrial presequence) at $A_{595} = 0.06$ and grow with aeration at 37°C.

2. At $A_{595} = 0.6$:

    a. Remove a 1.0-mL aliquot of uninduced cells to a microcentrifuge tube, pellet cells, aspirate supernatant, and resuspend cells in 200 μL of 1X SDS loading buffer. Use a 10-μL aliquot as control for SDS-PAGE.

    b. Induce target protein expression by adding IPTG to 0.3 mM final concentration and continue incubation at 37°C with aeration for 2 h.

3. Harvest cells:

    a. Save a 1.0-mL aliquot of induced cells as in **step 2a**. Use a 5-μL aliquot in SDS-PAGE.

    b. Harvest remaining cells in Sorvall GSA bottles by centrifugation at 3600g for 5 min at 4°C. Decant supernatant.

4. Resuspend cells in 1/10 volume of original culture in 50 mM Tris-HCl, pH 7.5, 10% sucrose. First, resuspend pellets in one-half of the total resuspension volume, transferring to a chilled plastic beaker. Wash GSA bottles with remaining one-half of volume and combine.

5. Aliquot washed cells into Sorvall SS-34 tubes, 200-mL cell equivalent per tube, and centrifuge cells at 3000g for 5 min at 4°C. Aspirate supernatant and freeze-dry cell pellets in liquid nitrogen. Store at –80°C.

3.2.2. Cell Lysis and Soluble Fraction Preparation

The following protocol uses 400-mL induced cell pellets ($A_{595} = 1$–1.2) as the starting material with an overall expression of 15 μg mtSSB/mL of induced cells. All buffers contain 1 mM PMSF, 10 mM sodium metabisulfite, 2 μg/mL leupeptin, and 2 mM DTT. The ionic strength of buffers is determined using a Radiometer conductivity meter.

All steps are performed at 0–4°C.

1. Thaw cell pellets (2 × 200 mL) on ice at least 30 min.

2. Resuspend cell pellets completely in 1/25 volume 50 mM Tris-HCl, pH 7.5, 10% sucrose. Use one-half of the total volume to resuspend the cells and the remaining one-half divided to wash the tubes, combining all samples.

3. Add 5X lysis solution to 1X final concentration. Incubate 30 min on ice, mixing every 5 min by inversion.

4. Freeze cell lysis suspension in liquid nitrogen. Thaw partially in ice water, then transfer to ice until thawed completely. Centrifuge in Sorvall SS-34 rotor at 17,500g for 30 min at 4°C.

5. Collect soluble fraction (supernatant) by pipetting into a fresh tube. Adjust the sample to 100 mM NaCl by dilution using 30 mM Tris-HCl, pH 7.5, 10% glycerol, 2 mM EDTA.

3.2.3. Blue Sepharose Chromatography and Glycerol Gradient Sedimentation

All chromatography steps are performed at 0–4°C. All column and gradient buffers contain 1 mM PMSF, 10 mM sodium metabisulfite, 2 μg/mL leupeptin, and 2 m*M* DTT. The following protocol uses a soluble bacterial fraction from 400 mL of induced cells loaded onto 10 mL of equilibrated Blue Sepharose.

1. Load salt-adjusted soluble fraction (approx 15–20 mg protein) at a flow rate of 1 CV per hour onto a 10-mL column of Blue Sepharose equilibrated with 10 column volumes (CV) 30 m*M* Tris-HCl, pH 7.5, 10% glycerol, 100 m*M* NaCl, 2 m*M* EDTA.
2. Wash column with 4 CV of 30 m*M* Tris-HCl, pH 7.5, 10% glycerol, 800 m*M* NaCl, 2 m*M* EDTA at a flow rate of 2 CV per hour, collecting 1 CV fractions.
3. Elute column with three steps of increasing NaSCN stringency at 2 CV per hour:
   a. 4 CV of 30 m*M* Tris-HCl, pH 7.5, 10% glycerol, 0.5 *M* NaSCN, 2 m*M* EDTA, collecting 1/2 CV fractions.
   b. 4 CV of 30 m*M* Tris-HCl, pH 7.5, 10% glycerol, 1 *M* NaSCN, 2 m*M* EDTA, collecting 1/3 CV fractions.
   c. 4 CV of 30 m*M* Tris-HCl, pH 7.5, 10% glycerol, 1.5 *M* NaSCN, collecting 1/4 CV fractions.
      mtSSB elutes in the 1. 5 *M* NaSCN step.
4. Analyze column fractions, 5 µL, by SDS-PAGE on three 17% gels as in **Subheading 3.1.3.**
5. Pool fractions containing mtSSB and spin concentrate sample in a Centricon-30 concentrator (pretreated with 5% Tween-20) to desalt. Exchange elution buffer for 8% glycerol, 50 m*M* phosphate buffer, pH 7.6, 200 m*M* $(NH_4)_2SO_4$, 2 m*M* EDTA.
6. Prepare three (10-mL) 12–30% glycerol gradients in 50 m*M* Tris-HCl, pH 7.5, 200 m*M* $(NH_4)_2SO_4$, 2 m*M* EDTA and chill on ice at least 1 h.
7. Layer 1–1.2 mL of dialyzed mtSSB fraction onto each gradient and spin in SW41 rotor at 170,000*g* for 62 h at 4°C.
8. Fractionate gradients by collecting 225-µL aliquots. Analyze 30 µL per fraction by TCA precipitation followed by SDS-PAGE on 17% gels and staining with silver nitrate.
9. Pool peak fractions of tetrameric mtSSB avoiding fractions that sediment slowly, trailing toward the top of the gradient, that are likely to contain dimers and monomers. Tetrameric mtSSB typically sediments in fractions 25–31. Concentrate gradient pool using a Centricon-30 spin concentrator (pretreated with 5% Tween-20).
10. Aliquot concentrated mtSSB, freeze in liquid nitrogen, and store at –80°C. Alternatively, mtSSB may be stabilized by the addition of an equal volume of stabilization buffer (20 m*M* phosphate buffer, pH 7.6, 80% glycerol, 2 m*M* EDTA) and stored at –20°C.

3.3. DNA Polymerase Stimulation Assay

This assay measures the stimulation of DNA synthesis by *Drosophila* pol γ on an oligonucleotide-primed single-stranded DNA substrate by *Drosophila* mtSSB.

A   **B**

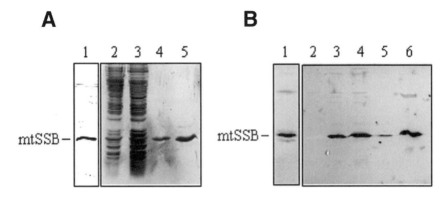

Fig. 1. Bacterial overexpression and purification of *Drosophila* mtSSB. Protein fractions were denatured and electrophoresed in a 17% SDS–polyacrylamide gel. Proteins were detected by silver staining (**A**) or by immunoblotting using the goat anti-rabbit IgG–alkaline phosphatase method with rabbit antiserum against *D. melanogaster* (*Dm*) mtSSB at a 1:1000 dilution (**B**). (**A**) Lane 1: *Dm* mtSSB (0.3 μg); lane 2: induced BL21 λDE3 cells; lane 3: soluble extract; lanes 4 and 5: Blue Sepharose pool (0.25 and 0.5 μg, respectively). (**B**) lane 1: *Dm* mtSSB (0.2 μg); lane 2: uninduced BL21 λDE3 cells; lane 3: induced cells; lane 4: soluble extract; lanes 5 and 6: Blue Sepharose pool (0.05 and 0.3 μg, respectively).

1. Each reaction (0.05 mL) contains 50 mM Tris-HCl, pH 8.5, 4 mM MgCl$_2$, 10 mM DTT, 30 mM KCl, 400 μg/mL bovine serum albumin, 20 μM each of dGTP, dATP, dCTP, and [^{3}H]-dTTP (1000 cpm/pmol), 10 μM (as nt) singly-primed recombinant M13 DNA, 0.8 μg of mtSSB, and 0.1 unit pol γ (sixfold excess of primer ends over pol γ molecules). Prepare the reaction mix in a microcentrifuge tube on ice. Add the radioactivity last. Vortex and centrifuge briefly in the microcentrifuge.
2. Dispense the mix, 47 μL, to prechilled and labeled 10 × 75-mm tubes on ice.
3. Add the mtSSB, 2 μL, to each tube, and then add pol γ, 1 μL, avoiding bubbles. Mix gently by flicking the tube three times.
4. Incubate the reaction for 30 min at 30°C, then transfer to ice.
5. Stop the reactions with 1 mL of 10% TCA, 0.1 mM NaPP$_i$, mix, and leave on ice at least 5 min.
6. Filter samples through glass fiber filters. Wash the tube twice with 1 M HCl, 0.1 mM NaPP$_i$, then wash the filtration funnel three times with 1 M HCl, 0.1 mM NaPP$_i$ and once with 95% ethanol.
7. Dry the filters under a heat lamp approx 5 min, then count in scintillation fluid in a liquid scintillation counter.
8. Spot 1–2 μL of mix directly onto filters, dry, and count in scintillation fluid without filtration, to calculate the specific radioactivity of the mix.

4. Note

1. The purity and yield of the final preparations are determined by SDS-PAGE followed by silver staining. **Figure 1** shows a typical purification of the recombinant form of mtSSB. We do not calculate a specific activity because mtSSB is used in excess in the DNA polymerase stimulation assay *(15)*. However, if one desires such a value, it can be determined as that amount required to saturate the assay.

References

1. Kornberg, A. and Baker, T. A. (1992) *DNA Replication*, 2nd ed., W. H. Freeman, New York.
2. Pavco, P. A. and Van Tuyle, G. C. (1985) Purification and general properties of the DNA-binding protein (P16) from rat liver mitochondria. *J. Cell Biol.* **100,** 258–264.
3. Mignotte, B., Barat, M., and Mounolou, J. C. (1985) Characterization of a mitochondrial protein binding to single-stranded DNA. *Nucleic Acids Res.* **13,** 1703–1716.
4. Van Dyck, E., Foury, F., Stillman, B., and Brill, S. J. (1992) A single-stranded DNA binding protein required for mitochondrial DNA replication in *S. cerevisiae* is homologous to *E. coli* SSB. *EMBO J.* **11,** 3421–3430.
5. Curth, U., Urbanke, C., Greipel, J., Gerberding, H., Tiranti, V., and Zeviani, M. (1994) Single-stranded DNA-binding proteins from human mitochondria and *Escherichia coli* have analogous physicochemical properties. *Eur. J. Biochem.* **221,** 435–443.
6. Thommes, P., Farr, C. L., Marton, R. F., Kaguni, L. S., and Cotterill, S. (1995) Mitochondrial single-stranded DNA-binding protein from *Drosophila* embryos. Physical and biochemical characterization. *J. Biol. Chem.* **270,** 21,137–21,143.
7. Ghrir, R., Lecaer, J. P., Dufresne, C., and Gueride, M. (1991) Primary structure of the two variants of *Xenopus laevis* mtSSB, a mitochondrial DNA binding protein. *Arch. Biochem. Biophys.* **291,** 395–400.
8. Tiranti, V., Rocchi, M., DiDonato, S., and Zeviani, M. (1993) Cloning of human and rat cDNAs encoding the mitochondrial single-stranded DNA-binding protein (SSB). *Gene* **126,** 219–225.
9. Stroumbakis, N. D., Li, Z., and Tolias, P. P. (1994) RNA- and single-stranded DNA-binding (SSB) proteins expressed during *Drosophila melanogaster* oogenesis: a homolog of bacterial and eukaryotic mitochondrial SSBs. *Gene* **143,** 171–177.
10. Baker, T. A., Sekimizu, K., Funnell, B. E., and Kornberg, A. (1986) Extensive unwinding of the plasmid template during staged enzymatic initiation of DNA replication from the origin of the *Escherichia coli* chromosome. *Cell* **45,** 53–64.
11. Sherman, L. A. and Gefter, M. L. (1976) Studies on the mechanism of enzymatic DNA elongation by *Escherichia coli* DNA polymerase II. *J. Mol. Biol.* **103,** 61–76.
12. LaDuca, R. J., Fay, P. J., Chuang, C. McHenry, C. S., and Bambara, R. A. (1983) Site-specific pausing of deoxyribonucleic acid synthesis catalyzed by four forms of *Echerichia coli* DNA polymerase III. *Biochemistry* **22,** 5177–5188.

13. Kunkel, T. A., Meyer, R. R., and Loeb, L. A. (1979) Single-strand binding protein enhances fidelity of DNA synthesis *in vitro*. *Proc. Natl. Acad. Sci. USA* **76,** 6331–6335.

14. Kunkel, T. A., Schaaper, R. M., and Loeb, L. A. (1983) Depurination-induced infidelity of deoxyribonucleic acid synthesis with purified deoxyribonucleic acid replication proteins *in vitro*. *Biochemistry* **22,** 2378–2384.

15. Farr, C. L., Wang, Y., and Kaguni, L. S. (1999) Functional interactions of mitochondrial DNA polymerase and single-stranded DNA-binding protein: template–primer DNA binding and initiation and elongation of DNA strand synthesis. *J. Biol. Chem.* **274,** 14,779–14,785.

16. Maier, D., Farr, C. L., Poeck, B., Alahari, A., Vogel, M., Fischer, S., et al. (2001) Mitochondrial single-stranded DNA-binding protein is required for mitochondrial DNA replication and development in *Drosophila* melanogaster. *Mol. Biol. Cell* **12,** 821–830.

17. Alberts, B. M. and Herrick, G. (1971) DNA–cellulose chromatography. *Methods Enzymol.* **21,** 198–217.

20

Expression, Purification, and In Vitro Assays of Mitochondrial Single-Stranded DNA-Binding Protein

Kang Li and R. Sanders Williams

1. Introduction

The mitochondrial single-stranded DNA-binding protein (mtSSB) is encoded by a single-copy nuclear gene in mammals. After *de novo* synthesis in the cytosol, mtSSB is transported into mitochondria, where it participates in mitochondrial DNA (mtDNA) replication, RNA transcription, and maintenance of mitochondrial DNA D-loop structure *(1–3)*. Mitochondrial SSB requires three different types of intermolecular interaction for its function: self-association to form homotetramers, binding to single-stranded DNA (ssDNA), and interaction with other proteins *(4–7)*.

Among different species, the amino acid sequence of mtSSBs is conserved among organisms as diverse as yeasts and humans *(8–12)*. Mitochondrial SSB has been purified from a variety of sources *(9,13)* and shown to form a homotetramer, similar to the *Escherichia coli* SSB *(3)*. The crystal structure for the human mtSSB has been solved and reveals a tightly associated tetramer that provides a hydrophobic path around which ssDNA can wrap *(14)*.

We have expressed and purified murine mtSSB from *E. coli* by utilizing glutathione-*S*-transferase (GST)/mtSSB fusion constructs and GST–affinity chromatography *(15)*. The experimental design for the expression and purification of mtSSB is presented in **Fig. 1**. This procedure permits rapid production and isolation of mtSSB to >90% purity, and monomers prepared in this manner undergo spontaneous tetramerization after the GST moiety is removed (**Fig. 2**). We have successfully expressed and purified a number of mutant mtSSBs as well as His-tagged mtSSB by using this system.

From: *Methods in Molecular Biology, vol. 197: Mitochondrial DNA: Methods and Protocols*
Edited by: W. C. Copeland © Humana Press Inc., Totowa, NJ

A Expression constructs

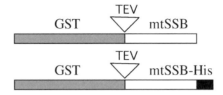

B Flow chart of experimental procedures

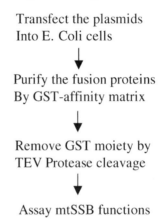

Transfect the plasmids
Into E. Coli cells

↓

Purify the fusion proteins
By GST-affinity matrix

↓

Remove GST moiety by
TEV Protease cleavage

↓

Assay mtSSB functions

Fig. 1. Experimental design. (**A**) Expression constructs including a GST moiety (hatched box) and mtSSB moiety (open box), polyhistidine tag (black box), and TEV cleavage site (triangle). (**B**) Flowchart of experimental procedure.

2. Materials

2.1. Expression Vector and PCR Primers

Amersham Pharmacia Biotech has more than ten different GST fusion vectors (the pGEX series of vectors) from which one can select to meet specific needs. These plasmids vectors vary with respect to restriction sites available for cloning and protease cleavage sites that are used to separate the protein of the interest from the GST moiety. We used pGEX-VP and pGEX-VH, which were modified by Dr. Yih-Sheng Yang from pGEX-cs (*16*). The pGEX-VH adds a 6× His tag to the C-terminus of the fusion proteins. The primers used for subcloning murine mtSSB cDNA into the expression vector bear a *Bam*HI restriction site. The primer sequences are as follows: 5′-CGAAGGATC CGAGTCTGAAGTAGCCAGCAGTTTG (forward), and 5′-GCTAGGATCCT ATGCCTTTTCTTTTGTCTGGTC (reverse).

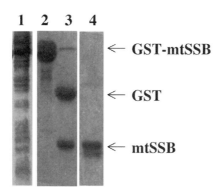

Fig. 2. Expression and purification of murine mtSSB. Coomasie Blue staining of protein preparations. Lane 1: total protein extract prepared from an *E. coli* strain transformed with a GST–mtSSB fusion gene construct; lane 2: affinity-purified GST–mtSSB fusion protein; lane 3: cleavage of GST–mtSSB fusion protein by rTEV protease; lane 4: purified mtSSB.

2.2. mtSSB Expression and Isolation

1. LB medium: dissolve 10 g Bacto-tryptone, 5 g Bacto-yeast extract, and 5 g NaCl in H_2O to a final volume of 1 L, adjust pH to 7.4–7.5.
2. 2X YT medium: dissolve 16 g Bacto-tryptone, 10 g Bacto-yeast extract, and 5 g NaCl in H_2O to a final volume of 1 L, adjust pH to 7.4–7.5.
3. 1 M IPTG (isopropylthio-β-galactoside): dissolve 2.38 g of IPTG in deionized water, make the final volume up to 10 mL, sterilize the IPTG stock solution by filtering.
4. Lysis buffer: 1X phosphate-buffered saline (PBS) with 5 mM EDTA, 0.5 mM phenylmethylsulfanyl fluoride (PMSF), and protease inhibitor mix (Roche Molecular Biochemicals).
5. TEK solution: 100 mM Tris-HCl, pH 7.8, 1 mM EDTA, 100 mM KCl.
6. Storage buffer: 50 mM Tris-HCl, pH 7.5, 200 mM KCl, 1 mM EDTA, 1 mM dithiolthreitol (DTT), and 10% glycerol.
7. rTEV protease from Gibco-BRL.
8. Glutathione–Sepharose 4B beads for GST affinity chromatography are from AmershamPharmacia Biotech.

2.3. ssDNA Binding Assay

1. Single-stranded oligonucleotide: a 34- to 40-mer oligonucleotide is preferred. Single-stranded oligonucleotide probes of at least this size are required to wrap around the mtSSB tetramer.
2. γ-P^{32}-ATP: 3000 Ci/mmol (Amersham).

2.4. Sucrose Gradient Assay

1. Sucrose buffers: 5%, 10%, 15%, and 20% are prepared in 0.4X storage buffer.
2. TK Buffer: 50 mM Tris-HCl, pH 7.5, 100 mM KCl.

3. Protein A agarose bead slurry: weigh 435 mg of protein A beads, add 2 mL of suspension buffer as suggested by the manufacturer (Sigma). Take 15 μL of protein A beads, wash twice with and resuspend in 35 μL of 0.4X storage buffer.

4. 1 X TBS-T: 20 mM Tris-HCl, pH 7.6, 136 mM NaCl, 0.1% Tween-20.

3. Methods

3.1. Expression Constructs

1. Amplify murine mtSSB cDNA by using the oligo primers designed in **Subheading 2.** and the High Fidelity PCR system (Roche Molecular Biochemicals). We typically use one nanogram of cDNA template and run 15 cycles of polymerase chain reaction (PCR) at an annealing temperature of 48°C.

2. Purify the PCR product by using a quick spin column (Qiagen).

3. Digest the PCR product and the expression vectors, pGEX-VP and pGEX-VH, with *Bam*HI.

4. Gel-purify the mtSSB/*Bam*HI DNA fragment by the Gel Extraction system (Qiagen). The necessity of gel purification depends on the outcome of PCR.

5. Ligate mtSSB/*Bam*HI and pGEX/*Bam*HI at 16°C for overnight.

6. Transform *E. coli* cells (DH5α) with the ligation mixture and grow the bacterial cells on LB medium with ampicillin.

7. Isolate plasmid from ampicillin-resistant colonies by Wizard mini-prep spin columns (Promega). Correct insertion of the mtSSB cDNA should be verified by restriction digestion and, subsequently, by sequencing.

3.2. Expression and Purification of mtSSB

1. Inoculate a single colony of mtSSB/DH5α in 10 mL of LB medium and grow at 37°C overnight.

2. Combine the overnight culture with 500 mL of 2X YT medium in a 2-L flask and grow with aerobic agitation at 37°C. Monitor the growth by checking OD_{600} reading. When OD_{600} reaches 0.4–0.5, add IPTG to a final concentration of 1 mM, continue to grow for 3–4 h until the OD_{600} reaches 1.0–1.1.

3. Harvest the bacterial cells by centrifugation at 2800g with a swinging bucket rotor. Wash the cells once with 100 mL of 1X PBS. Resuspend the cells in 10 mL of lysis buffer.

4. Break the cells by sonication. We typically place the tube containing the bacterial suspension in an ice bucket, sonicate five times for 30 s at 30% output.

5. Clear the cell debris by centrifuging at 12,000g at 4°C for 10 min.

6. Transfer the supernatant to a new tube, add 1.2 mL of 10% Triton X-100 (the final concentration should be 1%), stir for 10 min.

7. Rinse glutathione–Sepharose beads with 1X PBS, use low-speed spin to pellet the beads, and remove the solution. The beads should occupy 1–1.2 mL. Transfer the cell protein extract to the beads and rotate at 4°C for 2 h or overnight.

8. After binding of the GST–mtSSB fusion protein, wash the glutathione–Sepharose beads sequentially by using 5 mL of the following buffers: lysis buffer, once; 1X PBS/0.5 M NaCl, twice; 1X PBS/0.1 M KCl, once; TEK, twice.

9. Elute the GST–mtSSB fusion protein with TEK/5 m*M* glutathione. Use 1.5 mL of buffer for the first elution and 1 mL for the second elution.
10. Digest fusion protein with TEV protease (Gibco-BRL) at 30°C for 2 h (**Note 1**).
11. Dialyze against 1X PBS/0.1 *M* KCl/1 m*M* EDTA for 3 h by using a dialysis slide (Pierce).
12. Transfer the protein solution to a new tube and combine with 1 mL of glutathione–Sepharose beads prewashed with dialysis buffer. This removes GST moiety.
13. Dialyze the protein preparation against storage buffer, aliquot, and store at –20°C or –80°C. (*See* **Note 2**.)
14. Protein concentration may be assessed by various methods. We employed Coomasie Blue staining (Pierce).

3.3. Labeling of Single-Stranded DNA Probe

1. Label single-stranded oligonucleotide (*see* **Note 3**) by using T4 kinase. We typically set up a 25-µL kinase reaction containing 5 µL of γ-P^{32}-ATP, 8–10 pmol of oligonucleotide, 5X reaction buffer, and 2 µL of kinase (Gibco-BRL), and incubate the reaction mix at 37°C for 40–45 min.
2. Following the kinase reaction, add 125 µL of TE to the reaction mix and pass through a Sephadex G-50 quick spin column *(17)*. Free γ-P^{32}-ATP will be retained in the Sephadex G-50 matrix.
3. Add NaCl to 0.3 *M*, mix well; then add 2.2 volumes of ethanol and precipitate the ssDNA probe.
4. Dissolve the ssDNA probe in 100 µL of TE, removing 1 µL for scintillation counting to verify labeling by the radioisotope.

3.4. Single-Stranded DNA Binding Activity (see Note 4)

1. Set up single-stranded DNA binding reaction in 15 µL (final volume) containing 50,000 cpm (counts per minute) γ-P^{32}-labeled oligonucleotide probe, 0.2–1 µ*M* mtSSB, and 0.4X storage buffer. Allow reaction to proceed for 15 min at 30°C or room temperature.
2. To stop the reaction, add 1.5 µL of 1X native DNA gel loading buffer, chill on ice.
3. Resolve protein-bound and unbound oligonucleotide probe by electrophoresis through a 6% polyacrylamide gel containing 1X TBE buffer. With a 20-cm × 20-cm × 0.75-mm gel, we typically run the electrophoresis at 200 V (constant voltage) for 3 h.
4. After electrophoresis, transfer the gel to a Whatman filter paper and dry the gel.
5. Detect the radioisotope using x-ray film or a PhosphoImager cassette (Molecular Dynamics).

3.5. Assay for Monomeric and Tetrameric mtSSB by Sucrose Gradient Density Centrifugation

1. Make a step gradient by layering 0.8 mL of 20%, 15%, 10%, and 5% sucrose buffer into a 3.8-mL ultracentrifuge tube (Sorvall TST 60.1 rotor). Chill in an ice bucket for 20 min.

2. Treat mtSSB as desired, then load sample (at least 5 μ*M*) on top of the gradient.
3. Centrifuge at 340,000*g* in 4°C for 4 h.
4. Take fractions (150 μL per fraction) from top to bottom.
5. Protein concentration of each fraction may be assessed by using Coomasie Blue staining (Pierce).

3.6. Assessment of Heterotetrameric Forms of mtSSB (see Note 5)

We have adopted two methods to assess formation and binding activity of heterotetrameric forms of mtSSB. Both methods utilize mtSSB tagged with six histidine residues at the C-terminus (mtSSB-His, see **Fig. 1**). The methods depend on a denaturation–renaturation step to form heterotetramers.

1. Mix mtSSB-His with a variant mtSSB (e.g., carrying an amino acid substitution) in a total volume of 20 μL with 1X storage buffer, heat the protein mixture to 95°C, then gradually cool down to 20°C using a thermal cycler instrument.
2. Assess formation of heterotetramers by immunoprecipitation. Add 1 μL of anti-His-tag antibody (Sigma) to the mtSSB mixture, make up final volume to 50 μL with TK buffer. Rotate at 4°C for 1.5–2 h. Add 50 μL of protein A–agarose beads, rotate at 4°C for 1.5–2 h. Wash the beads four times, each with 400 μL of TBS-T buffer. Remove supernatant, add 9 μL of 50 m*M* Tris-HCl, pH 6.8, and 11 μL of 2X protein gel-loading buffer, immediately heat to 95°C for 5 min. Proceed to protein gel eletrophoresis and Western blot analysis using anti-mtSSB antibody *(15)*.
3. Assess ssDNA binding activity of heterotetramers by gel mobility shift assays. The ssDNA binding reaction condition is the same as described in **Subheading 3.4.** Gel eletrophoresis should be extended longer to allow a distinction among different forms of heterotetrameric complexes.

4. Notes

1. The TEV protease cleavage reactions should be carried out after eluting the GST fusion protein from glutathione–Sepharose beads. We have tried to cut the GST–mtSSB fusion protein when it is bond to the affinity beads, but this results in incomplete cleavage at the TEV site inserted between the GST and mtSSB moieties.
2. MtSSB protein is stable in high-salt buffer. Purified protein may precipitate in low-salt solutions. Adding NaCl or KCl can help dissolve the protein precipitate.
3. It is known that mtSSB tetramers bind ssDNA in each of several different binding modes *(1–3)*. Therefore, it is essential to use the proper length of oligonucleotide that allows mtSSB to interact with ssDNA in different modes. The minimum length of oligonucleotide needed to wrap completely around a rat mtSSB tetramer is 32 nucleotides *(9)*.
4. When performing binding reactions, we usually assess the specificity of ssDNA binding activity by including parallel reactions that include either cold double-stranded DNA (plasmid DNA) or ssDNA (M13mp18) as competitors *(15)*.

5. Formation of heterotetrameric forms of SSB is facilitated by heating mixtures of mtSSB variants to 95°C and by cooling down slowly to room temperature. We have used a slightly larger mtSSB variant, mtSSB-His, to aid the detection of heterotetramer binding activity. We typically run the electrophoresis for extended time-periods to separate heterotetramer and homotetramer bands clearly in gel mobility shift assays.

References

1. Chase, J. W. and Williams, K. R. (1986) Single-stranded DNA binding proteins required for DNA replication. *Annu. Rev. Biochem.* **55,** 103–136.
2. Meyer, R. R. and Laine, P. S. (1990) The single-stranded DNA-binding protein of Escherichia coli. *Microbiol. Rev.* **54,** 342–380.
3. Lohman, T. M. and Ferrari, M. E. (1994) *Escherichia coli* single-stranded DNA-binding protein: multiple DNA-binding modes and cooperativities. *Annu. Rev. Biochem.* **63,** 527–570.
4. Van Tuyle, G. C. and Pavco, P. A. (1985) The rat liver mitochondrial DNA-protein complex: displaced single strands of replicative intermediates are protein coated. *J. Cell. Biol.* **100,** 251–257.
5. Williams, K. R., Spicer, E. K., LoPresti, M. B., Guggenheimer, R. A., and Chase, J. W. (1983) Limited proteolysis studies on the *Escherichia coli* single-stranded DNA binding protein. Evidence for a functionally homologous domain in both the *Escherichia coli* and T4 DNA binding proteins. *J. Biol. Chem.* **258,** 3346–3355.
6. Chase, J. W., L'Italien, J. J., Murphy, J. B., Spicer, E. K., and Williams, K. R. (1984) Characterization of the *Escherichia coli* SSB-113 mutant single-stranded DNA-binding protein. Cloning of the gene, DNA and protein sequence analysis, high pressure liquid chromatography peptide mapping, and DNA-binding studies. *J. Biol. Chem.* **259,** 805–814.
7. Chase, J. W., Flory, J., Ruddle, N. H., Murphy, J. B., Spicer, E. K., and Williams, K. R. (1985) A monoclonal antibody that recognizes the functional domain of *Escherichia coli* single-stranded DNA binding protein that includes the ssb-113 mutation. *J. Biol. Chem.* **260,** 7214–7218.
8. Mahoungtou, C., Ghrir, R., Lecaer, J. P., Mignotte, B., and Barat-Gueride, M. (1988) The amino-terminal sequence of the *Xenopus laevis* mitochondrial SSB is homologous to that of the *Escherichia coli* protein. *FEBS Lett.* **235,** 267–270.
9. Hoke, G. D., Pavco, P. A., Ledwith, B. J., and Van Tuyle, G. C. (1990) Structural and functional studies of the rat mitochondrial single strand DNA binding protein P16. *Arch. Biochem. Biophys.* **282,** 116–124.
10. Tiranti, V., Barat-Gueride, M., Biji, J., DiDonato, S., and Zeviani, M. (1991) A full-length cDNA encoding a mitochondrial DNA-specific single-stranded DNA binding protein from *Xenopus laevis*. *Nucleic Acids Res.* **19,** 4291.
11. Van Tuyle, G. C. and Pavco, P. A. (1981) Characterization of a rat liver mitochondrial DNA–protein complex. Replicative intermediates are protected against branch migrational loss. *J. Biol. Chem.* **256,** 12,772–12,779.

12. Tiranti, V., Rocchi, M., DiDonato, S., and Zeviani, M. (1993) Cloning of human and rat cDNAs encoding the mitochondrial single-stranded DNA-binding protein (SSB). *Gene* **216,** 219–225.
13. Van Dyck, E., Foury, F., Stillman, B., and Brill, S.J. (1992) A single-stranded DNA binding protein required for mitochondrial DNA replication in *S. cerevisiae* is homologous to *E. coli* SSB. *EMBO J.* **11,** 3421–3430.
14. Yang, C., Curth, U., Urbanke, C., and Kang, C. (1997) Crystal structure of human mitochondrial single-stranded DNA binding protein at 2.4 A resolution. *Nat. Struct. Biol.* **4,** 153–157.
15. Li, K. and Williams, R. S. (1997) Tetramerization and single-stranded DNA binding properties of native and mutated forms of murine mitochondrial single-stranded DNA-binding proteins. *J. Biol. Chem.* **272,** 8686–8694.
16. Parks, T. D., Leuther, K. K., Howard, E. D., Johnston, S. A., and Dougherty, W. G. (1994) Release of proteins and peptides from fusion proteins using a recombinant plant virus proteinase. *Anal. Biochem.* **216,** 413–417.
17. Sambrook, J., Fritsch, E. F., and Maniatis, T. (1989) *Molecular Cloning: A Laboratory Manual*, Cold Spring Harbor Laboratory Press, Cold Spring Harbor, NY, pp. E37–E38.

21

Recombinant Yeast mtDNA Helicases

Purification and Functional Assays

Silja Kuusk, Tiina Sedman, and Juhan Sedman

1. Introduction

At least two DNA helicases are important for maintenance of the yeast *Saccharomyces cerevisiae* mitochondrial DNA (mtDNA). Pif1 protein is a $5'{\rightarrow}3'$ DNA helicase that affects the stability of mtDNA at higher temperatures but is not essential at 28°C. Pif1 is probably involved in mtDNA repair and recombination *(1,2)*, but it also affects telomeres *(3)* and nuclear ribosomal DNA *(4)*. Hmi1 protein is the second mitochondrial enzyme that has single-stranded DNA (ssDNA) stimulated ATPase and $3'{\rightarrow}5'$ DNA helicase activities. Hmi1 is essential for mtDNA maintenance in regular growth conditions *(5)*. Pif1 and Hmi1 proteins are normally not abundant, thus, their overexpression is necessary for biochemical studies. In this chapter, we describe purification protocols for recombinant Hmi1 and Pif1 proteins expressed in *Escherichia coli*.

We isolate Hmi1 as a glutathione-*S*-transferase (GST) fusion protein. The GST tag of the Hmi1 fusion protein can be safely removed with thrombin cleavage without significant loss in enzymatic activity. Our attempts to use recombinant His-tagged Hmi1 protein have not been successful as the protein appears to be largely insoluble and tightly associated with some host proteins. The expression construct pGEX 4T1–HMI1Δ15 used here lacks the last 15 codons of HMI1 open reading frame (ORF). The C-terminal peptide encoded by this part of the ORF is apparently removed during import into mitochondria *(6)*. The fusion protein encoded by the full-length HMI1 ORF is expressed at a significantly lower level compared to the HMI1Δ15 deletion version. The same purification strategy has been applied for several different Hmi1 mutants. As a

From: *Methods in Molecular Biology, vol. 197: Mitochondrial DNA: Methods and Protocols*
Edited by: W. C. Copeland © Humana Press Inc., Totowa, NJ

rule, enzymatically inactive helicase mutants are expressed at a lower level in comparison with the wild-type protein.

Purification schemes for Pif1 protein have been described previously. Overexpressed Pif1 was first purified from yeast mitochondrial "inclusion body-like" aggregates *(7)* and later from recombinant baculovirus-infected Sf9 cells *(8)*. We have obtained Pif1 protein from *E. coli* in relatively large quantities using a C- or N-terminal His tag. The enzymatic activity of the recombinant protein appears to be comparable to the native Pif1 protein purified from yeast *(7)*.

We also describe versions of two standard functional assays–ATPase assay and helicase unwinding assay–that are adapted for mtDNA helicases. We use this ATPase assay mainly for detecting rapidly ssDNA-dependent ATPase activity during purification of helicase proteins and for estimating the activity of immobilized Hmi1–GST and Pif1–GST fusion proteins. Helicase unwinding assay with different substrates can be applied for examination of biochemical function of a given DNA helicase.

2. Materials

For reagents and buffers not stored at room temperature, the appropriate temperature is indicated. Standard molecular biology methods and materials will not be described here; *see* **ref. 9**.

2.1. Purification of Recombinant Hmi1, Pif1–CHis, and Pif1–NHis Proteins

1. Competent cells of *E. coli* strains BL21 and BL21(DE3) (Novagen, Darmstadt, Germany) *(10)*.
2. Expression constructs: pGEX 4T1–HMI1Δ15, pET19b–PIF1Δ40, pET24d–PIF1Δ40 (*see* **Note 1**).
3. 2X YT medium: 16 g/L Bacto-tryptone (Difco, Detroit, MI), 10 g/L Bacto-yeast extract (Difco), 5 g/L NaCl.
4. Ampicillin 100 mg/mL, kanamycin, 45 mg/mL. Store at –20°C.
5. 1 *M* isopropylthio-β-galactoside (IPTG) (Fermentas, Vilnius, Lithuania). Store at –20°C.
6. Lysozyme (Amresco, Solon, OH). Store at 4°C.
7. Calf thrombin (cat. no. T9681, Sigma, St. Louis, MO) 10 U/μL, dissolved in 10 m*M* Tris-HCl, pH 8.0. Store at –70°C.
8. Glutathione (Sigma). Store at 4°C.
9. Glutathione–agarose (cat. no. G4510, Sigma), Ni-NTA resign (Qiagen, Hilden, Germany), MonoQ Sepharose and MonoS Sepharose (Amersham Pharmacia Biotech, Piscataway, NJ). Store at 4°C.

10. Components for different buffers:
    a. Sucrose (Amresco).
    b. Glycerol (Amresco).
    c. Urea (Serva GmbH, Heidelberg, Germany).
    d. 5 M NaCl.
    e. 1 M Tris-HCl, pH 8.0.
    f. 1 M Tris-HCl, pH 6.8.
    g. 100 mM unbuffered Tris.
    h. 0.5 M Mes-NaOH, pH 6.5, (store at 4°C).
    i. 0.5 M EDTA, pH 8.0.
    j. 1 M Dithiothreitol (DTT) (Boehringer-Mannheim, Mannheim, Germany) (store at –20°C).
    k. 100 mM Phenylmethylsulfonyl fluoride (PMSF) (Serva GmbH) dissolved in isopropanol.
    l. 0.5 M EGTA, pH 8.0.
    m. 100 mM CaCl$_2$ (store at 4°C).
    n. 0.5 M Imidazole-HCl, pH 6.0, (Amresco) (store at 4°C).
    o. 10% (v/v) NP-40.
    All buffers used for protein purification should be made directly before use (the same day).
11. Buffer A: 50 mM Tris-HCl, pH 8.0, 300 mM NaCl, 10% (w/v) sucrose, 1 mM EDTA, pH 8.0, 1 mM DTT, 1 mM PMSF.
12. Buffer B: 20% (v/v) glycerol, 20 mM Tris-HCl, pH 8.0, 300 mM NaCl, 0.1 mM EDTA, pH 8.0, 1 mM DTT.
13. Buffer Q: 20 mM Tris-HCl, pH 8.0, 20% (v/v) glycerol, 0.1 mM EDTA, pH 8.0, 1 mM DTT, 1 mM PMSF.
14. Buffer S: 25 mM Mes-NaOH, pH 6.5, 20% (v/v) glycerol, 0.1 mM EDTA, pH 8.0, 1 mM DTT, 1 mM PMSF.
15. Buffer C: 25 mM Tris-HCl, pH 8.0, 300 mM NaCl, 10% (v/v) glycerol, 1 mM PMSF.
16. Buffer D: buffer C + 10 mM imidazole-HCl, pH 6.0.
17. Buffer E: 10% (v/v) glycerol, 25 mM Tris-HCl, pH 6.8, 300 mM NaCl, 1 mM PMSF.
18. Buffer F: 10% (v/v) glycerol, 25 mM Tris-HCl, pH 8.0, 0.1 mM EDTA, pH 8.0, 0.2 mM DTT, 0.5 mM PMSF.
19. Buffer G: 2 M urea, 25 mM Tris-HCl, pH 8.0, 300 mM NaCl, 0.1 mM EDTA, pH 8.0, 0.2 mM DTT, 0.2 mM PMSF.
20. Buffer H: 25 mM Tris-HCl, pH 8.0, 1 M NaCl, 0.1 mM EDTA, pH 8.0, 0.2 mM DTT, 0.2 mM PMSF.
21. Ultrasonic homogenizer, model CP300 (Cole-Parmer Instrument Co., Chicago, IL), end-over-end mixer (Leo, Tartu, Estonia), disposable plastic columns, dialysis membrane with volume of 2 mL/cm and molecular-weight cutoff of 12,000–14,000 Daltons (Spectrum Medical Industries, Inc., Los Angeles, CA).

22. Bradford reagent: 0.01% (w/v) Coomassie Brilliant Blue G-250, 5% (v/v) ethanol, 4.25% (v/v) ortophosphoric acid *(11)*.
23. Liquid nitrogen.
24. Reagents for denaturing protein gel electrophoresis according to Laemmli *(12)*.

2.2. ATPase Assay

1. 10X Reaction buffer: 70 mM MgCl$_2$, 300 mM Tris-HCl, pH 7.5, 10 mM DTT. Store at –20°C.
2. Acetylated bovine serum albumin (BSA), 1 mg/mL (Promega, Madison, WI). Store at –20°C.
3. 1 mM ATP (Amersham Pharmacia Biotech). Store at –20°C.
4. [γ-^{32}P] ATP (5000 Ci/mmol, cat. no. AA0018, Amersham Pharmacia Biotech). Store at 4°C.
5. Single-stranded oligonucleotide DNA. Store at –20°C.
6. Stop solution: 0.8% (w/v) charcoal Norit A 100 mesh (cat. no. 26,001-0, Aldrich Chemical Company, Inc., Milwaukee, WI), 2 mM KH$_2$PO$_4$, 0.036% (v/v) HCl.
7. Scintillation counter.

2.3. Preparation of Helicase Substrates

1. DNA oligos 10 pmol/μL:
   3OVER: 5′ AGTGAGTCGTATTACAATTG 3′
   5OVER: 5′ GATCCAACCGCAGGTCTA 3′
   LONG: 5′ GGGCAATTGTAATACGACTCACTATAGTAGACCTGCGGTTG GATC 3′ (*see* **Note 2**).
2. 10X Annealing: 333 mM Tris-acetate, pH 7.9, 10 mM Mg-acetate, 66 mM K-acetate. Store at –20°C.
3. Acetylated BSA 1 mg/mL (Promega). Store at –20°C.
4. 2 mM dATP, 2 mM dTTP, 2 mM dCTP (Amersham Pharmacia Biotech). Store at –20°C.
5. [α-^{32}P] dCTP (3000 Ci/mmol; Amersham Pharmacia Biotech).
6. Klenow fragment 2 U/μL (Fermentas).
7. Substrate loading dye (SLD): 500 μL 50% (v/v) glycerol, 500 μL 5X TBE, 100 μL 10% (w/v) sodium dodecyl sulfate (SDS), 100 μL 0.5 M EDTA, pH 8.0, 0.05% (w/v) bromophenol blue.
8. Reagents for nondenaturing polyacrylamide TBE gel electrophoresis.
9. S$_{100}$T$_{10}$E$_{0.1}$ buffer: 100 mM NaCl, 10 mM Tris-HCl, pH 8.0, 0.1 mM EDTA, pH 8.0.

2.4. Helicase Unwinding Assay

1. 10X Reaction buffer: 70 mM MgCl$_2$, 300 mM Tris-HCl, pH 7.5, 10 mM DTT. Store at –20°C.
2. Acetylated BSA 1 mg/mL (Promega). Store at –20°C.
3. 100 mM ATP (Amersham Pharmacia Biotech). Store at –20°C.

4. Helicase substrate, prepared as described in **Subheading 2.3.**
5. SLD, prepared as described in **Subheading 2.3.**
6. Reagents for nondenaturing polyacrylamide TBE gel electrophoresis.

3. Methods
3.1. Purification of Recombinant GST–Hmi1 Protein

1. Transform competent *E. coli* BL21 cells with pGEX 4T1–HMI1Δ15 expression construct (*see* **Note 3**).
2. Pick several colonies from the fresh plate and inoculate a liquid culture in 2X YT + 0.1 mg/mL ampicillin. Incubate the flasks with vigorous shaking at 30°C until OD_{600} reaches 0.3–0.35.
3. Cool the flasks on ice until the temperature of the culture is 21–23°C and add IPTG to 0.5 mM (120 mg/L).
4. Continue to grow at 21–23°C for 8–12 h. OD_{600} of the induced culture will reach 0.75–1.1 by this time.
5. Collect the cells 2700g for 10 min in a Sorvall GS3 rotor.

All of the subsequent steps are performed on ice or at 4°C. The following protocol is used for a 4-L culture.

6. Resuspend the bacterial pellet in 20 mL buffer A.
7. Add 20 mg lysozyme resuspended in 1 mL buffer A just before use. Incubate for 1 h.
8. Snap-freeze in liquid nitrogen (*see* **Note 4**).
9. Thaw the cells and add buffer A to the final volume of 50 mL.
10. Sonicate three times for 20 s 50% duty cycle, output control 0.7 with the CP300 sonicator to decrease the viscosity of the lysate. Between the cycles, cool the lysate on ice for 30–60 s.
11. Centrifuge 17,000g (15,000 rpm for 20 min in a Sorvall SS34 rotor).
12. Recover the supernatant and add 0.8 mL glutathione–agarose beads (lyophilized glutathion agarose has been rehydrated, washed extensively with double distilled water (ddH$_2$O) and equilibrated with buffer A). Incubate on the end-over-end mixer for 3–5 h.
13. Let the beads sediment by gravity for 1 h.
14. Remove most of the supernatant. Make sure that the beads are not transferred.
15. Add a little buffer A to the agarose beads and transfer the suspension into a disposable plastic column.
16. Wash the column five times with 1 mL buffer B, twice with 1 mL buffer B + 0.1% (v/v) NP-40, once with 1 mL buffer B, twice with 1 mL buffer B + 1 M NaCl, and twice with 1 mL buffer B.
17. Elute GST–Hmi1 protein with buffer B + 10 mM glutathione (1 mL buffer B + 100 μL glutathione, 30.7 mg/mL in 100 mM unbuffered Tris). Collect 150-μL fractions. Check the fractions for protein content by mixing 2–5 μL with 100 μL Bradford reagent.

18. Pool the peak fractions containing 0.2–2 mg/mL protein (*see* **Note 5**). Total volume of the pooled fractions is usually about 1.5 mL.
19. Add $CaCl_2$ to the final concentration of 2.6 mM and 20–30 U of calf thrombin. Incubate overnight on ice (*see* **Note 6**).
20. Add 1 mM PMSF to inactivate thrombin, and 4 mM EGTA to chelate Ca^{2+}.
21. Prepare a 0.75-mL MonoQ–Sepharose column and equilibrate with buffer Q + 100 mM NaCl.
22. Dilute the protein fraction five times with buffer Q + 48 mM NaCl and load immediately onto the MonoQ–Sepharose column.
23. Wash the column four times with 1 mL buffer Q + 100 mM NaCl.
24. Most of the Hmi1 protein is in the flow-through and 100 mM NaCl wash. (*see* **Note 7**).
25. Prepare a 200-µL MonoS–Sepharose column and equilibrate with buffer S + 100 mM NaCl.
26. Load the MonoQ flow-through and 100 mM NaCl wash containing Hmi1 protein onto the MonoS–Sepharose column.
27. Wash the column three times with 700 µL buffer S + 100 mM NaCl and three times with 700 µL buffer S + 150 mM NaCl.
28. Elute Hmi1 protein with buffer S + 400 mM NaCl.
29. Freeze the protein in aliquots using liquid nitrogen and keep at –70°C. The protein is stable at least for 1 yr (*see* **Note 8**).
30. Check the enzymatic activity of Hmi1 protein using ATPase and helicase unwinding assays as described in **Subheadings 3.3.** and **3.5.**

The final concentration of Hmi1 protein is 0.1–0.3 mg/mL and the yield is approximately 0.2 mg protein from a 4-L culture (*see* **Note 5**). The purified protein is shown in **Fig. 1**. The results of ATPase and helicase unwinding assays done with this protein are shown in **Fig. 2** and **Fig. 3**, respectively.

3.2. Purification of Recombinant Pif1–CHis and Pif1–NHis Proteins

1. Transform competent *E. coli* BL21(DE3) cells with expression constructs pET19b–PIF1Δ40 to purify Pif1–NHis protein, and pET24d–PIF1Δ40 to purify Pif1–CHis protein (*see* **Note 9**).
2. Pick several colonies from a fresh plate and inoculate a liquid culture in 2X YT + 0.1 mg/mL ampicillin for Pif1–CHis, or in 2X YT + 0.045 mg/mL kanamycin for Pif1–NHis. Incubate the flasks with vigorous shaking at 21–23°C until OD_{600} reaches 0.3–0.35.
3. Add IPTG to 1 mM (240 mg/L).
4. Continue to grow at 21–23°C for 8–12 h. OD_{600} of the induced culture will reach 0.75–1.1 by this time.
5. Collect the cells 2700g for 10 min in a Sorvall GS3 rotor.

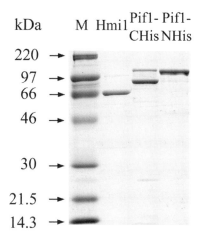

Fig. 1. Sodium dodecyl sulfate–polyacrylamide gel with the purified Hmi1, Pif1–CHis, and Pif1–NHis proteins. The M lane shows protein molecular size marker with molecular weights (kDa) as indicated (Rainbow mix, Amersham Pharmacia Biotech). All other lanes contain 3 μg of indicated protein each. The gel is stained with Coomassie Blue.

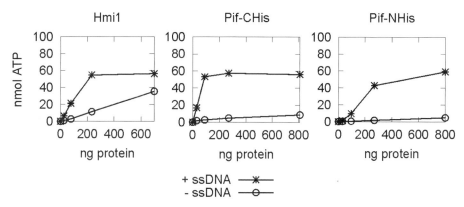

Fig. 2. Results of an ATPase assay performed with Hmi1, Pif1–CHis, and Pif1–NHis proteins. Four reactions were done with gradually increasing amounts of each protein. The amounts were 26 ng, 77.78 ng, 233.3 ng, and 700 ng of Hmi1 and 30 ng, 90 ng, 270 ng, and 810 ng of Pif1–CHis and Pif1-NHis. With each protein, two sets of reactions were made: one in the presence of 150 ng single-stranded DNA oligo and the other one without. The reactions were carried on for 5 min at 30°C. The amount of ATP hydrolyzed was calculated as described in **Note 17** and plotted against the amount of protein in the reaction.

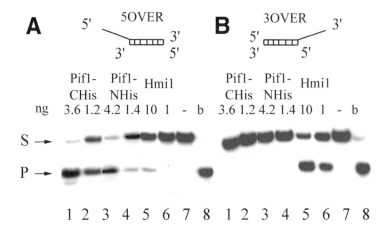

Fig. 3. Autoradiogram of helicase unwinding assay gels. The assays were performed with Hmi1, Pif1–CHis, and Pif1–NHis proteins using amounts of proteins (ng) as indicated. Gel A shows the assays done with 5OVER substrate, and gel B shows the assays done with 3OVER substrates. Lane 7 is negative control where the helicase substrates are incubated in reaction mixture for 15 min at 30°C without protein. Lane 8 shows helicase substrates denatured by heating for 3 min at 100°C. S indicates the helicase substrate and P indicates the unwound single-stranded oligo.

All of the subsequent steps are performed on ice or at 4°C. The following protocol is used for a 1- to 2-L culture.

6. Resuspend the bacterial pellet in 20 mL buffer C.
7. Add 20 mg lysozyme resuspended in 1 mL buffer C just before use. Incubate for 1 h.
8. Snap-freeze in liquid nitrogen (*see* **Note 4**).
9. Thaw the cells and add buffer C to the final volume of 50 mL.
10. Sonicate three times 20 s 50% duty cycle, output control 0.7 with the CP300 sonicator to decrease the viscosity of the culture. Between the cycles, cool the lysate on ice for 30–60 s.
11. Centrifuge 27,000*g* for 20 min in a Sorvall SS34 rotor.
12. Recover the supernatant and add 1 mL Ni-NTA beads (lyophilized Ni-NTA resign has been rehydrated, washed extensively with ddH$_2$O and equilibrated with buffer C). Incubate on the end-over-end mixer for 3–5 h.
13. Let the beads sediment by gravity for 1 h.
14. Remove most of the supernatant. Make sure that the beads are not transferred.
15. Add a little buffer C to the beads and transfer the suspension into a disposable plastic column.
16. Wash the column five times with 1 mL buffer D (buffer C + 10 m*M* imidazole-HCl, pH 6.0), twice with 1 mL buffer D + 0.1% (v/v) NP-40, once with 1 mL buffer D, twice with 1 mL buffer D + 1 *M* NaCl, and twice with 1 mL buffer D.

17. Wash the Pif1–NHis column twice with 1 mL buffer E + 100 m*M* imidazole-HCl, pH 6.0, (*see* **Note 10**).

18. Elute Pif1–CHis from the column with a 10-mL gradient of imidazole-HCl, pH 6.0, from 0 m*M* to 500 m*M* in buffer E, and Pif1–NHis with a 10-mL gradient of imidazole-HCl, pH 6.0, from 100 m*M* to 500 m*M* in buffer E. Collect 0.5-mL fractions. Check the fractions for protein content by mixing 2–5 µL with 100 µL Bradford reagent.

19. Pool the peak fractions containing 0.2–2 mg/mL protein (*see* **Note 5**). Usually, the total volume of the pooled fractions is about 5 mL for Pif1–CHis and 10 mL for Pif1–NHis.

20. Dialyze against 500 mL buffer F + 100 m*M* NaCl for 15 h, with one change.

21. Centrifuge at 10,550*g* for 30 min in a Sorvall HB4 rotor. Continue with Pif1–CHis as described in **Subheading 3.2.1.** and with Pif1–NHis as described in **Subheading 3.2.2.** (*see* **Note 11**).

3.2.1. Pif1–CHis

1. Prepare a 0.8-mL MonoQ–Sepharose column and equilibrate with buffer F + 100 m*M* NaCl.
2. Load the supernatant from the 10,550*g* spin onto the MonoQ–Sepharose column.
3. Wash the column four times with 1 mL buffer F + 100 m*M* NaCl. Collect the flow-through and 100 m*M* NaCl wash that contain most of the Pif1–CHis protein.
4. Freeze the protein in aliquots using liquid nitrogen and keep at –70°C.
5. Check the enzymatic activity of Pif1–CHis protein using ATPase and helicase unwinding assays as described in **Subheadings 3.3.** and **3.5.**

The final concentration of Pif1–CHis protein is 0.1–0.3 mg/mL and the yield is approximately 1 mg protein from a 1-L culture (*see* **Note 5**). The purified protein is shown on **Fig. 1** (*see* **Note 12**). The results of ATPase and helicase unwinding assays done with this protein are shown in **Fig. 2** and **Fig. 3**, respectively.

3.2.2. Pif1–NHis

1. Resuspend the pellet from the 10,550*g* spin in buffer G.
2. Incubate on the end-over-end mixer for 6 h.
3. Dialyze against 100 mL buffer H for 6 h and then against buffer H + 50% (v/v) glycerol for 20 h with one change.
4. Centrifuge at 13000*g* (13,000 rpm) for 15 min to remove any insoluble matter.
5. Freeze the protein in aliquots using liquid nitrogen and keep at –70°C.
6. Check the enzymatic activity of Pif1–NHis protein using ATPase and helicase unwinding assays as described in **Subheadings 3.3.** and **3.5.**

The final concentration of Pif1–NHis protein is 0.5–0.6 mg/mL and the yield is approximately 1 mg protein from a 1-L culture (*see* **Note 5**). The purified protein is shown in **Fig. 1**. The results of ATPase and helicase unwinding assays done with this protein are shown in **Fig. 2** and **Fig. 3** respectively.

3.3. ATPase Assay

1. Set up a master mix as follows: 10 μL of 10X buffer, 10 μL of 1 mg/mL BSA, 10 μL of 1 m*M* ATP, 10 μL diluted [γ-^{32}P] ATP, and 60 μL ddH$_2$O (*see* **Notes 14** and **15**). The indicated volumes are given for 10 reactions, 10 μL each. Mix the components, remove half of the master mix, and add 0.75 μg single-stranded oligonucleotide DNA. Dispense the master mixes into 10-μL aliquots.
2. Prewarm the reaction mixtures for 2–5 min at 30°C, add enzyme, mix carefully, and start incubation. Include control samples without added protein for background estimation.
3. We usually take samples for analysis at 2-, 5-, and 10-min time-points.
4. Stop the reaction by adding immediately 150 μL stop solution.
5. Mix on a vortex for 10 min.
6. Spin the samples at 10,000g (13,000 rpm for 1 min in an Eppendorf centrifuge).
7. Remove 100 μL of the supernatant to a new tube and add 150 μL stop solution.
8. Mix on a vortex for 10 min.
9. Spin the samples at 10,000g (13,000 rpm for 1 min in an Eppendorf centrifuge).
10. Remove 150 μL of the supernatant to a new tube and count the amount of [^{32}P] according to Cerenkov with a scintillation counter (*see* **Note 15**). Include 3.75 μL of the original reaction mixture that has not been subjected to charcoal extraction (*see* **Note 16**) in order to calculate the amount of ATP hydrolyzed in one reaction (*see* **Note 17**). The results of an ATPase assay done with Hmi1, Pif1–CHis, and Pif1–NHis proteins purified as described in **Subheadings 3.1.** and **3.2.** are shown in **Fig. 2**.

3.4. Preparation of Helicase Substrates

1. Combine the following in an Eppendorf tube: 1 μL of 10 pmol/μL 3OVER oligo and 1 μL of 10 pmol/μL LONG oligo, or 1 μL of 10 pmol/μL 5OVER oligo and 1 μL of 10 pmol/μL LONG oligo (*see* **Notes 2** and **18**).
2. Add 2.5 μL of 10X annealing buffer and 15 μL ddH$_2$O.
3. Heat the tube for 5 min at 68°C and cool slowly down to room temperature (30 min or longer).
4. Spin the tubes briefly and add 2.5 μL of 1 mg/mL BSA, 2 μL of [α-^{32}32P] dCTP, and 1 μL of 2 U/μL Klenow to the 3OVER substrate tube, or 2.5 μL of 1 mg/mL BSA, 2 μL of [α-^{32}P] dCTP, 1 μL of 2 m*M* dATP, 1 μL of 2 m*M* dTTP, and 1 μL of 2 U/μL Klenow to the 5OVER substrate (*see* **Note 2**).
5. Incubate at room temperature for 30 min.
6. Add 1 μL of 2 m*M* dCTP and incubate at room temperature for 10 min.
7. Stop the reaction with 3 μL SLD.
8. Purify the substrates on a 12% TBE-PAGE (29 : 1 acrylamide : bisacrylamide), 12 cm × 10 cm × 1.0 mm. We run these gels for 3 h at 80 V in 1X TBE.

9. Wrap the gel in plastic wrap and expose briefly (1–5 min) to Fuji SuperRX x-ray film. Cut out the band corresponding to the expected size of the substrate.
10. Cut the gel slice into small pieces (approximately 1 mm × 1 mm).
11. Add 200 µL $S_{100}T_{10}E_{0.1}$ and freeze–thaw the mixture (*see* **Note 19**).
12. Centrifuge at 18°C for 30 min at 40,000g (30,000 rpm in a Beckman Ti70.1 rotor).
13. Recover the supernatant and store at 4°C. It should contain 10,000–20,000 cpm/µL or approx 5–10 fmol/µL of the oligo substrate. Routinely, we use 0.5–1 µL per unwinding assay reaction.

3.5. Helicase Unwinding Assay

1. Set up a master mix as follows: 10 µL of 10X reaction buffer, 5 µL helicase substrate prepared as described in **Subheading 3.4.**, 10 µL of 1 mg/mL BSA, 4 µL of 100 mM ATP, and 66 µL ddH$_2$O. The indicated volumes are given for 10 reactions, 10 µL each. Divide the master mix into 9.5-µL aliquots and put on ice. Add the helicase protein (0.2–200 ng) and mix gently (*see* **Note 20**).
2. Incubate for 15 min at 30°C.
3. Put the tubes on ice and add 1 µL of SLD.
4. Spin briefly and load onto 12% TBE-PAGE (29 : 1 acrylamide : bisacrylamide), 10 cm × 10 cm × 0.75 mm.
5. Run the gel at 4°C 10 V/cm (100–120 V) until the bromophenol dye has moved approximately 6 cm or two-thirds of the gel length.
6. Dry the gel under vacuum at 80°C onto Whatman 3MM paper (*see* **Note 21**). Expose with a screen to Fuji SuperRX x-ray film at –70°C. Optimal exposure times vary from 2 h to overnight. A typical helicase assay done with Hmi1, Pif1–CHis, and Pif1–NHis proteins purified as described in **Subheadings 3.1.** and **3.2.** and using 3OVER and 5OVER substrates prepared as described in **Subheading 3.4.** is shown in **Fig. 3**.

4. Notes

1. To make pGEX 4T1–HMI1Δ15 expression construct, HMI1 ORF was cloned into *Bam*HI site of pGEX 4T1 plasmid (Amersham Pharmacia Biotech) as described in **ref. 5**. To make pET19b–PIF1Δ40 expression construct, PIF1 ORF was cloned into pET19b vector (Novagen) between *Nco*I and *Sac*I sites. The first codon of PIF1 ORF in the construct corresponds to the *Met*39 of the Pif1 preprotein. The next codon in our construct encodes for Ala and not for Ser as in the wild-type protein. To make pET24d–PIF1Δ40 expression construct, PIF1 ORF starting with the 39th codon was cloned into pET24d vector (Novagen) between *Nde*I and *Bam*HI sites.
2. The 3OVER oligo is designed to be complementary with the 5′ half of the LONG oligo, leaving unpaired the three G nucleotides in the 5′ end. This end is filled with [α-^{32}P] dCTP and Klenow fragment. The 5OVER oligo is designed to be complementary with the 3′ half of the LONG oligo. With Klenow fragment

[α-^{32}P] dCTP, dTTP and dATP, five nucleotides are added to make the double-stranded region of the same length (23 base pairs) as with the 3OVER oligo.

3. We have also successfully used other *E. coli* strains including DH5α and XL1Blue.

4. The cells can be kept at –70°C until needed. We usually take 1 mL of the lysate and check for protein expression before large-scale purification.

5. Protein concentration is estimated using BSA as a standard.

6. We try not to incubate longer than 10–12 h in order to minimize losses in enzymatic activity.

7. Next, 150–1000 m*M* NaCl gradient in buffer Q can be applied to the MonoQ–Sepharose column, and some Hmi1 protein also appears in later fractions. We have not used these fractions because a prominent contaminating 65- to 70-kDa protein and most of the uncut GST-Hmi1 that stick to MonoQ in 100 m*M* NaCl buffer are recovered there.

8. We do not recommend repeated freeze–thaw cycles of the frozen Hmi1 aliquots because the activity of the protein drops significantly.

9. The Pif1 protein can be also purified as a GST fusion protein using pGEX-4T1-based expression constructs. The GST-Pif1 fusion protein displays ssDNA-dependent ATPase and helicase activities and we have used it as an immobilized version of Pif1 protein in DNA-binding assays. The fusion protein can be cleaved with thrombin to obtain the mature Pif1 protein. However, we prefer to use His tag because the yield of the GST fusion protein is approximately 10-fold lower.

10. This extra wash helps to get rid of a major contaminating 60- to 70-kDa protein.

11. Pif1–CHis protein remains soluble during dialysis against buffer F + 100 m*M* NaCl. In contrast, Pif1–NHis protein precipitates. We have taken advantage of the fact that practically all of the contaminating proteins remain soluble and recovered the Pif1–NHis protein using 2 *M* urea treatment followed by specific dialysis steps.

12. Although Pif1–NHis protein appears as a dominant approx 100-kDa band on a denaturing SDS-PAGE, the Pif1–CHis protein migrates as a doublet of full-length approx 100-kDa protein and a truncated approx 85-kDa protein. A similar doublet has been described for Pif1 protein overexpressed in baculovirus system *(8)*.

13. [γ-^{32}P] ATP used in these assays should be of good quality. Crude preparations of [γ-^{32}P] ATP often contain free inorganic phosphate that is the cause of high background. The Amersham pharmacia Biotech Redivue [γ-^{32}P] ATP has been proven satisfactory for us.

14. Usually, we take about 50,000 cpm of [γ-^{32}P] ATP per reaction.

15. Charcoal binds only compounds containing aromatic nitrogen bases but not inorganic phosphate.

16. A 3.75-µL aliquot of the original master mix corresponds to the fraction counted ($V = 150$ µL) after two consecutive charcoal extractions.

17. The cpm value of the 3.75-µL aliquot/cpm value of 150 µL of stopped reaction is *x*; the total amount of ATP in a 10-µL reaction (100 nmol)/*x* = apparent amount of ATP hydrolyzed (nmol); the apparent amount of ATP hydrolyzed – amount of ATP hydrolyzed in control sample without protein = amount of ATP hydrolyzed by the added protein.
18. The same strategy can also be applied for making forked substrates. Also, we have used basically the same protocol with three oligos for preparing substrates with a nick or a gap in one DNA strand as well as more complicated forked substrates. With three oligos, we recommend taking two shorter oligos in twofold molar excess. This ensures that the annealed reaction mixture will contain very little hybrids consisting of only two oligonucleotides. Often, these molecules are not separated well from hybrids consisting of three oligonucleotides in the gel system we use. When making triplet forked substrates, we label the shorter oligos with [γ-^{32}P] ATP in a polynucleotide kinase reaction before hybridizing to the LONG oligo.
19. We use this method to recover the substrate from the gel because, in our hands, it is milder than other purification methods. Thus, we can avoid any denaturation of the hybrid molecules that would otherwise generate unnecessary background in the helicase assay. Recovery of the procedure is approx 50%.
20. Include one reaction without enzyme as a negative control. To get a completely unwound substrate, denature one sample for 3 min at 100°C and put on ice.
21. As the helicase substrates used in these assays are based on quite short oligo-nucleotides, it might be helpful to dry the gels onto Whatman DE81 paper placed between the gel and a layer of 3MM paper, so that the losses during vacuum drying are minimal. This step is absolutely necessary for quantitative experiments.

Acknowledgments

We are grateful to Annely Kukk for excellent technical support and to Aare Abroi for critical reading of the manuscript. This work was funded by research grant from Estonian Science Foundation to J. S. (grant 4474).

References

1. Foury, F. and Lahaye, A. (1987) Cloning and sequencing of the *PIF* gene involved in repair and recombination of yeast mitochondrial DNA. *EMBO J.* **6,** 1441–1449.
2. Lahaye, A., Stahl, H., Thines-Sempoux, D., and Foury, F. (1991) Pif1: a DNA helicase in yeast mitochondria. *EMBO J.* **10,** 997–1007.
3. Schulz, V. P. and Zakian, V. A. (1994) The *Saccharomyces* PIF1 DNA helicase inhibits telomere elongation and *de novo* telomere formation. *Cell* **76,** 145–155.
4. Ivessa, A. S., Zhou, J.-Q., and Zakian, V. A. (2000) The *Saccharomyces* Pif1p DNA helicase and highly related Rrm3p have opposite effects on replication fork progression in ribosomal DNA. *Cell* **100,** 479–489.

5. Sedman, T., Kuusk, S., Kivi, S., and Sedman, J. (2000) A DNA helicase required for maintenance of the functional mitochondrial genome in *Saccharomyces cerevisiae*. *Mol. Cell. Biol.* **20,** 1816–1824.

6. Lee, C. M., Sedman, J., Neupert, W., and Stuart, R. A. (1999) The DNA helicase, Hmi1p, is transported into mitochondria by a C-terminal cleavable targeting signal. *J. Biol. Chem.* **274,** 20,937–20,942.

7. Lahaye, A., Leterme, S., and Foury, F. (1993) PIF1 DNA helicase from *Saccharomyces cerevisiae*. *J. Biol. Chem.* **268,** 26,155–26,161.

8. Zhou, J.-Q., Monson, E. K., Teng, S.-C., Schulz, V. P., and Zakian, V. A. (2000) Pif1p helicase, a catalytic inhibitor of telomerase in yeast. *Science* **289,** 771–774.

9. Sambrook, J., Fritsch, G. F., and Maniatis, T. (1989) *Molecular Cloning: A Laboratory Manual*, 2nd ed., Cold Spring Harbor Laboratory Press, Cold Spring Harbor, NY.

10. Studier, F. W. and Moffatt, B. A. (1986) Use of bacteriophage T7 RNA polymerase to direct selective high-level expression of cloned genes. *J. Mol. Biol.* **189,** 113–130.

11. Bradford, M. M. (1976) A rapid and sensitive method for the quantitation of microgram quantities of protein utilizing the principle of protein–dye binding. *Anal. Biochem.* **72,** 248–254.

12. Laemmli, U. K. (1970) Cleavage of structural proteins during the assembly of the head of bacteriophage T4. *Nature* **227,** 680–685.

22

Isolation of DNA Topoisomerases I and II Activities from Mitochondria of Mammalian Heart

Robert L. Low

1. Introduction

DNA topoisomerases constitute a ubiquitous class of enzymes that engage chromosomal DNA and promote the passage of the two strands of the DNA helix through one another (reviewed in **refs.** *1* and *2*). This occurs while preserving the phosphodiester backbone of each strand. In some instances, the DNA topoisomerase promotes passage of one helical segment of duplex DNA through another segment. The ability to transiently break DNA, pass DNA through the break, and then rejoin the DNA gives the DNA topoisomerases the ability to change the linking number of DNA (the number of times the two strands of DNA in a constrained chromosome revolve around one another). In vivo, the DNA topoisomerase activities are needed to alleviate torsional strain imposed on the DNA helix. They eliminate, for example, the overwinding or underwinding of the DNA helix that occurs during DNA replication, transcription, and at the end of the DNA replication cycle when daughter chromosome helices need to be disentangled from one another. In vitro, this topoisomerase activity is reflected in the enzyme's capacity to remove supercoils from supercoiled DNA templates, as well as to knot/unknot and catenate/decatenate circular, covalently closed DNA circular substrates.

Genetic, biochemical, and structural studies of DNA topoisomerases in bacteria (in particular, *Escherichia coli*), yeasts, and vertebrates have identified three families of DNA topoisomerase activity (reviewed in **ref.** *3*). These families are denoted type 1A, 1B, and II, respectively. The type 1A, 1B group includes the procaryotic and eucaryotic DNA topoisomerase I (topo I) enzymes, respectively, as well as the procarytic topo I-like DNA topoisomerase III

From: *Methods in Molecular Biology, vol. 197: Mitochondrial DNA: Methods and Protocols*
Edited by: W. C. Copeland © Humana Press Inc., Totowa, NJ

activities. The enzymes of this 1A, 1B group break only one of the DNA strands during catalysis, change the linking number of the DNA in steps of one, and do not require ATP. The enzymes in the type II family, in contrast, produce 5′-staggered breaks in both DNA strands during catalysis and generate an enzyme–DNA covalent intermediate in which the active-site tyrosine of each monomer of the dimeric enzyme is covalently attached via the DNA phosphate of each breakpoint. In a subsequent step, this intermediate acts as a "gate" for passage of another segment of duplex DNA through the DNA, prior to resealing and restoring the phosphodiester linkage of the DNA as the enzyme dissociates. These type II activities consequently change the linking number of the DNA in steps of two and also require ATP *(4,5)*. Although *Saccharomyces cerevisiae* and *Drosophila* only contain one type II DNA topoisomerase gene, mammalian cells contain two genes, for enzyme isoform homologs, called DNA topoisomerase IIα and IIβ (reviewed in **refs.** *6* and *7*). In humans, these genes map to chromosomes 17q21–22 and 3p24, respectively. Mammalian cells also contain two isoforms of DNA topoisomerase III, named topoisomerase IIIα and IIIβ *(8,9)*.

Compared to the remarkable progress made in the study of nuclear DNA topoisomerases in the past several years, the identification and study of the mitochondrial DNA topoisomerase activities have met limited success. The presence of potent endonuclease activity (Endonuclease G) *(10)* and ATPase in crude fractions of mitochondrial protein has been especially problematic. These activities effectively prevent detection of any ATP/MG^{2+}-dependent topoisomerase activity in vitro (e.g., DNA topoisomerase II). In addition, topoisomerase activities are present in mitochondria at very low levels, perhaps in amounts 0.1–1% to that of nuclear types II and I DNA topoisomerases (estimated to be 10^5–10^6 enzyme molecules/nucleus, respectively). This had made it very difficult to purify enzymes to near-homogeneity. Furthermore, the presence of small fragments of nuclear DNA in standard preparations of mitochondria has also raised concerns that activities attributed to mitochondria could, in fact, represent nuclear contaminants.

Thus far, the only DNA topoisomerase activity conclusively demonstrated to be mitochondrial in origin is a type II enzyme, called DNA topoIImt, found within the mitochondrion of the trypanosomatid *Crithidia fasciculata* *(11)*. Unlike mtDNA in vertebrates and fungi, the mtDNA in trypanosomes exists as a huge, catenated network of minicircle and maxicircle DNAs termed kinetoplast DNA *(12)*. DNA topoIImt likely plays a role in kinetoplast DNA replication and in maintaining the structure of the kinetoplast network, promoting decatenation of parental circles from the mtDNA network, segregating catenated dimers, and recatenating newly replicated daughter circles to the network. Evidence for the existence of a DNA topoisomerase II activity in trypanosome mitochondria also emerged from the discovery that the epipodophyllotoxins, a class of drugs that promote cleavage of DNA by type II enzymes, promote cleavage of kinetoplast DNA minicircles *(13)*. The trypanosome DNA topoIImt was subsequently purified to near-homogeneity and shown to a homodimer of a 132-kDa polypeptide *(14)*. The purified enzyme was shown to possess enzymatic properties of a classic eucaryotic type II enzyme and, as deduced from its gene sequence, to share partial, amino acid-sequence identity with that of other eucaryotic type II enzymes. Antibodies prepared against the purified protein have been used to demonstrate both that the DNA topoIImt is physically located at the periphery of the kinetoplast network and that the enzyme becomes covalently crosslinked to the kinetoplast DNA circles when cells are treated with etoposide *(15)*.

Although the existence of a type II activity in mammalian mitochondria was proposed over 20 yr ago *(16)*, and still seems likely, the evidence for such an activity has remained inconclusive. Several years ago, a putative type II topoisomerase activity was identified in protein fractions of mitochondria isolated from both human leukemic cells *(17)* and calf thymus *(18)*. Unfortunately, these activities were never purified to homogeniety. Whether the human and calf thymus activities identified were truly mitochondrial in origin or derived from trace fragments of nuclear DNA that still contaminated the mitochondrial membrane fractions still remains unresolved. These activities were identified and partially purified using ATP-dependent catenation/decatenation and unknotting assays, respectively *(17,18)*. Using these assays, this activity was demonstrated to be sensitive to topo II inhibitors such as VP16 and VP26. Unfortunately, neither activity was shown to be capable of relaxing a supercoiled DNA template in an ATP-dependent fashion. The existence of a type II topoisomerase in mammalian mitochondria has also been suggested from the recognition that a type II activity may produce a common deletion found in human mtDNA. Nucleotide sequences where the mtDNA is deleted seem to resemble a nucleotide consensus sequence often targeted by vertebrate type II DNA topoisomerases *(19,20)*.

Although DNA topoisomerase II activity has been difficult to find in mitochondria, a DNA topoisomerase I activity with enzymatic properties similar to those of the nuclear topo I activity has been identified in isolated mitochondria from a number of cell sources, including oocytes of *Xenopus laevis (21)*, rat liver *(22)*, calf thymus *(23)*, human leukemic cells *(24)*, bovine liver *(25)*, *S. cerevisiae (26)*, and human platelets *(27)*. The presence of this activity in mitochondria of purified human platelets that are effectively devoid of nuclear contamination provides strong evidence that mammalian mitochondria indeed contain a type I enzyme. The platelet mitochondrial topo I activity reportedly shows a strong preference for Ca^{2+} over other divalent cations. Its activity has been shown to reside with a 60-kDa polypeptide, a size significantly smaller than that of the 100-kDa nuclear enzyme *(27)*. Of interest, an "autoimmune" antibody that reacts with the nuclear topo I also recognizes this polypeptide. This finding raises the possibility that the same gene, or a very closed related gene, encodes the mitochondrial activity *(28)*. In contrast, a larger, 78-kDa, polypeptide has been found to be associated with the purified calf thymus mitochondrial topo I *(23)*. Although the latter activity exhibits sensitivity to camptothecin and ATP inhibition similar to that of the nuclear topo I, the mitochondrial topo I reportedly shows a distinct pH profile, thermal stability, and response to the inhibitor Berenil. This has suggested that the mitochondrial activity is a different enzyme than that in the nucleus.

In yeast, the mitochondrial topo I activity is associated with a 79-kDa polypeptide that, as found with the platelet enzyme, crossreacts with antiserum against its nuclear topo I protein *(26)*. This mitochondrial activity appears absent from yeast strains in which the *TOP1* gene has been inactivated *(29)*. The mitochondrial enzyme may be encoded by the *TOP1* gene, but this still remains uncertain because the mitochondrial enzyme shows sensitivity to 2,2,5,5-tetramethyl-4-imidazolidinone and alkaline pH and a strict requirement for a divalent cation for activity that the nuclear topo I does not. In addition to this prototypic type I activity, an unusual ATP(ADP)-activated topoisomerase-like activity has also been found in yeast mitochondria. Of interest, this latter enzyme is not apparently encoded by either *TOP1*, *TOP2*, or *TOP3 (30)*.

The identity of any of the mitochondrial topo I enzymes, whether a processed form of the nuclear topo I or a protein specific to the mitochondrion, should become evident once amino acid sequences of peptides, prepared from the purified enzyme, can be obtained. A method to isolate the mitochondrial topo I associated with complexes of mtDNA and protein recovered from bovine heart mitochondria is described in **Subheading 3.1.** In addition, in the past several months, we successfully purified a type II DNA topoisomerase activity from bovine heart mitochondria to near-homogeneity. This activity was detected using an assay that follows the enzyme's ability to carry out ATP-dependent relaxation of a supercoiled plasmid DNA substrate. A method to identify this activity is presented in **Subheading 3.2.** The topoII activity copurifies with mitochondria, which are isolated free of nuclear contamination. Identification of tryptic peptides prepared from the purified protein using mass spectrometry reveals that this mitochondrial activity is, in fact, a truncated form of DNA topo IIβ (R. Low and S. Orton, unpublished results).

2. Materials

2.1. Isolation of Mitochondria

All reagents can be purchased from Sigma-Aldrich (St. Louis, MO), except as noted.

1. Two adult bovine hearts, obtained fresh from a local meat processing plant, packed in ice and transported directly to the laboratory.
2. Cheesecloth (Softwipe; American Fiber and Finishing, Albemarle, NC), double layered, cut into 20×15-cm-length strips.
3. Eight-inch surgical knife, assorted spatulas.
4. Commercial meat grinder, operated in cold room, 4°C.
5. Stainless-steel (5 L), three-speed, commercial Waring blender.
6. Tissue wash buffer (buffer A): Prepare 12 L. Per liter: 15 mL of 2 M Tris-HCl (pH 7.9), 8.7 g of NaCl, and 10 mL of 0.5 M ethylenediaminetetraacetic acid (EDTA), pH 8.0. Store at 4°C.
7. Tissue homogenization buffer (buffer B): Prepare 12 L. Per liter: 15 mL of 2 M Tris-HCl, pH 7.9, 10 mL of 0.5 M EDTA, pH 8.0, 8.45 g of sodium glutamate, and 103 g of sucrose. Store at 4°C.
8. 250- and 500-mL Centrifuge bottles.
9. 15- and 40-mL Dounce homogenizers.
10. Sterile plastic centrifuge tubes, 50 mL.
11. Liquid-nitrogen storage container.

2.2. Isolation of mtDNA–Protein Complex

1. Mitochondria dilution buffer (buffer C): Prepare 1 L. Per liter: 15 mL of 2 M Tris-HCl, pH 7.9, 0.05 mL of 0.5 M EDTA, pH 8.0, 5.8 g of NaCl, 3.7 g of potassium glutamate, and 100 g of glycerol. Store at 4°C.
2. 1 M DTT (1-mL aliquots, stored at –20°C).
3. 10% (w/v) Triton X-100 and 10% (w/v) n-octylglucoside (prepared fresh).
4. Protease inhibitors: PMSF (100 mM in isopropyl alcohol, stored at –20°C), pepstatin and leupeptin (each 10-mg/mL stocks stored at –20°C), sodium metabisulfite.

2.3. Isolation of mtDNA Topoisomerases I and II

1. Chromatography buffer (Buffer D): Prepare 1 liter. Per liter: 15 ml of 2 M Tris-HCl, pH 7.9, 5.8 g of NaCl, 3.7 g of potassium glutamate, and 200 g of glycerol. Store at 4°C.
2. DEAE-Sepharose (10 ml), washed extensively with 500 ml of 1 M NaCl, followed by 1 liter of H_2O, stored at 4°C, up to one week.
3. Hydroxylapatite (HTP grade, BioRad).
4. Centricon 10 filtration concentrators (Amicon).

5. Chromatography supplies (Pharmacia).
6. Electrophoretic grade agarose (Gibco-BRL).

3. Methods

All steps should be carried out at 4°C unless otherwise stated.

3.1. Isolation of Mitochondria from Adult Bovine Heart Ventricular Myocardium

1. Open the heart chambers using surgical-type scissors. First, open the right and left ventricles by cutting along the pulmonic and aortic outflow tracts. Extend the incisions by cutting across the tricuspid and mitral valves into the right and left atria, respectively. Trim the ventricular myocardial tissue free of epicardial fat and heart valve tissue. Slice the right and left ventricular muscle into $1 \times 1 \times 2$-cm strips, using a surgical knife. Weigh the myocardial tissue. The combined weight usually averages about 2400 g. After weighing, mince the tissue by passing it once through an electric, commercial meat grinder.
2. Wash the minced tissue to remove blood (*see* **Note 1**). Place one-half (1200 g) of the tissue into a 4-L plastic beaker and fill with buffer A. Stir gently using a large spatula to suspend the tissue. Incubate at 4°C for 5 min. Stir several more times before filtering the suspension through a colander, lined with double-layered cheesecloth. Pour an additional liter of buffer A over the tissue and stir gently. Once the sanguineous fluid has drained, wrap the cheesecloth around the tissue and squeeze out the remaining liquid. Distribute the tissue among six 500-mL centrifuge bottles. Fill each bottle with buffer A and mix by gentle shaking. Centrifuge at $500g$ for 5 min. Discard the supernatant. Refill each bottle with buffer A, shake, and recentrifuge at $500g$ for 5 min. The washed tissue should appear pale tan–brown. Transfer the tissue to a clean 4-L beaker. Repeat the washing procedure with the remaining 1200 g of the minced tissue.
3. In this step, homogenize the washed tissue in a Waring blender. To each 1200-g portion of tissue, add 2500 mL of buffer B. Stir the suspension with a large spatula and quickly pour one-third of the volume into the Waring blender chamber. Blend at the lowest speed setting for 20 s. Wait 10 min and repeat blending for another 20 s (*see* **Note 2**). Transfer the tan–pink, liquid homogenate into a clean 4-L beaker. Repeat this procedure until all of the tissue has been processed.
4. Next, isolate the mitochondria by differential centrifugation. Centrifuge the homogenate at $500g$ for 10 min in 500-mL centrifuge bottles. Carefully decant the turbid supernatant ("cell lysate") from each bottle through a double-layer sheet of cheesecloth, stretched across the top of a clean 4-L beaker. The cheesecloth serves to strain fat and loose clumps of cellular debris. Discard the loose tan–brown pellet of cellular debris. The cheesecloth tends to clog after about 600 mL of lysate has been filtered. Replace the cheesecloth as needed. Once all of the lysate has been collected, recentrifuge it at $500g$ for 10 min. Carefully decant the supernatant to avoid carrying over the loose, brown particulate material that

has pelleted at the bottom of the bottle. This material contains nuclear DNA fragments and cell debris. Repeat this low-speed centrifugation (500*g*, 10 min) step six more times. After the final clarification, collect the mitochondria at 17,000*g* for 20 min. Discard the supernatant.

5. Wash the pelleted mitochondria. To do this, resuspend the pellet and dilute to 1000 mL using buffer B. Recentrifuge the suspension at 17,000*g* for 20 min. Collect the "washed" mitochondrial pellets and resuspend them in a total of 160 mL of buffer B using a Dounce homogenizer. Four to five passes of the pestle should be sufficient.

6. Distribute the mitochondrial suspension among four 50-mL plastic tubes; freeze and store the tubes in liquid nitrogen.

3.2. Isolation of the Insoluble Complex of mtDNA and Its Associated Proteins

1. Thaw two 50-mL tubes of the frozen mitochondria at 4°C. Dilute the thawed mitochondria with 200 mL of buffer C and supplement with 0.5 m*M* PMSF, 1 m*M* sodium metabisulfite, and 0.5 µg/mL of pepstatin and leupeptin. Add 10 mL of Triton X-100 in drops, with stirring, to promote lysis of the mitochondria. Incubate on ice for 30 min.

2. Next, centrifuge the lysate (fraction I) at 150,000*g* for 60 min. Discard the supernatant fraction. Save the pellets. Resuspend the pellets in 70-mL of buffer C. Use a Dounce homogenizer. Perform several passes of the pestle until no particles are evident. Douncing takes several min.

3. After 60 min at 4°C, centrifuge the suspension at 3000*g* for 10 min. Carefully decant the supernatant (#1). Wash the pellets with 35 mL total of buffer B. Recentrifuge this resuspension for 3000*g* for 10 min. Save the supernatant (#2). Discard the pellet. This pellet contains fragments of nuclear DNA.

4. Combine the supernatants (#1 and #2) and centrifuge at 150,000*g* for 60 min. Resuspend the dark brown pellets with 20 mL of buffer B using a Dounce homogenizer. Add 6 mL of 10% (w/v) *n*-octylglucoside to this suspension (*see* **Note 3**). Incubate on ice for 30 min.

5. Centrifuge the suspension at 150,000*g* for 60 min. Discard the supernatant and save the pellet, which contains the mtDNA–protein complex (fraction II).

3.3. Isolate the mtDNA Topoisomerase I and II Activities

1. Recover the DNA topoisomerase activities and other mtDNA replicative proteins from the mtDNA–protein complex. Resuspend the mtDNA–protein pellet with 15 mL of 300 m*M* Tris-HCl (pH 7.9), 1 *M* NaCl, 20 m*M* EDTA, 20 m*M* DTT, using a 15-mL Dounce homogenizer. Use multiple, repeat passes of the pestle. Incubate the suspension at least 2 h at 4°C.

2. Centrifuge the suspension at 150,000*g* for 60 min. Carefully decant the supernatant (fraction III). Measure the protein concentration of the fraction III using the Bradford assay *(31)*. This fraction is stable at 4°C for at least 2 d.

3. Before proceeding, assess whether there is significant contamination of the mtDNA pellet by nuclear DNA fragments. To do this, resuspend the pellet in 10 mL of 10 m*M* Tris-HCl, pH 7.5, 1 m*M* EDTA (TE). Add SDS to 1%, and add proteinase K to 0.1 mg/mL and incubate 3 h at ambient temperature to hydrolyze proteins. Next, extract the DNA successively with an equal volume of phenol:chloroform (1:1) followed by chloroform, prior to precipitating the DNA in 70% ethanol. Resuspend the DNA in 1 mL of TE and dialyze overnight against 2 L of TE. Digest a 50-μL aliquot of the DNA with 1 μg of RNAse A for 30 min at 37°C, followed by 20 U of *Eco*R1 restriction endonuclease for 4 h at 37°C, in buffer supplied by the manufacturer (New England Biolabs). Phenol-extract the DNA digest and run the DNA through a 0.8% (w/v) agarose gel in a TAE buffer system at 1.5 V/cm for 10 h. Stain the gel in 1 μg/mL of ethidium bromide for 30 min, destain in H_2O for 5 min, and photograph the band patterns under ultraviolet (UV) (300 mm) illumination. The *Eco*R1 fragments of mtDNA should be the predominant species (**Fig. 1**) (*see* **Note 4**).

4. Prepare to fractionate the mt DNA topoisomerase I and II activities present in fraction III by hydroxylapatite chromatography. To prepare for the hydroxylapatite step, add 1 mL of washed DEAE–Sepharose to fraction III (approx 12 mL), mix by inversion, and slowly add 36 mL of cold H_2O containing 10 m*M* DTT. Mix further by inversion, and after 10 min, centrifuge at 8000*g* for 10 min. Carefully decant the supernatant (fraction IV). This DEAE treatment removes any residual contaminants of nucleic acids in fraction III.

5. Directly apply fraction IV to a 10-mL (5×2-cm^2) column of hydroxylapatite equilibrated with buffer D. Once the fraction is applied, wash the column with 20 mL of buffer D and elute with a 100-mL linear gradient of 0–1.5 *M* potassium phosphate, pH 8.0, containing buffer D, plus 10 m*M* dithiothreitol, 0.5 m*M* PMSF, and 0.5 μg/mL pepstatin and leupeptin. Collect 1.5-mL fractions.

6. Assay even fractions, starting with fraction 14 for DNA topoisomerase II activity. Each 40-μL reaction contains 0.5 μL of 2 *M* Tris-HCl, pH 7.9, 1 μL of 5 *M* NaCl, 0.3 μL of 1 *M* Mg(OAc)$_2$, 0.25 μL of 20 mg/mL bovine serum albumin, 0.1 μL of 1 *M* DTT, 0.5 μL of 1 mg/mL pUC18 plasmid DNA, 1 μL of the fraction being assayed, and none or 1 μL of 100 m*M* ATP. Reactions are incubated 60 min at 37°C and stopped by the addition of 2 μL of 20% (w/v) SDS.

7. The terminated reactions are then applied to 150 mL (12×15 cm) 0.8% (w/v) agarose gel. Electrophoresis is carried out at 1.5 V/cm for 10 h in a TAE buffer system. After electrophoresis, the gel is stained in ethidium bromide and photographed under UV illumination as detailed in **step 3**. One unit of DNA topoisomerase II activity is the amount of enzyme that relaxes 50% of the input DNA in 60 min. Assay each even fraction plus and minus ATP. The ATP-dependent topo II activity usually elutes starting at about fraction 28.

8. Pool fractions 28–34 (fraction V) and concentrate the pool to 200 μL in a Centricon 10 filtration concentrator, at 5000*g*, as recommended by the manufacturer.

9. Assay even fractions from the hydroxylapatite column for DNA topoisomerase I starting at fraction 30. The DNA topoisomerase I reaction mixture is identical to

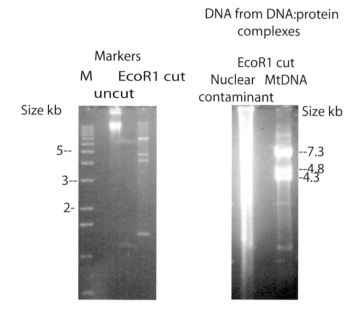

Fig. 1. Assessment of whether the isolated complexes of mtDNA and protein are contaminated with fragments of nuclear DNA. See text for details. Molecular size markers are shown in the gel on the left. Lane M contains a commercial 1-kb ladder (Gibco-BRL); lane 2, labeled uncut, corresponds to an approx 1 µg sample of bovine mtDNA; lane 3 contains an *Eco*R1 digest prepared from a second sample of the mtDNA shown in lane 2. The ethidium-stained gel on the right shows restriction enzyme analysis of the DNAs, which were phenol extracted from the pellet collected at 3000*g* for 10 min (the "nuclear contaminant"), and the pellet of mtDNA–protein collected from the subsequent centrifugation at 150,000*g* for 60 min. Each DNA fraction was dialyzed, treated with RNase A, digested with *Eco*R1 restriction endonuclease, and samples of the digests analyzed by agarose gel electrophoresis. The left lane, labeled nuclear contaminant, corresponds to DNA extracted from the 3000*g* at 10 min pellet; the intense 7.3-, 4.8-, and 4.3-kb *Eco*R1 fragments of bovine mtDNA in the right lane correspond to bands obtained with the DNA extracted from the mtDNA–protein complex and indicate that this fraction is relatively free of nuclear contamination.

that used to assay topo II, except that ATP is omitted and the incubation time is shortened to 5 min. The extent of DNA template relaxation is assessed by agarose gel electrophoresis, as detailed in **step 7**. The peak of DNA topoisomerase I activity usually elutes between fractions 40 and 44.

10. Pool the topo I peak fractions (fraction V, topo I) and similarly concentrate the pool to 200 µL in a Centricon 10 filtration concentrator. One unit of DNA topoisomerase I activity is the amount of enzyme that relaxes 50% of the input DNA in 5 min at 37°C.

A   B

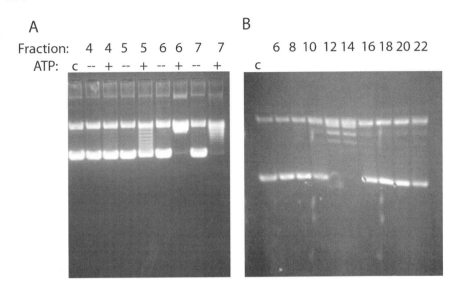

Fig. 2. Further purification of mt DNA topoisomerases I and II by glycerol-gradient velocity sedimentation. See text for details of assay and sedimentation conditions. Photographs of the ethidium stained agarose gel assays are shown. (**A**) Assays of DNA topoisomerase II (fraction V) recovered from its glycerol gradient. DNA topoisomerase II assays of glycerol gradient fractions 4–7, minus and plus ATP, are as indicated. (**B**) Assays of DNA topoisomerase I (fraction V) recovered from its glycerol gradient. DNA topoisomerase I assays (minus ATP) of glycerol gradient even fractions from 6 to 22 are as indicated.

11. Purify the DNA topoisomerase II activity further by glycerol-gradient velocity sedimentation. In order to remove trace contamination of the DNA topoisomerase II fraction V by endonuclease and topoisomerase I activities, the fraction V topo II pool should be further purified on a glycerol gradient. Dilute a 100-μL aliquot of the fraction V (topo II) pool with an equal volume of H_2O and carefully layer the sample on top of a 4-mL linear 15–42% (w/v) glycerol gradient containing 300 mM NaCl, 30 mM Tris-HCl, pH 7.9, 10 mM Mg(OAc)$_2$, 10 mM DTT, and 0.1% (w/v) *n*-octylglucoside. Carry out the sedimentation at 300,000g in a swinging-bucket rotor for 20 h at 3°C. Collect the gradient by drops from the bottom of the tube. Collect 22–24 five-drop fractions. The topoisomerase II activity should be assayed as described in **step 6**, using 2-μL aliquots of fractions being assayed. The topoisomerase II activity usually peaks at about fraction 6 (**Fig. 2A**). Pool peak fractions. This activity is stable for about 10 d at 4°C. A residual contaminate of topo I activity in the topo II fraction V pool usually elutes around fraction 12–14 (not shown).

12. The DNA topoisomerase I fraction V pool should be similarly further purified on a glycerol gradient. Similarly dilute the fraction V topoisomerase I concentrate

1:1 with H_2O and layer onto a 4-mL linear 12–42% glycerol prepared as in **step 11**. Carry out the sedimentation and collect fractions as described in **step 9** for topo II. Assay fractions for topoisomerase I as detailed in **Subheading 3.3.**

13. Pool peak fractions (e.g., fractions 12–14 [fraction VI] (**Fig. 2B**). This activity is stable (90% retention of activity) for at least 2 mo at 4°C.

4. Notes

1. The heart is a highly vascular tissue and contains a large amount of blood. Once the heart tissue is minced, most of the blood should be washed away, as much as possible. We have observed that failure to adequately remove the red blood cells from the disrupted tissue promotes significant oxidation of mtDNA replicative proteins and crosslinking of proteins to the DNA. Many of the DNA replicative activities bound to the mtDNA–protein complexes cannot be released from the DNA by a high-salt (1 M NaCl) treatment unless reducing agents such as 10 mM DTT or 30 mM of 2-mercaptoethanol are added.

2. Blending the heart tissue at the highest speed setting or for longer times (>20 s) in the Waring blender appears to only slightly improve the yield of mitochondria recovered per gram of tissue and leads to mitochondrial damage and significantly more contamination of the mitochondria with fragments of nuclear debris and nuclear DNA.

3. Washing the suspension of the mtDNA–protein complexes with 2–3% *n*-octylglucoside prior to treating the complexes with 1 M NaCl eliminates a number of detergent-soluble, membrane-associated proteins that otherwise contaminate the fraction III. This *n*-octylglucoside wash removes most, although not all, of the Endonuclease G activity from the mtDNA–protein complex fraction but does not release DNA polymerase γ or mt DNA topoisomerase I and II activities.

4. If the isolated mtDNA–protein complex is contaminated with unacceptable levels of nuclear DNA, shorten the blending time to 10 s when disrupting the minced myocardial tissue and extend the 3000g centrifugation step to remove nuclear DNA fragments from 10 min to 20 min. If contamination is still a problem, try purifying the unbroken mitochondria on linear 0.5–2 M sucrose gradient prior to disruption of the mitochondria with Triton and isolation of the mtDNA–protein complex.

References

1. Cozzarelli, N. R. (1980) DNA topoisomerases. *Cell* **22(Pt. 2)**, 327–328.
2. Wang, J. C. (1985) DNA topoisomerases. *Annu. Rev. Biochem.* **54**, 665–697.
3. Wang, J. C. (1996) DNA topoisomerases. *Annu. Rev. Biochem.* **65**, 635–692.
4. Brown, P. O. and Cozzarelli, N. R. (1979) A sign inversion mechanism for enzymatic supercoiling of DNA. *Science* **206**, 1081–1083.
5. Liu, L. F., Liu, C.-C., and Alberts, B. M. (1980) Type II DNA topoisomerases: enzymes that can unknot a topologically knotted DNA molecule via a reversible double-strand break. *Cell* **19**, 697–707.

6. Watt, P. M. and Hickson, I. D. (1994) Structure and function of type II DNA topoisomerases. *Biochem. J.* **303**, 681–695.

7. Austin, C. A. and Marsh, K. L. (1998) Eucaryotic DNA topoisomerase IIβ. *Bioessays* **20**, 215–226.

8. Hanai, R., Caron, P. R., and Wang, J. C. (1996) Human TOP3: a single-copy gene encoding DNA topoisomerase III. *Proc. Natl. Acad. Sci. USA* **93**, 3653–3657.

9. Ng, S.-W., Liu, Y., Hasselblatt, K. T., Mok, S. C., and Berkowitz, R. S. (1999) A new human topoisomerase III that interacts with SGS1 protein. *Nucleic Acids Res.* **27**, 993–1000.

10. Cummings, O. W., King, T. C., Holden, J. A., and Low, R. L. (1987) Purification and characterization of the potent endonuclease in extracts of bovine heart mitochondria. *J. Biol. Chem.* **62**, 2005–2015.

11. Melendy, T., Sheline, C., and Ray, D. S. (1988) Localization of a type II DNA topoisomerase to two sites at the periphery of the kinetoplast DNA of *Crithidia fasciculata*. *Cell* **55**, 1083–1088.

12. Simpson, L. (1986) Kinetoplast DNA in trypanosomid flagellates. *Intl. Rev. Cytol.* **99**, 119–179.

13. Shapiro, T. A., Klein, V. A., and Englund, P. A. (1989) Drug-promoted cleavage of kinetoplast DNA minicircles. *J. Biol. Chem.* **264**, 4173–4178.

14. Melendy, T. and Ray, D. S. (1989) Novobiocin affinity purification of a mitochondrial type II topoisomerase from the trypanosomatid *Crithidia fasciculata*. *J. Biol. Chem.* **264**, 1870–1876.

15. Ray, D. S., Hines, J. C., and Anderson, M. (1992) Kinetoplast-associated DNA topoisomerase in *Crithidia fasciculata*: crosslinking of mitochondrial topoisomerase II to both minicircles and maxicircles in cells treated with the topoisomerase inhibitor VP16. *Nucleic Acids Res.* **20**, 3353–3356.

16. Castora, F. J. and Simpson, M. V. (1979) Search for a DNA gyrase in mammalian mitochondria. *J. Biol. Chem.* **254**, 11,193–11,195.

17. Castora, F. J., Lazarus, G. M., and Kunes, D. (1985) The presence of two mitochondrial DNA topoisomerases I human acute leukemia cells. *Biochem. Biophys. Res. Commun.* **130**, 854–866.

18. Lin, J.-H. and Castora, F. J. (1991) DNA topoisomerase II from mammalian mitochondria is inhibited by the antitumor drugs, m-AMSA and VM-26. *Biochem. Biophys. Res. Commun.* **176**, 690–697.

19. Mita, S., Rizzuto, R., Moraes, C. T., Shanske, S., Arnaudo, E., Fabrizi, G. M., et al. (1990) Recombination via flanking direct repeats is a major cause of large-scale deletions of human mitochondrial DNA. *Nucleic Acids Res.* **18**, 561–567.

20. Blok, R. B., Thorburn, D. R., Thompson, G. N., and Dahl, H.-H. M. (1995) A topoisomerase II cleavage site is associated with a novel mitochondrial DNA deletion. *Hum. Genet.* **95**, 75–81.

21. Brun, G., Vannier, P., Scovassi, I., and Callen, J. C. (1981) DNA topoisomerase I from mitochondria of Xenopus laevis oocytes. *Eur. J. Biochem.* **118**, 407–415.

22. Fairfield, F. R., Bauer, W. R., and Simpson, M. V. (1985) Studies on mitochondrial type I topoisomerase and on its function. *Biochim. Biophys. Acta* **824**, 45–57.

23. Lazarus, G. M., Henrich, J. P., Kelly, W. G., Schmitz, S. A., and Castora, F. J. (1987) Purification and characterization of a type I DNA topoisomerase from calf thymus mitochondria. *Biochemistry* **26,** 6195–6203.
24. Castora, F. J. and Lazarus, G. M. (1984) Isolation of a mitochondrial DNA topoisomerase from human leukemia cells. *Biochem. Biophys. Res. Commun.* **121,** 77–86.
25. Lin, J. H. and Castora, F. J. (1995) Response of purified mitochondrial DNA topoisomerase I from bovine liver to camptothecin and m-AMSA. *Arch. Biochem. Biophys.* **324,** 293–299.
26. Wang, A. T., Kulpa, V., and Wernette, C. M. (1997) Mitochondrial DNA topoisomerase I of *Saccharomyces cerevisiae. Biochimie* **79,** 341–350.
27. Kosovsky, M. J. and Soslau, G. (1991) Mitochondrial DNA topoisomerase I from human platelets. *Biochim. Biophys. Acta* **1078,** 56–62.
28. Kosovsky, M. J. and Soslau, G. (1993) Immunological identification of human platelet mitochondrial DNA topoisomerase I. *Biochim. Biophys. Acta* **1164,** 101–107.
29. Wang, J., Kearney, K., Derby, M., and Wernette, C. M. (1995) On the relationship of the ATP-independent, mitochondrial associated DNA topoisomerase of *Saccharomyces cerevisiae* to the nuclear topoisomerase I. *Biochem. Biophys. Res. Commun.* **214,** 723–729.
30. Ezekiel, U. R., Towler, E. M., Wallis, J. W., and Zassenhaus, H. P. (1994) Evidence for a nucleotide-dependent topoisomerase activity from yeast mitochondria. *Curr. Genet.* **27,** 31–37.
31. Bradford, M. M. (1976) A rapid and sensitive method for the quantitation of microgram quantities of protein utilizing the principle of protein–dye binding. *Anal. Biochem.* **72,** 248–254.

23

Endonuclease G Isolation and Assays

Robert L. Low and Mariana Gerschenson

1. Introduction

Mitochondria in mammalian cells contain a nuclearly encoded, Mg^{+2} (Mn^{+2})-dependent DNA endonuclease activity (reviewed in **ref. 1**), that has been termed Endonuclease G (or Endo G) *(2)*. This enzyme activity, an approx 60-kDa homodimer of a 29-kDa polypeptide, accounts for nearly all of the potent DNA endonuclease activity identified in crude protein extracts of isolated mitochondria *(3)*. The 1133-bp (base pairs) cDNA for the human Endonuclease G gene has been mapped to chromosome 9q34.1 *(4)*.

Endonuclease G is readily detected in standard nuclease assays performed with mitochondrial extracts of soluble protein. The activity only becomes evident once the membranes of isolated mitochondria are disrupted by the addition of detergent. In the intact organelle, the endonuclease activity appears quiescent in the sense that samples of purified intact mitochondria can be incubated in the presence of Mg^{+2}-containing buffers for extended periods (>1 h) at 37°C and the mitochondrial DNA (mtDNA) remains supercoiled and unfragmented. However, with the addition *n*-octylglucoside or Triton X-100, the mtDNA is degraded extensively (**Fig. 1**). So far, the factor(s) that normally prevent Endonuclease G from attacking the mtDNA in vivo remain unknown. In vitro, Endonuclease G extensively degrades both single-stranded and duplex DNA templates in the presence of low-millimolar concentrations of Mg^{+2} or Mn^{+2}, producing 5'-phosphoryl, 3'-hydroxyl scissions. These activities are highest when the ionic strength of the reaction mixture is low (<20 mM KCl). Also with extended incubations in vitro, the purified enzyme can extensively degrade plasmid DNA substrates into DNA fragments less than about 10 nucleotides in length, with some products being dinucleotides and trinucleotides

From: *Methods in Molecular Biology, vol. 197: Mitochondrial DNA: Methods and Protocols*
Edited by: W. C. Copeland © Humana Press Inc., Totowa, NJ

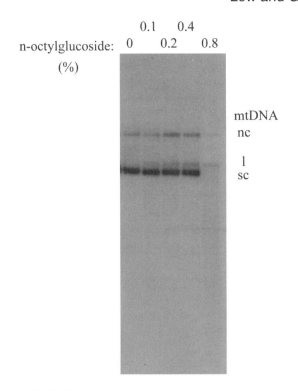

Fig. 1. Mitochondrial DNA remains undegraded when samples of intact bovine heart mitochondria are incubated at 37°C in the presence of Mg^{2+}, but the DNA becomes extensively fragmented when mitochondrial membranes are disrupted. A 3.5-µL sample of bovine heart mitochondria (20 mg protein/mL) was suspended in 100 µL of 20 mM Tris-HCl, pH 8.0, 100 mM NaCl, 2.5 mM MgAc$_2$, without or with 0.1%, 0.2%, 0.4%, or 0.8% (w/v) n-octylglucoside, as indicated. After 60 min at 37°C, 1% (w/v) SDS and 10 mM EDTA were added and the DNA was extracted, successively once each with phenol, phenol–chloroform (1:1), and chloroform. Following precipitation in 70% (v/v) ethanol, the DNA was collected at 30,000g for 15 min and resuspended in 40 µL of 40 mM Tris-acetate, pH 8.0, 2 mM EDTA, and electrophoresed in a 0.8% (w/v) agarose gel. The gel was run overnight at 1.5 V/cm in a 40 mM Tris-acetate, pH 8.0, 2 mM EDTA buffer system. After electrophoresis, a high-stringency Southern Blot analysis *(40)* was done and probed with a d(N)$_6$-labeled [^{32}P] probe (approx 2 × 10^8 dpm/µg) 1-kb Pst1–Hpa1 restriction fragment of the bovine mtDNA D-loop region sequence. A photograph of the autoradiogram is shown. The supercoiled (sc), nicked circular (nc), and full-length linear (l) forms of mtDNA are as indicated.

(3). Although most studies have focused on the capacity of the enzyme to hydrolyze DNA, the endonuclease can also degrade homopolymeric and single-stranded RNAs as well in vitro. Although Endonuclease G can degrade

any natural single-stranded and double-stranded DNA template extensively, the enzyme does not attack the nucleotide sequence randomly but shows an unusually strong preference to initially nick within tracts of consecutive guanine residues (hence the name Endo G) *(2,5–7)*. Despite the name, the enzyme preferentially nicks as avidly within the opposing runs of complementary cytosine residues and next to residues other than guanine.

This Endonuclease G activity has been identified in mitochondria of mammalian cells; mitochondria of *Neurospora (1,8–10)* and yeast *(1,11–16)* also possess a potent endonuclease activity. The *Neurospora* and yeast enzyme activities also only become evident once mitochondria are disrupted by a treatment with detergent, and they extensively degrade single-stranded and double-stranded DNA and also RNA templates in vitro. Furthermore, both the fungal mitochondrial nuclease and Endonuclease G appear to be associated with mitochondrial membranes; yet, once released, they are quite soluble in aqueous buffers. As often found for membrane-associated enzymes, the Endonuclease G activity in vitro is significantly enhanced by mitochondrial phospholipids *(17)*. Despite these similarities, the *Neurospora* and yeast activities differ from Endonuclease G in two interesting respects. First, unlike Endonuclease G, the fungal mitochondrial nucleases possess an intrinsic 5'-exonuclease activity that expands nicks introduced within double-stranded DNA templates into long single-stranded gaps in vitro *(10,16,18)*. Second, neither fungal mitochondrial endo-exonucleases reportedly nicks preferentially within tracts of consecutive guanine residues. Instead, the *Neurospora* mitochondrial activity has been shown to nick preferentially midway within an AGCACT nucleotide sequence motif that has been suggested to function analogously to that of the "chi site" sequence targeted by the *Escherichia coli recBC nuclease (19)*. The yeast enzyme, in contrast, has apparently not yet been found to exhibit a particular preference to attack specific sequences but to nick DNA templates randomly.

Despite differences between the fungal mitochondrial endo-exonucleases and Endonuclease G activities, the cloning and sequencing of the yeast enzyme's gene (named *NUC1) (20)*, and the subsequent cloning and sequencing of the bovine Endonuclease G gene *(21)* revealed a striking degree of identity (42%) between the amino acid sequences of the enzymes. Additionally, inactivation of the *NUC1* gene in a yeast-strain construct abolished nearly all of the nuclease activity detected in the mitochondria *(22)*. This demonstrated that each of the various single-stranded and double-stranded specific DNA and RNA nuclease activities previously reported to be in yeast mitochondria could be attributed to a single enzyme, as the purification studies in bovine and *Neurospora* implied. All of these data suggested that mitochondria contain a single major nuclease activity that has been highly conserved during evolution. Also, recent genome-sequencing projects have identified the Endonuclease G gene in *Homo sapiens,* *Mus musculus, Caenorhabditis elegans, Rattus norvegicus, Xenopus laevis,* and *Drosophila melanogaster.* Of interest, although many bacterial species such as *E. coli* lack Endonuclease G, *Serratia marcesens* and *Anabaeba* sp. each contain a nuclease gene that shows 42% and 28% identity, respectively, with that of the deduced amino acid sequence of bovine Endonuclease G *(21,23).* This striking conservation likely comes from the evolutionary origin of mitochondria from endosymbiontic alpha proteobacteria.

Although Endonuclease G accounts for most of the endonuclease activity identified in protein extracts of mitochondria prepared from many vertebrate tissues, additional types of mitochondrial nuclease activity likely exist. Consequently, some caution should be exercised when assigning a detectable nucleolytic mitochondrial activity to Endonuclease G. Mitochondria of plasmacytoma cells have been reported to contain a 37-kDa single-strand-specific endonuclease *(24).* This activity is probably not Endonuclease G. This plasmacytoma cell mitochondrial activity extensively degrades single-stranded DNA but does not extensively degrade; it only nicks a supercoiled plasmid duplex DNA template. In addition to the plasmacytoma cell activity, a 55-kDa Endonuclease G activity with enzymatic properties identical to those of the 29-kDa Endonuclease G has more recently been identified in bovine mitochondria using nuclease activity gels *(25).* Whether this 55-kDa species seen represents a novel enzyme or an oxidatively crosslinked dimeric (or otherwise modified) form of the Endonuclease G enzyme remains to be determined. Also, two additional Endonuclease G-like enzyme activities have also been inferred from the recently available human genome sequence, named "Endonuclease G-like 1" and "Endonuclease G-like 2," at human chromosome 3p21.3 *(26).* The Endonuclease G-like 1- and Endonuclease G-like 2-deduced protein sequences share about 57% and 38% identity with that of Endonuclease G, respectively. Northern blot analysis has revealed two splice variants that are expressed in most tissues. The enzymatic properties of these activities and a role of the enzymes in mtDNA metabolism still remain uncharacterized.

Some of the early studies of Endonuclease G had initially reported the activity to be nuclear in location *(2,7).* However, almost all (>95%) of the Endonuclease G is located in mitochondria *(27).* In addition, the unprocessed precursor form of the enzyme displays a mitochondrial targeting sequence, and immunohistochemical staining using antibodies prepared against peptide segments of the enzyme shows the protein to be almost entirely distributed among the mitochondria *(21).* A low level of Endonuclease G activity is often detected in isolated nuclei; however, no splice variants of the gene's mRNA have been identified to indicate that some of the enzyme is targeted to cellular compartments other than mitochondria *(28).* The enzymatic and biochemical properties of this nuclear activity appear indistinguishable from those of the Endonuclease G purified from mitochondria *(29)*. The detected amount of this nuclear activity has been shown to be attributed to contamination of standard preparations of isolated nuclei with mitochondria *(29,30)*.

Endonuclease G's biological role appears to be to eliminate oxidatively damaged mtDNA circles or to participate in a pathway for the repair of oxidative damage in mtDNA. Levels of Endonuclease G measured in mitochondrial extracts in vitro have been observed to vary over 200-fold among different rat tissues, with the highest values being found in heart mitochondria *(27)*. The specific activity of the enzyme (relative to mitochondrial protein or mtDNA) seems to reflect the tissue's relative rate of oxygen consumption and, by inference, the rate of oxidative injury to mtDNA. Although Endonuclease G will degrade undamaged DNA in vitro, treatment of DNA with oxidants such as L-ascorbic acid and peplomycin, or with cisplatin has been found to enhance the susceptibility of the DNA to nucleolytic attack in vitro, further implicating the enzyme in degradation of damaged mtDNA in vivo *(31)*.

The striking preference of Endonuclease G to nick within tracts of consecutive guanine residues in vitro is intriguing because mtDNA contains a single, evolutionarily conserved sequence tract of consecutive guanine residues that provides a strong Endonuclease G target *(5)*. This sequence tract resides in the noncoding displacement (D)-loop region of the mitochondrial genome close to where both transcription and DNA replication begin. In bovine mtDNA, this Endonuclease G target comprises a run of 12 consecutive guanine and complementary cytosine residues near one end of the D-loop region adjacent to the gene for tRNA[phe] *(32)*. In the mouse and human mtDNAs, the analogous guanine : cytosine tract, called conserved sequence box II (CSB II) *(33)*, resides a short distance further upstream within the D-loop region, near the site where mtDNA replication begins. Because Endonuclease G is membrane associated and the mtDNA appears to be attached to the inner membrane in this region of the DNA *(34)*, it is possible that oxidant injury to the CSB II site by an increased burst of electrons straying from the OXPHOS pathway could be what senses ongoing oxidant injury to the mtDNA. Although the evidence that Endonuclease G functions at this site remains sketchy, it is tempting to speculate that the enzyme acts at this site in vivo.

In human and mouse mtDNAs, the CSB II has also been identified as the site where endonucleolytic processing of an RNA transcript generates the primer needed to initiate mtDNA replication *(35)*. Although an RNase H activity (called RNase MRP) has been previously implicated in this process *(36)*, it has been proposed that Endonuclease G, acting as a CSB II-directed RNase H activity, and not the MRP activity, serves this role *(21)*. Whether Endonuclease G could serve as an RNase H in vivo remains uncertain. In vitro, at least, the

Endonuclease G can avidly nick DNA and does not exhibit the strict substrate specificity for the RNA of a DNA : RNA heteroduplex that typically defines an RNase H activity.

Endonuclease G's presence in crude extracts of mitochondria and in fractions of mitochondrial protein interfers with a variety of mtDNA replication and repair assays in vitro. Even low levels of Endonuclease G activity in mitochondrial fractions are sufficient to quantitatively nick supercoiled DNA templates, making it difficult to detect Mg^{+2}-dependent DNA topoisomerase II activity and activities required to produce stable D-loop replication intermediates in vitro. Even slight degradation of single-stranded and double-stranded DNA templates by Endonuclease G promotes vigorous foldback/hairpin priming of single-stranded DNA and a repair-type DNA synthesis in double-stranded DNA. As yet, there are no known specific inhibitors of Endonuclease G activity or available antibodies against the enzyme. We recommend that during the purification of any mtDNA replication/repair activity, active fractions be assessed for Endonuclease G contamination. The following protocols for the isolation of and assaying of Endonuclease G activity should be useful.

2. Materials
2.1. Stock Solutions

All reagents can be purchased from Sigma–Aldrich (Boston, MA) except where noted.

1. 1 M N-Hydroxyethylpiperazine-N-2-ethanesulfonic acid (HEPES), adjust pH with sodium hydroxide to pH 7.5.
2. 0.5 M Magnesium acetate [$Mg(OAc)_2$].
3. 10 mg/mL Bovine serum albumin. Store aliquots at –20°C.
4. 0.2 M Dithiothreitol (DTT) in water. Store aliquots at –20°C.
5. 2 M Potassium chloride.
6. 0.5 M Ethylenediaminetetraacetic acid (EDTA), pH 8.0.
7. 80% (w/v) Glycerol in water.
8. 10% Sodium dodecyl sulfate (SDS).
9. 1 M Tris-hydrochloric acid (Tris-HCl), pH 7.4, 8.0, and 8.8.
10. 5 M Sodium chloride (NaCl).
11. Liquid nitrogen.
12. 10% N-Octlyglucoside.
13. Triton X-100.
14. Tris-buffered phenol.
15. Chloroform.
16. 100% Ethanol.
17. Ethidium bromide.

2.2. Isolation of Mitochondria from Animal Tissues for Endonuclease G Studies

1. Homogenization buffer: 30 m*M* Tris-HCl (pH 7.4), 150 m*M* NaCl, 20 m*M* EDTA, and 10% (w/v) sucrose. Store at 4°C for up to 4 mo. Prior to use, add 2 m*M* DTT and 1 m*M* phenylmethylsulfonyl fluoride (PMSF).
2. A Polytron tissue processor (Brinkmann Instruments) at 4°C.

2.3. Preparation of Mitochondrial Fraction IIs

1. Ammonium sulfate.

2.4. Isolation of Endonuclease G Using a Glycerol Gradient

1. Potassium glutamate.
2. Hydroxylapatite.
3. Potassium phosphate.
4. Centricon 10 Filtration Units (Amicon).
5. Buffer A: 500 mL of 30 m*M* Tris-HCl, pH 8.0, 20 m*M* potassium glutamate, 100 m*M* NaCl, 5 m*M* DTT, 0.1 m*M* EDTA, 10% (v/v) glycerol. Store at 4°C for 24 h.

2.5. Agarose Gel Assay

1. 1–2 µg/mL Supercoiled plasmid. Any commercial DNA plasmid, whether it contains a long guanine tract or not, can be used.
2. TAE: 100 m*M* Tris-acetate, pH 7.5, and 1 m*M* EDTA.
3. High-quality agarose (SeaKem, Pharmacia) for 0.8% gels prepared with TAE.
4. Endonuclease diluent: 30 m*M* Tris-HCl, pH 8.0, 300 m*M* NaCl, 0.2 mg/mL bovine serum albumin, 2 m*M* DTT, and 30% (v/v) glycerol. Store in 500-µL aliquots at –20°C for 6 mo.

2.6. Acid Solubilization Assay

1. 1–2 µg/mL of [^{3}H] single-stranded M13 DNA (40 disintegrations per min [dpm]/ng) or heat denatured *E. coli* DNA.
2. Trichloroacetic acid.
3. Scintillation cocktail (Ready Safe, Beckman) and vials.

2.7. Sequencing Gel Assay

1. A 100- to 300-bp DNA singly end-labeled [5'-^{32}P] containing cytosines or guanines (100,000 dpm/µL) *(29)*.
2. Loading dye: 90% (v/v) formamide, 10 m*M* EDTA, 0.2% (w/v) bromphenol blue and xylene cyanol. Store in 500-µL aliquots at –20°C for 6 mo.
3. A 1.5 mm thick 12% (w/v) polyacrylamide gel (acrylamide : bisacrylamide, 19 : 1) containing 7 *M* urea, 100 m*M* Tris-borate, pH 8.3, and 1 m*M* EDTA.
4. TBE: 100 m*M* Tris-borate (pH 8.3) and 1 m*M* EDTA.

2.8. Endonuclease G Activity Gel

1. Salmon testes DNA at 2 mg/mL in 10 mM Tris-HCl, pH 7.5, 0.5 mM EDTA (TE buffer). The DNA should be phenol-extracted, chloroform-extracted twice, and dialyzed extensively against TE buffer. Store at −20°C for 6 mo.
2. A 12% (w/v) acrylamide (0.4% [w/v] bisacrylamide) resolving gel with 0.1% (w/v) SDS, 0.4 mM Tris-HCl, pH 8.8, 20 µg/mL of *Bam*H1 endonuclease cut salmon testes DNA that has been heat denatured 5 min at 94°C, 0.005% (v/v) TEMED, and 0.05% (w/v) ammonium persulfate. Pour a stacking gel with 4% (w/v) acrylamide (0.13% [w/v] bisacrylamide), 0.1% SDS, 0.125 M Tris-HCl (pH 6.8), 0.005% (w/v) TEMED, and 0.005% (w/v) ammonium persulfate. The gel is "aged" overnight at ambient temperature prior to being used.
3. Molecular-weight standards for proteins ("broad range set," Bio-Rad).
4. 40 mM Tris-glycine.
5. Buffer B (1 L): 50 mM Tris-HCl (pH 8), 0.1 mM EDTA, 5 mM DTT, 0.25 mM Mg(OAc)$_2$, 0.02% (v/v) Tween-20. Store at 4°C for 6 mo.

3. Method

3.1. Isolation of Mitochondria from Animal Tissues for Endonuclease G Studies

1. All steps for isolation of mitochondria are performed at 4°C.
2. Mince 10 g of tissue using a scalpel blade into millimeter-sized pieces.
3. Add 100 mL of homogenization buffer.
4. Homogenize for three intervals of 30 s each.
5. Centrifuge the tissue homogenate twice at 1000g for 5 min to remove cellular debris and nuclei.
6. Collect the mitochondrial pellet at 20,000g for 20 min and gently resuspended in homogenization buffer.
7. Freeze the mitochondria in liquid nitrogen and stored at −70°C.

3.2. Preparation of Mitochondrial Fraction IIs

1. Thaw a mitochondrial suspension on ice and dilute with an equal volume of homogenization buffer.
2. Increase the NaCl concentration to 300 mM and add 1% (w/v) N-octlyglucoside or 0.5% (w/v) Triton X-100. Incubate on ice for 20 min.
3. Clarify the lysate by centrifugation at 30,000g for 2 h at 4°C.
4. Fractionate the supernatant (so-called fraction I or lysate) using ammonium sulfate (AS), 0–0.16 g/mL (0–30% saturation). Incubate on ice for 1 h. Spin for 30 min at 20,000g at 4°C. Keep the supernatant.
5. Fraction II is fractionated using ammonium sulfate, 0.16–0.36 g/mL (30–60% saturation). Again, incubate on ice for 1 h. Spin for 30 min at 20,000g. The protein precipitated using the 60% saturated AS is resuspended in 30 µL of 100 mM

HEPES, pH 7.5, 150 mM NaCl, 5 mM DTT, 5 mM EDTA, 10% (v/v) glycerol, and 1% (w/v) N-octylglucoside (so-called mitochondrial fraction II).
6. The resuspended fraction II is frozen in liquid nitrogen and stored at –70°C.

3.3. Isolation of Endonuclease G Using a Glycerol Gradient

Endonuclease G activity is commonly contaminated with DNA polymerase gamma activity during chromatographic purification on hydroxylapatite, heparin agarose, and DNA–cellulose columns. In the described technique, glycerol-gradient velocity sedimentation offers an effective step to prepare nuclease free of DNA polymerase gamma (*see* **Fig. 3**).

1. Dilute a 40-mL sample (20–30 mg/mL) of mitochondria in buffer A at 4°C. All steps are done at 4°C unless otherwise indicated.
2. Disrupt the mitochondria by addition of 0.5% Triton X-100. Incubate for 30 min.
3. Centrifuged twice at 3000g for 10 min at 4°C to remove trace contaminants of nuclear and insoluble debris. Keep the supernatant. Centrifuge the supernatant in a fixed-angle rotor at 150,000g for 60 min. Keep the pellet; it constitutes the insoluble complex of mtDNA and protein. Resuspend the mtDNA–protein complex in 3 mL of 300 mM Tris-HCl, pH 8.8, 1 M NaCl, 5 mM DTT, and keep on ice overnight.
4. Centrifuge the suspension in a fixed angle rotor at 200,000g for 60 min. Keep the supernatant.
5. Prepare a 2-mL hydroxylapatite column and equilibrate with buffer A. Load the supernatant and then run a 20-mL linear gradient of 0–1 M potassium phosphate, pH 8.0, in buffer A and collect fractions.
6. Assay the fractions for DNA polymerase γ activity as previously described *(37)*.
7. Active DNA polymerase fractions are pooled and concentrated to 200 μL using a Centricon 10 (Amicon) centrifugation filter.
8. This concentrate is then sedimented through a 4-mL linear 15–42% (v/v) glycerol gradient containing 30 mM Tris-HCl, pH 8.0, 5 mM DTT, 300 mM NaCl, 10 mM Mg(OAc)$_2$, 0.1 % (w/v) n-octylglucoside. Sedimentation is carried out at 300,000g in a swinging-bucket rotor for 20 h. Fractions are collected, dropwise from the bottom of the tube.
9. Fractions are assayed for DNA polymerase γ and Endonuclease G activities (*see* **Fig. 3** for an example).

3.4. Agarose Gel Assay

The agarose gel assay is the most sensitive and easiest method for identifying Endonuclease G activity in fractions of mitochondrial protein. This is an excellent assay to determine whether enzyme fractions contain sufficient levels of Endonuclease G to damage DNA templates used in DNA replication and repair assays. In this assay, Endonuclease G activity is detected from the enzyme's endonucleolytic nicking of a negatively supercoiled plasmid DNA substrate as followed by agarose gel electrophoresis. In this nicking reaction, a single scission introduced into either DNA strand converts the supercoiled form of the plasmid DNA substrate into a nicked, circular DNA product, which, with the loss of superhelicity, migrates more slowly during agarose gel electrophoresis. More extensive degradation of the duplex DNA by Endonuclease G is typically seen using higher levels of enzyme *(3)*.

Examples of typical agarose gel assay of bovine heart Endonuclease G activity are shown in **Figs. 2** and **3**. In **Fig. 2A**, a titration of the Endonuclease G nicking activity present in a crude mitochondrial lysate is shown. In **Fig. 2B**, an agarose gel assay of Endonuclease G activity eluting off a phosphocellulose column at about 350 mM potassium phosphate during application of a linear 0–1 M potassium phosphate gradient is shown. In **Fig. 3**, Endonuclease G activity is monitored using a single-stranded DNA substrate after glycerol gradient velocity sedimentation.

1. The reactions are prepared on ice. Each reaction contains in 40 µL: 30 mM Tris-HCl, pH 8.0, 20 mM KCl, 1 mM Mg(OAc)$_2$, 0.2 mg/mL of bovine serum albumin, 2 mM DTT, 500 ng of negatively supercoiled plasmid DNA (3–6 kb), and nanogram quantities of enzyme. Mix gently using a vortex mixer.
2. The reactions are carried out for 30 min at 37°C and stopped by the addition of sodium dodecyl sulfate to 1% (w/v).
3. The DNA reaction products are then resolved by gel electrophoresis, through a 0.8% (w/v) agarose gel. Electrophoresis is typically carried out at 1.5–3 V/cm for 6 h in TAE, depending on the size of the substrate.
4. Stain the gel in 1 µg/mL ethidium bromide for 30 min at ambient temperature and destain briefly in water.
5. DNA band patterns are visualized under UV (300 nm) illumination and photographed.

One unit of activity is defined as the amount of enzyme that nicks one-half of the input DNA template in 30 min at 37°C (*see* **Note 1**).

3.5. Acid Solubilization Assay

This assay follows the endonucleolytic production of short, acid-soluble single-stranded DNA fragments from a single-strand [^{3}H]DNA template. This assay, though not as sensitive as the agarose gel assay, is quick and quantitative. It is a good assay to use when Endonuclease G itself is being purified or when there is a need to accurately measure Endonuclease G activity in different preparations of mitochondrial extracts.

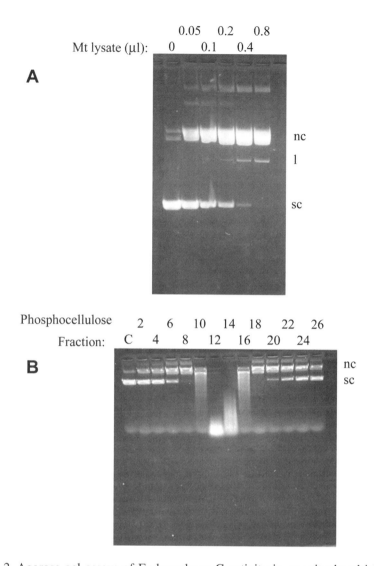

Fig. 2. Agarose gel assays of Endonuclease G activity in an mitochondrial lysate and during enzyme purification by phosphocellulose chromatography. (**A**) Titration of clarified lysate of bovine heart mitochondria (mt). Reactions were prepared as described in the text and included 0, 0.05, 0.1, 0.2, 0.4, or 0.8 µL of lysate (approx 3 mg protein/mL), as indicated. A photograph of the agarose gel is shown. The nicked circular (nc) and full-length linear (l) DNA products and supercoiled (sc) substrate DNA are seen, as labeled. (**B**) Activity profile of Endonuclease G during purification of the enzyme during phosphocellulose chromatography. A hydroxylapatite fraction was dialyzed for 2 h at 0°C against 30 mM Tris-HCl, pH 8.0, 20 mM potassium glutamate, 2 mM DTT, 20% (v/v) glycerol (buffer C), and applied to a 1-mL (2-cm × 0.5-cm^2) phosphocellulose column equilibrated in buffer C. Bound Endonuclease G activity was eluted from the column using a 20-mL, linear 0–1 M potassium phosphate, pH 8.0, in buffer C. Active fractions (8–18, peak at fraction 12), eluted near 300 mM phosphate, are as shown in the photograph of the agarose gel.

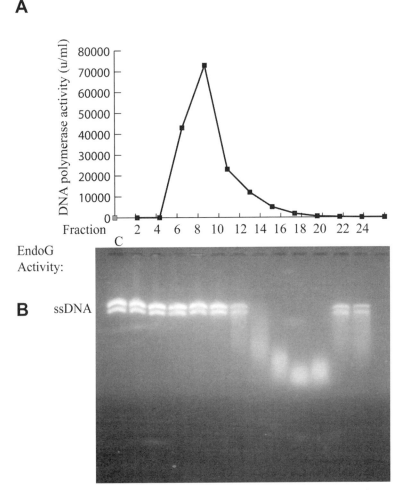

Fig. 3. Endonuclease G and DNA polymerase γ activities after glycerol-gradient velocity sedimentation. **(A)** Activity profile for DNA polymerase γ activity, with a peak at fraction 8. **(B)** The agarose gel assay of Endonuclease G activity in even fractions using an M13 viral single-stranded DNA substrate is shown. Endo G activity is in fractions 14–20, with a peak in fractions 16–20. The doublet band of the substrate includes the full-length linear and circular forms of the M13 DNA. Chromatography, endonuclease, and polymerase assays were carried out as detailed in the text.

1. The 40-μL reaction is prepared on ice with 30 mM Tris-HCl, pH 8.0, 1 mM Mg(OAc)$_2$, 0.2 mg/mL bovine serum albumin, 2 mM DTT, 1 μg of sonicated, heat-denatured *E. coli* [^{3}H] DNA (40 dpm/ng) or M13 ss [^{3}H] DNA (26,000 cpm/μg), and 0.01–0.10 μg of fraction II. This fraction precipitate contains > 80% of the

endonuclease activity identified in the mitochondrial lysate *(27)*. Mix gently using a vortex mixer.

2. Reactions are carried out for 30 min at 37°C and stopped by the addition of 4 µL of 100% (w/v) trichloroacetic acid. After 15 min at 0°C, [^{3}H] DNA precipitates are pelleted at 30,000g for 15 min.

3. A 20-µL aliquot of the supernatant (containing acid-soluble radioactivity) is carefully removed and counted in 5 mL of scintillant in a liquid scintillation counter.

One unit of acid solubilization activity is defined as the amount of enzyme that produces 1 ng of acid soluble [^{3}H] DNA fragments in 30 min at 37°C.

3.6. Sequencing Gel Assay

This assay monitors preferential nicking by Endonuclease G of the $(G)_{12}$:$(C)_{12}$ sequence tract found in mtDNA. The prominent cleavage of the $(G)_{12}$ sequence tract of bovine mtDNA by Endonuclease G is indicated in the sequencing gel assay shown in **Fig. 4**. Although more labor-intensive than other assays, this assay provides the best test of whether an endonuclease activity is, in fact, Endonuclease G. In this assay, nicking within the guanine (or complementary cytosine) sequence tract is assessed from the pattern of cleavages seen occurring within a singly 5′-end-labeled substrate. The cleavages produced in the restriction fragment substrate are detected by sequencing gel electrophoresis and autoradiography.

1. The 40-µL reaction sequencing gel assay, prepared at 4°C, contains 30 mM Tris-HCl, pH 8.0, 2.5 mM Mg(OAc), 0.2 mg/mL bovine serum albumin, 2 mM DTT, 100 ng of 100- to 300-bp DNA singly end-labeled [5′-^{32}P] containing cytosines or guanines (approx 10,000 dpm) and picogram to nanogram fraction II Endonuclease G. Mix gently using a vortex mixer.

2. Incubate the reaction at 37°C for 30 min and terminate by the addition of 0.5% (w/v) sodium dodecyl sulfate.

3. The [^{32}P] DNA products are phenol and chloroform extracted and ethanol precipitated, and the precipitants are collected at 30,000g for 15 min.

4. Wash the pellets briefly with 70% (v/v) ethanol and resuspend the [^{32}P] DNA products in loading dye.

5. Prerun the 1.5-mm 12% polyacrylamide gel for 60 min at 40 W.

6. Heat the [^{32}P] DNA for 2 min at 90°C and load the samples onto the prerun polyacrylamide gel.

7. Load the [^{32}P] DNA and in adjacent lanes: G and G+A Maxam–Gilbert sequencing ladders, prepared from the restriction fragment substrate. These are necessary to assign cleavage sites to the $(G)_{12}$ locus.

8. Electrophoresis is carried out at 40 W for 4–6 h at 60–70°C in TBE.

9. The gel is then placed on a piece of x-ray film for support and covered with Saran Wrap. Autoradiography is carried out on x-ray film at –80°C for 12 h using an intensifying screen.

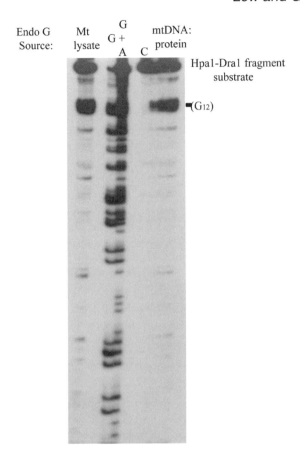

Fig. 4. Sequencing gel assay of partially purified Endonuclease G activity recovered from the mitochondrial lysate versus isolated complexes of mtDNA and protein. The mitochondrial lysate and isolated complexes of mtDNA and protein were prepared as described in the text. The Endonuclease G activity in each fraction was further purified by hydroxylapatite and DNA cellulose chromatography *(3)*. The DNA–cellulose fractions were assayed for site specific nicking of the singly end-labeled [5′-^{32}P]HpaI–DraI fragment that harbors the bovine mtDNA CSB II locus *(3)*. Reactions and sequencing gel electrophoresis were carried out as described in the text. The $(G)_{12}$ target site is as indicated in the photograph of the autoradiogram.

10. Optional: The endonucleolytic cleavages (the bands) can also be quantitated using the PhosphorImager.

3.7. Endonuclease G Activity Gel

This assay also provides an excellent way to assess whether an endonuclease activity encountered in mitochondrial preparations is likely Endonuclease G.

Molecular size
markers (kD)

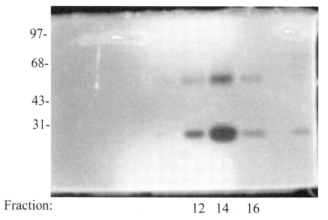

97-

68-

43-

31-

Fraction: 12   14   16

Fig. 5. Activity gel assay to detect Endonuclease G activity recovered from a glycerol gradient velocity sedimentation. Endonuclease G activity was recovered from a high-salt eluate of the mtDNA–protein complex and further purified by hydroxylapatite chromatography, concentrated, and then fractionated by velocity sedimentation, essentially as detailed in **Subheading 3.** An activity gel was prepared and run as detailed in the methods, using 10-µL aliquots of even-numbered glycerol gradient fractions. Positions of the phosphorylase b, bovine serum albumin, ovalbumin, and carbonic anhydrase molecular-weight markers are as indicated.

Both the molecular weights and endonuclease activities are resolved in a SDS–polyacrylamide gel with salmon testes DNA by electrophoresis. The results of a typical activity gel assay performed on fractions from a glycerol gradient velocity sedimentation step during purification of Endonuclease G is shown in **Fig. 5.** As seen, a prominent 29-kDa band and a minor 58-kDa band of Endonuclease G activity are evident in the assays of fractions 12, 14, and 16. The following protocol is essentially as described Ikeda et al. *(25)* with minor modifications.

1. Samples of mitochondrial protein fractions (10–40 µL of fraction II; *see* **Note 2**) and molecular-weight markers should be denatured with 68 mM Tris-HCl, pH 7.0, 10% (v/v) glycerol, 2% SDS, 0.7 M of 2-mercaptoethanol, 0.05% (w/v) bromophenol blue prior to electrophoresis.
2. The mitochondrial proteins and molecular-weight markers (load the markers either in the exterior lanes) are loaded on a 12% (w/v) acrylamide (0.4% [w/v] bisacrylamide]) gel that includes heat-denatured fragments of salmon testes DNA. Electrophoresis is carried at 12 mA, at ambient temperature in a 40 mM Tris-glycine, 0.1% SDS buffer system (Laemmli). Once the bromophenol blue marker has reached the bottom of the gel, electrophoresis is stopped.

3. Remove the gel from the apparatus and use a razor blade to slice the lane with molecular-weight markers away from the gel.

4. Take the gel piece with the molecular-weight markers and stain with 0.1% (w/v) Commassie Blue R-250 in 40% methanol and 10% acetic acid for 1 h. Destain the gel with repeated changes of 40% methanol and 10% acetic acid. The molecular-weight markers should be visible and can be used to determine the size of the endonuclease activity band.

5. Simultaneously wash the gel with the endonuclease fractions with 400 mL of 50 mM Tris-HCl, pH 8.0, 0.1 mM EDTA, 1 mM DTT, 0.5% (w/v) Triton X-100 at ambient temperature while gently rocking for 30 min. This step removes the SDS and allows the proteins in the gel to slowly renature (*see* **Note 3**).

6. Remove the wash in **step 5** and replace with 400 mL of buffer B at 4°C for 3 h.

7. Remove the wash in **step 6** and replace with 250 mL of fresh buffer B with occasional rocking at 37°C for 8 h. This incubation allows the endonuclease to locally degrade the DNA cast within the gel.

8. After this incubation, the gel is soaked in 0.5 µg/mL of ethidium bromide for 1 h at ambient temperature and subsequently washed in 50 mM Tris-HCl, pH 8.0, 0.1 mM EDTA for 30 min.

9. A clear band outline should be visible under ultraviolet light box illumination because of endonuclease activity and can be photographed.

4. Notes

1. If the endonuclease activity being tested is suspected to be Endonuclease G, then the activity should extensively degrade the double-stranded DNA substrate at high enzyme levels, and extensively fragment M13 single-stranded DNA. These activities should be strongly inhibited by either 120 mM NaCl or 5 mM N-ethylmaleimide (>95% inhibition).

2. The activity gel assay is suitable when analyzing concentrates of mitochondrial protein fractions prepared from the mitochondrial lysate by ammonium sulfate precipitation or, during column chromatography by insoluble complexes of mtDNA and protein that are treated with 600 mM NaCl to release proteins bound to the mtDNA. Unfortunately, the levels of Endonuclease G in mitochondrial lysates are typically too low to detect the Endonuclease G directly.

3. In some activity gel protocols, the SDS is removed from the gel following electrophoresis using a wash with 20% (v/v) isopropyl alcohol *(38)*. After removal of the SDS, renaturation of activity is often promoted by first transiently treating the gel with guanidinium HCl *(39)*. In our experience, we have not been able to recover much Endonuclease G activity using this latter method; we prefer to use the Triton X-100 exchange method.

References

1. Fraser, M. J. and Low, R. L. (1993) Fungal and mitochondrial nucleases, in *Nucleases*, 2nd ed. (Roberts, R. J., Linn, S. M., and Lloyd, S., eds.), Cold Spring Harbor Laboratory Press, Cold Spring Harbor, NY, pp. 171–207.

2. Ruiz-Carrillo, A. and Renaud, J. (1987) Endonuclease G: a $(dG)_n \cdot (dC)_n$-specific DNase from higher eukaryotes. *EMBO J.* **6,** 401–407.

3. Cummings, O. W., King, T. C., Holden, J. A., and Low, R. L. (1987) Purification and characterization of the potent endonuclease in extracts of bovine heart mitochondria. *J. Biol. Chem.* **62,** 2005–2015.

4. Tiranti, V., Rossi, E., Ruiz-Carrillo, A., Rossi, G., Rocchi, M., DiDonato, S., et al. (1995) Chromosomal localization of mitochondrial transcription factor A (TCF6), single-stranded DNA-binding protein (SSBP), and endonuclease G (ENDOG), three human housekeeping genes involved in mitochondrial biogenesis. *Genomics* **25,** 559–564.

5. Low, R. L., Cummings, O. W., and King, T. C. (1987) The bovine mitochondrial endonuclease prefers a conserved sequence in the displacement loop region of mitochondrial DNA. *J. Biol. Chem.* **262,** 16,164–16,170.

6. Low, R. L., Buzan, J. M., and Couper, C. L. (1988) The preference of the mitochondrial endonuclease for a conserved sequence block in mitochondrial DNA is highly conserved during mammalian evolution. *Nucleic Acids Res.* **16,** 6427–6445.

7. Cote, J., Renaud, J., and Ruiz-Carrillo, A. (1989) Recognition of $(dG)_n \cdot (dC)_n$ sequences by Endonuclease G. *J. Biol. Chem.* **264,** 3301–3310.

8. Linn, S. and Lehman, I.R. (1966) An endonuclease from mitochondria of *Neurospora crassa. J. Biol. Chem.* **241,** 2694–2699.

9. Fraser, M. J. and Cohen, H. (1983) Intracellular localization of neurospora crassa endo-exonuclease and its putative precursor. *J. Bacteriol.* **154,** 460–470.

10. Chow, T. Y.-K. and Fraser, M. J. (1983) Purification and properties of single strand DNA-binding endo-exonuclease of *Neurospora crassa. J. Biol. Chem.* **258,** 12,010–12,018.

11. Jacquemin-Sablon, H., Jacquemin-Sablon, A., and Paoletti, C. (1979) Yeast mitochondrial deoxyribonuclease stimulated by ethidium bromide. 1. Purification and properties. *Biochemistry* **18,** 119–127.

12. Morosoli, R. and Lusena, C. V. (1980) An endonuclease from yeast mitochondrial fractions. *Eur. J. Biochem.* **110,** 431–437.

13. Rosamond, J. (1981) Purification and properties of an endonuclease from the mitochondrion of *Saccharomyces cerevisiae. Eur. J. Biochem.* **120,** 541–546.

14. von Tigerstrom, R.G. (1982) Purification and chacteristics of a mitochondrial endonuclease from the yeast *Saccharomyces cerevisiae. Biochemistry* **21,** 6397–6403.

15. Foury, F. (1982) Endonucleases in yeast mitochondria: apurinic and manganese-stimulated deoxyribonuclease activities in the inner mitochondrial membrane of *Saccharomyces cerevisiae. Eur. J. Biochem.* **124,** 253–259.

16. Dake, E., Hofmann, T. J., McIntire, S., Hudson, A., and Zassenhaus, H. P. (1988) Purification and properties of the major nuclease from mitochondria of *Saccharomyces cerevisiae. J. Biol. Chem.* **263,** 7691–7702.

17. Parks, W. A., Couper, C. L., and Low, R. L. (1990) Phosphatidylcholine and phosphatidylethanolamine enhance the activity of the mammalian mitochondrial endonuclease in vitro. *J. Biol. Chem.* **265,** 3436–3439.

18. Koa, H., Fraser, M. J., and Kafer, E. (1990) Endo-exonuclease of *Aspergillus nidulans*. *Biochem. Cell Biol.* **68,** 387–392.

19. Fraser, M. J., Hatahet, Z., and Huang, X. (1989) The actions of *Neurospora* endo-exonuclease on double strand DNAs. *J. Biol. Chem.* **264,** 13,093–13,101.

20. Vincent, R. D., Hofmann, T. J., and Zassenhaus, H. P. (1988) Sequence and expression of NUC1, the gene encoding the mitochondrial nuclease in *Saccharomyces cerevisiae*. *Nucleic Acids Res.* **16,** 3297–3312.

21. Cote, J. and Ruiz-Carrillo, A. (1993) Primers for mitochondrial DNA replication generated by Endonuclease G. *Science* **261,** 765–769.

22. Zassenhaus, H. P., Hofmann, T. J., Uthayashanker, R., Vincent, R. D., and Zona, M. (1988) Construction of a yeast mutant lacking the mitochondrial nuclease. *Nucleic Acids Res.* **16,** 3283–3296.

23. Muro-Pastor, A. M., Flores, E., Herrero, A., and Wolk, C. P. (1992) Identification, genetic analysis and characterization of a sugar non-specific nuclease from the cyanobacterium *Anabaena* sp. PCC7120. *Mol. Microbiol.* **6,** 3021–3030.

24. Tomkinson, A. E. and Linn, S. (1988) Purification and properties of a single-specific endonuclease from mouse cell mitochondria. *Nucleic Acids Res.* **14,** 9579–9593.

25. Ikeda, S., Tanaka, T., Hasegawa, H., and Ozaki, K. (1996) Identification of a 55-kDa endonuclease in rat liver mitochondria with nucleolytic properties similar to Endonuclease G. *Biochem. Mol. Biol. Int.* **38,** 1049–1057.

26. Daigo, Y., Isomura, M., Nishiwaki, T., Tamari, M., Ishikawa, S., Kai, M., et al. (1999) Characterization of a 1200-kb genomic segment of chromosome 3p22–p21.3. *DNA Res.* **6,** 37–44.

27. Houmiel, K. L., Gerschenson, M., and Low, R. L. (1991) Mitochondrial endo-nuclease activity in the rat varies markedly among tissues in relation to the rate of tissue metabolism. *Biochim. Biophys. Acta* **1079,** 197–202.

28. Prats, E., Noel, M., Letourneau, J., Tiranti, V., Vaque, J., Debon, R., et al. (1997) Characterization and expression of the mouse Endonuclease G gene. *DNA Cell Biol.* **16,** 1111–1122.

29. Gerschenson, M., Houmiel, K., and Low, R. L. (1995) Endonuclease G from mammalian nuclei is identical to the major endonuclease of mitochondria. *Nucleic Acids Res.* **23,** 88–97.

30. Meng, X. W., Fraser, M. J., Ireland, C. M., Feller, J. M., and Ziegler, J. B. (1998) An investigation of a possible role for mitochondrial nuclease in apoptosis. *Apoptosis* **3,** 395–406.

31. Ikeda, S. and Ozaki, K. (1997) Action of mitochondrial Endonuclease G on DNA damaged by L-ascorbic acid, peplomycin, and *cis*-diamminedichloroplatinum (II). *Biochem. Biophys. Res. Commun.* **235,** 291–294.

32. Anderson, S., Bankier, A. T., Barrell, B. G., de Bruijn, M. H. L., Coulson, A. R., Drouin, J., et al. (1981) Complete nucleotide sequence of bovine mitochondrial DNA. *Nature* **290,** 457–465.

33. Walberg, M. W. and Clayton, D. A. (1981) Sequence and properties of the human KB cell and mouse L cell D-loop regions. *Nucleic Acids Res.* **9,** 5411–5421.

34. Albring, M., Griffith, J., and Attardi, G. (1977) Association of a protein structure of probable membrane derivation with HeLa cell mitochondrial DNA near its origin of replication. *Proc. Natl. Acad. Sci. USA* **74,** 1348–1352.

35. Chang, D. D., Fisher, R. P., and Clayton, D. A. (1987) Roles for a promoter and RNA processing in the synthesis of mitochondrial displacement-loop synthesis. *Biochim. Biophys. Acta* **909,** 85–91.

36. Lee, D. Y. and Clayton, D. A. (1998) Initiation of mitochondrial DNA replication by transcription and R-loop processing. *J. Biol. Chem.* **273,** 30,164–30,621.

37. Naviaux, R. K., Markusic, D., Barshop, B. A., Nyhan, W. L., and Haas, R. H. (1999) Sensitive assay for mitochondrial DNA polymerase gamma. *Clin. Chem.* **45,** 1725–1733.

38. Blank, A., Sugiyama, R. H., and Dekker, C. A. (1982) Activity staining of nucleolytic enzymes after sodium dodecyl sulfate–polyacrylamide gel electrophoresis: use of aqueous isopropanol to remove detergent from gels. *Anal. Biochem.* **120,** 267–275.

39. Emoto, Y., Manome, Y., Meinhardt, G., Kisake, H., Kharbanda, S., Robertson, M., et al. (1995) Proteolytic activation of protein kinase C by an ICE-like protease in apoptotic cells. *EMBO J.* **14,** 6148–6156.

40. Ausubel, F. M. (1999) *Current Protocols in Molecular Biology.* Wiley, New York.

24

Targeting DNA Repair Proteins to Mitochondria

Allison W. Dobson, Mark R. Kelley, Glenn L. Wilson, and Susan P. LeDoux

1. Introduction

Mitochondrial DNA is a very vulnerable target for both alkylating and oxidizing agents *(1–5)*. Recent studies from our laboratory demonstrated an association between inefficient mtDNA repair capacity and an increased sensitivity to both types of DNA damaging agent *(5–9)*. To determine if a cell's sensitivity to a DNA damaging agent can be altered by changing the mtDNA repair capacity, experiments that involve targeting of repair enzymes to mitochondria were undertaken *(10)*. The methods utilized are described in detail within this chapter.

The experimental design is divided into three phases: First, the protein will be targeted to mitochondria by adding an N-terminal mitochondrial localization signal to the gene of interest in a construct for transfection (**Subheadings 3.1.–3.3.**); second, the presence of higher amounts of active enzyme in mitochondria of transfected cells will be confirmed (**Subheadings 3.4.–3.5.**); finally, the effects of this enzyme on mtDNA repair and the overall sensitivity of the cells to oxidative stress will be studied (**Subheadings 3.6.–3.8.**).

In order to target any protein to mitochondria, a localization signal must be added. The classical mitochondrial localization signal is an N-terminal stretch of positively charged amino acids that form an amphipathic α-helix. Manganese superoxide dismutase has a very strong localization signal *(11,12)*, and we have used it successfully to target repair proteins to mitochondria *(10)*. One of the greatest challenges in this procedure is in confirming mitochondrial localization, because many cells must be harvested in order to isolate sufficient mitochondria for analysis. The fact that some cell types have very few mitochondria per cell complicates the issue further.

From: *Methods in Molecular Biology, vol. 197: Mitochondrial DNA: Methods and Protocols*
Edited by: W. C. Copeland © Humana Press Inc., Totowa, NJ

A variety of agents can be used to induce oxidative damage to DNA. For the studies described within this chapter, menadione was utilized because it redox cycles with complex I of the electron transport chain within mitochondria and causes enhanced mtDNA damage *(13,14)*. This drug induces oxidative damage by generating the superoxide radical. In addition to reactive oxygen species, a variety of other agents have been found to damage mtDNA *(4,5)* and can be used if another form of damage is to be studied.

There are many ways to deliver genetic material into eukaryotic cells. However, because of our success with it, only one transfection procedure is described here. The most important aspect of transfection is that a high percentage of the cells contain and express the trans-gene of interest. To determine specific targeting of recombinant proteins, there are methods available for extensive purification of organelles, which, unfortunately, result in very low yields. To overcome this problem, we use enriched organelle fractions instead and these have been used to demonstrate organelle-specific repair enzyme activity *(10)*. Western blots can be used as an additional confirmation of protein localization, but those procedures are not described here.

A variety of methods can be used to assess changes in cell viability resulting from expression of recombinant DNA repair proteins. The clonogenic assay is particularly useful for determining both cell survival and proliferation. However, it is not applicable for some cell types. Transformed cells will generally form colonies when plated very sparsely, but primary cultures and certain neighbor-sensitive cell lines may not form colonies at all. Other commercial viability assays can be used in such cases.

2. Materials

2.1. Preparation of Constructs

1. Oligonucleotides designed to serve as primers to amplify the DNA repair enzyme of choice.
2. pcDNA3.0neo vectors or other vector of choice.

2.2. Transfections

1. Fugene 6 transfection reagent (Roche)
2. Serum-free media.
3. Complete media.
4. Plasmid DNA.

2.3. Selection for Stable Clones

1. G4-18 geneticin antibiotic or other appropriate selection antibiotic.
2. Complete media.

2.4. Isolation of Organelles

1. Digitonin solution: 325 mM digitonin, 2.5 mM EDTA, 250 mM mannitol, and 17 mM 3-[N-morpholino]propanesulfonic acid (MOPS), pH 7.4.
2. 2.5X Mannitol–sucrose buffer: 525 mM mannitol, 175 mM sucrose, 12.5 mM EDTA, 12.5 mM Tris-HCl, pH 7.5.
3. Organelle buffer: 20 mM HEPES, pH 7.6, 1 mM EDTA, 5 mM dithiothreitol (DTT), 300 mM KCl, 5% glycerol.
4. Protease inhibitors (for mammalian cell extracts): 100 mM 4-(2-aminoethyl) benzenesulfonylfluoride (AEBSF), 4 mM bestatin, 1.4 mM E-64, 2.2 mM leupeptin, 1.5 mM pepstatin, and 80 mM aprotinin (Sigma).
5. At least 10 million cells per sample.

2.5. Repair Enzyme Activity Assays

1. Single-stranded oligonucleotide with specific damage at one position.
2. γ-^{32}P or γ-^{33}P (ATP).
3. T4 polynucleotide kinase and 10X buffer (Promega).
4. Complementary oligonucleotide (to damaged one).
5. 10X Reaction buffer (may differ depending on targeted enzyme): 1 M NaCl, 250 mM phosphate buffer, pH 7.5, 20 mM Na–EDTA, 10 mM MgCl$_2$.

2.6. In Vivo Repair Assay

1. Menadione (Sigma) or other DNA damaging agent.
2. Serum-free media.
3. Hanks' balanced salt solution (HBSS).
4. Lysis buffer: 10 mM Tris-HCl, pH 8.0, 1 mM EDTA, 0.5% sodium dodecyl sulfate (SDS), 0.3 mg/mL proteinase K (Roche).
5. 5 M NaCl.
6. SEVAG (24 parts chloroform : 1 part isoamyl alcohol).

2.7. Quantitative Southern Blots

1. 10 M Ammonium acetate.
2. 95–100% Ethanol and 75% ethanol.
3. RNase (Roche).
4. Xho I restriction endonuclease (Roche).
5. TE: 10 mM Tris-HCl, pH 8.0, 1 mM EDTA.
6. 6X Alkaline loading dye: 0.5 M NaOH, 10 mM EDTA, 25% Ficoll, 0.25% bromocresol purple.
7. Large horizontal gel electrophoresis system and transfer apparatus.
8. HCl wash: 0.25 M HCl.
9. Tris wash: 0.5 M Tris-HCl, 1.5 M NaCl, pH 7.5.
10. NaOH wash: 0.5 M NaOH, 1.5 M NaCl.
11. Zeta-probe GT nylon membrane (Bio-Rad).

12. 10X SSC: 3 *M* NaCl, 0.3 *M* sodium citrate, pH 7.0.
13. Hybridization solution: 0.25 *M* sodium phosphate buffer, pH 7.2, 7% SDS.
14. 5% SDS wash: 20 m*M* sodium phosphate buffer, pH 7.2, 5% SDS.
15. 1% SDS wash: 20 m*M* sodium phosphate buffer, pH 7.2, 1% SDS.

2.8. Clonogenic Assay

1. Hemocytometer.
2. 1X Phosphate-buffered saline (PBS).
3. Fixing solution: 3 parts methanol : 1 part acetic acid.
4. Hematoxylin or methylene blue.

3. Methods
3.1. Preparation of Constructs

1. Design oligonucleotides to serve as primers to amplify the DNA repair enzyme from a cDNA plasmid.
2. Prepare the 5′ primer, containing an EcoRI site (underlined in the sequence below), or another restriction enzyme site that will be used for subcloning into vector of choice, the mitochondrial targeting sequence (MTS from MnSOD in bold, <u>GGAATTC</u>**ATGTTGAGCCGGGCAGTGTGCGGCACCAGCAGGC AGATGCCTGCCGCGCGCTTCTG**-aa1-aa2-aa3-aa4) and the sequence corresponding to the first three to four amino acids (aa1–aa4) of the enzyme/ protein you are planning to clone. The MTS sequence is 24 amino acids in length *(10)*.
3. Prepare a 3′ primer appropriate to the enzyme sequence that contains another restriction enzyme site of choice at the 5′-end of this primer.
4. Amplify the cDNA using a high-fidelity thermostable DNA polymerase by polymerase chain reaction (PCR) in a thermal cycler under the following conditions: 30 s denaturation (94°C), 1 min annealing (55°C), 2 min extension (72°C). The resulting PCR fragment should consist of an *Eco*RI site, the MTS, the enzyme coding region, and another restriction site.
5. Use the PCR product in a double restriction enzyme digest with *Eco*RI and the other restriction enzyme overnight at 37°C.
6. Subclone the restriction fragment into the *Eco*RI and other enzyme sites of pcDNA3.0neo or another vector of choice.
7. Sequence to confirm fidelity.

3.2. Transfections

1. Grow cells in 75-cm² flasks until they reach 75% confluence.
2. Add Fugene 6 reagent dropwise to serum-free media (ratio of 3 μL Fugene for 1 μg DNA to be used), flick tube to mix gently, and incubate for 10 min.
3. Add DNA dropwise to mixture in **step 2**, flick tube to mix gently, and incubate for 30 min.
4. Add the entire mixture (media, Fugene, and DNA) dropwise to cells in normal culture media, swirl vessel, and return cells to incubator.

3.3. Selection for Stable Clones

1. Twenty-four h after transfection, replenish cultures with fresh media containing 0.6 mg/mL G4-18.
2. Passage cells normally, including G4-18 (7–10 d). During the selection period, significant cell death should occur. Reduce the culture vessel size if necessary. Keep 0.6 mg/mL G4-18 in the media.
3. After 2 wk of selection, maintain cells in 0.4 mg/mL G4-18.
4. Isolate individual cells for clonogenic expansion (*see* **Note 1**).

3.4. Isolation of Organelles

1. Grow at least 10 million cells for each cell type. The number of cells required will depend on the number of mitochondria per cell. For HeLa cells, this is approximately three 75-cm^2 flasks at near-confluence.
2. Dislodge cells from monolayers with trypsin and collect each cell type into a single pellet by centrifugation at 1000g for 10 min.
3. Resuspend each cell pellet in ice-cold digitonin solution for 80 s. Pipet cells on ice until no clumps are observed.
4. Add the lysed cell mixture to 2.5X mannitol–sucrose buffer for a final strength of 1X (210 mM mannitol, 70 mM sucrose, 5 mM EDTA, 5 mM Tris-HCl, pH 7.5).
5. Centrifuge the ice-cold suspension at 4°C for 10 min at 800g to pellet nuclei.
6. Save the supernatant, resuspend the pelleted nuclear material in 1X mannitol–sucrose, and repeat the centrifugation.
7. Repeat **step 6** three additional times (*see* **Note 2**).
8. Combine the saved supernatants and centrifuge at 800g for 10 min to pellet any remaining nuclei, and very carefully draw off the resulting supernatant. Save the nuclei on ice.
9. Centrifuge the new supernatant at 10,000g for 20 min to pellet mitochondria. You should see a small pellet.
10. Decant supernatant carefully and save (this is cytosolic fraction). Add 5 µL/mL protease inhibitors (*see* **Note 3**).
11. Resuspend isolated mitochondria and nuclei in organelle buffer + 5 µL/mL protease inhibitors (add just before use).
12. Pulse-sonicate (1 s) the resuspended organelles on ice (*see* **Note 4**).
13. Centrifuge the sonicated organelles at 5000g for 10 min to pellet remaining cell debris, and carefully draw off the supernatants.
14. Determine protein concentrations for each sample (Bio-Rad/Bradford method) (*see* **Note 5**).

3.5. Repair Enzyme Activity Assays (Fig. 1)

1. End-label the damage-containing oligonucleotide at 37°C for 30 min in the following reaction: 20 pmol oligonucleotide, 20 pmol γ-^{32}P or γ-^{33}P ATP, 1 µL (10 U) T4 polynucleotide kinase, and 2 µL 10X buffer in a total volume of 20 µL. Heat to 90°C for 2 min to inactivate the kinase (*see* **Note 6**).

1 2 3 4 5 6 7 8 9 10 11 12 13 14 15

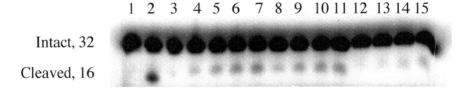

Intact, 32

Cleaved, 16

Fig. 1. Repair enzyme activity assay. An oligonucleotide with 8-oxoguanine at position 16 was end-labeled, annealed to complementary oligonucleotide, and incubated with nuclear HeLa fraction (lanes 4–7) or mitochondrial fraction (lanes 8–15) for 2, 10, 30, or 60 min. Lanes 8–11 are mitochondrial extracts from HeLa cells stably transfected with 8-oxoguanine glycosylase targeted to mitochondria, and lanes 12–15 are mitochondrial extracts from HeLa cells stably transfected with the pcDNA3 vector only. Removal of the damaged base by extract protein results in an abasic site and a strand break at the modified base. Lane 1: substrate oligonucleotide only; lane 2 reaction: 0.5 U *E. coli* Fpg (positive control) for 60 min; lane 3 reaction: dH$_2$O (negative control) for 60 min. All reactions except lane 1 were heated to 95°C.

2. Add 20 pmol complementary oligonucleotide, heat to 70°C, and cool slowly to anneal.
3. In 0.5-mL microtubes on ice, put equal amounts of organelle protein from samples to be compared. Protein extracts should contribute less than 30% of the 20-μL total reaction volume. Purified enzyme, if available, should be used as a positive control, and distilled water (dH$_2$O) as a negative control (*see* **Note 7**).
4. Prepare a "Master Mix" of labeled oligonucleotide and 10X reaction buffer (*see* **Note 8**).
5. On ice, add equal amounts of the Master Mix to each protein sample and incubate at 37°C for 30 min (*see* **Note 9**).
6. Stop reactions and add bromophenol blue loading dye (*see* **Note 10**).
7. Prepare 20% polyacrylamide, 8 *M* urea, 1X TBE vertical gels in a Bio-Rad Mini-Protean II apparatus. Use 1X TBE for electrophoresis buffer.
8. Flush urea from wells before loading samples. Subject to electrophoreses for 1 h at 100 V (*see* **Note 11**).
9. Leave gel on one glass plate and wrap in clear plastic. Expose to film overnight (*see* **Note 12**).

3.6. In Vivo Repair Assay

1. Dissolve menadione in sterile serum-free media to make 10 m*M* stock solution. Menadione is light sensitive, so work quickly in low light or keep solutions covered.
2. Prepare serial dilutions of menadione from the stock solution and additional serum-free media (*see* **Note 13**).
3. Rinse cultured cells with HBSS and replace media with menadione solution for 1 h (5% CO$_2$ at 37°C). Control cells should be treated with serum-free media only (*see* **Note 14**).

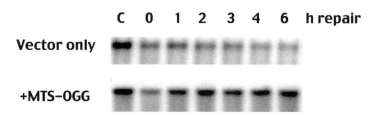

Fig. 2. Quantitative Southern blot for DNA damage and repair. HeLa cells stably transfected with vector only or MTS-OGG were drugged with 400 μ*M* menadione for 1 h and either lysed immediately or allowed repair time in their normal media and then lysed. Control samples were exposed to the drug diluent only. Total DNA was isolated from the lysates and analyzed in quantitative alkaline Southern blots with a probe corresponding to part of the human mitochondrial genome. Representative autoradiographs are shown here.

4. Remove menadione and lyse cells immediately on the culture surface with lysis buffer. Be sure to tilt the culture vessel such that lysis buffer spreads over the entire surface. It will become viscous very rapidly.
5. Return lysed cells to incubator for 5 min and then use a cell lifter to transfer lysate into 15-mL conical tubes. Incubate at 37°C overnight.

3.7. Quantitative Southern Blots (Fig. 2)

1. Add 0.2 volumes of 5 *M* NaCl to each cell lysate (*see* **Note 15**).
2. To extract DNA, add an equal volume of SEVAG, invert the tube and gently agitate for 10 min to mix thoroughly, and centrifuge 10 min at 2500*g* to separate the aqueous phase from the organic phase. The aqueous phase will be on the top.
3. Using a wide-mouth pipet, carefully draw off the aqueous phase, avoiding the protein at the interface (white) as much as possible (*see* **Note 16**).
4. Repeat **steps 2** and **3** two to three times (*see* **Note 17**).
5. Precipitate total DNA with 0.2 volumes of 10 *M* ammonium acetate, followed by 2.2 volumes of ethanol. Centrifuge for 10 min at 3000*g* or higher to pellet DNA.
6. Rinse DNA pellet with 75% ethanol and centrifuge again. Pour off ethanol and allow pellet to dry inverted for 5 min.
7. Resuspend DNA in dH$_2$O and restrict with *Xho*I overnight (this cuts the human mitochondrial genome once, but other restriction enzymes may be used). Treat with RNase (1 mg/mL).
8. Precipitate digested samples with ammonium acetate and ethanol as in **step 4**.
9. Prepare a horizontal 0.6% alkaline agarose gel: Weigh 1.2 g agarose into 200 mL dH$_2$O and boil to dissolve agarose. Cool to 55°C and add 400 μL of 0.5 *M* EDTA and 600 μL of 10 *N* NaOH. Allow at least 2 h for this gel to set and remove the comb very slowly. Use extreme caution handling this gel in later steps, as it slides and breaks very easily!

10. Resuspend the DNA in TE buffer and determine a precise and accurate concentration. We use a Hoefer DyNA Quant 200 Fluorometer and Hoechst 33258 dye.
11. For each sample, bring up 5 μg of total DNA to 20 μL in TE.
12. Heat samples for 15 min at 70°C and cool at room temperature.
13. Add NaOH to a final concentration of 0.1 N and incubate at 37°C for 15 min.
14. Add alkaline loading dye to each sample.
15. Load samples and subject to electrophoreses at 30 V (1.5-V/cm gel length) for 16 h.
16. In Tris-HCl wash, stain the gel with ethidium bromide to confirm even loading, DNA integrity, and take a picture for a permanent record.
17. Wash the gel two times for 10 min each with HCl wash.
18. Wash the gel two times for 10 min each with the NaOH wash.
19. Wash the gel two times for 10 min each with Tris-buffer wash (*see* **Note 18**).
20. Soak Zeta-probe membrane in 10X SSC.
21. Transfer the DNA to Zeta-probe membrane with vacuum transfer in 10X SSC (approximately 1 h). Add SSC periodically to top of gel during transfer (*see* **Note 19**).
22. Rinse membrane in 2X SSC briefly after transfer is complete.
23. Crosslink the membrane.
24. Hybridize membrane overnight with ^{32}P-labeled human mtDNA-specific PCR-generated probe. This 672-bp probe corresponding to part of the cytochrome-*c* oxidase, subunit 3 gene and part of the ATP synthase F0, subunit 6 gene is made by PCR amplification (94°C, 5 min followed by 30 cycles of 95°C, 30 s; 60°C, 30 s; 72°C, 1 min) from human genomic DNA with the following primers: 5′ CACAACTAACCTCCTCG 3′ and 5′ CTTTTTGGACAGGTGGTG 3′.
25. Wash membrane twice with 5% SDS wash and once with 1% SDS wash (5–10 min per wash).
26. Expose to film overnight.

3.8. Clonogenic Assays (Fig. 3)

1. Carefully count cells with a hemocytometer and plate 400 cells on each 60-mm plate.
2. Allow cells 24 h in culture to adhere and acclimate.
3. Treat plates with increasing doses of menadione as described in **Subheading 3.6.** The actual doses should be much lower, however, because cells plated this sparsely are far more sensitive than cells at near-confluence.
4. Replace normal culture media and return cells to the incubator for 10 d.
5. Rinse cells with PBS and fix in fixing solution at 37°C for 10 min.
6. Remove fixing solution and cover the bottom of each plate with hematoxylin (or methylene blue) for 10 min.
7. Pour off hematoxylin and rinse plates very gently in a large container of water.

Clonogenic Assay

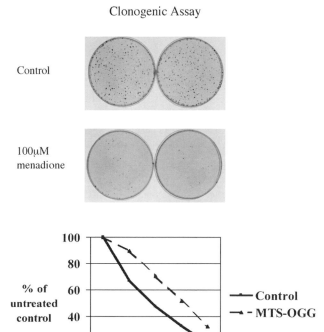

Fig. 3. Clonogenic assay. Cells were plated sparsely (400 cells per 60-mm dish) and allowed 24 h to adhere. They were then treated with 100 μ*M* menadione or diluent (control) for 1 h and then placed in their normal media for 10 d. Colonies were then fixed, stained, and counted. The graph shows results for HeLa cells stably transfected with vector only (control) or 8-oxoguanine glycosylase targeted to mitochondria (MTS-OGG).

8. Calculate survival/proliferation rates as follows:

$$\frac{\text{Average number of colonies}}{(\text{Number of cells plated}) \, (\text{PE [plating efficiency]})}$$

where

$$PE = \frac{\text{Number of colonies of untreated cells}}{\text{Number of cells plated}}$$

4. Notes

1. Isolation of single cells can be accomplished several ways. A cell sorter can be used to deposit one cell in each well of a 96-well plate, for example.
2. These additional centrifugations are only necessary to optimize the mitochondrial yield. If a very large number of cells is used, omit this step.
3. The cytosolic components may need to be concentrated. There are many commercial concentrators available if necessary.
4. Sonication is one of several ways to rupture organelle membranes. We use a Branson Sonifier Cell Disruptor 185 with the output control set at 6, and sonicate for 1 s. Some foam is then visible at the surface of the sample. Alternatively, Triton-X-100 can be added to the organelle lysis buffer to 0.2%, and after resuspension, the organelles can be incubated at 37°C for 30 min. This may interfere with protein quantitation, however.
5. This protein may also be used in Western blots to confirm localization of the protein of interest and proper organelle isolation *(10)*.
6. The total amount of labeled oligonucleotide required will depend on the enzyme. Some DNA repair enzymes are optimally active at higher substrate concentrations than others. Human 8-oxoguanine glycosylase (OGG) activity, for example, is best detected when the substrate concentration is 100 nM or higher.
7. The total reaction volume can be varied, but in general smaller volumes tend to yield better results.
8. The use of a master mix minimizes the human error introduced by additional pipetting steps; that is, it ensures more even distribution of 10X buffer and substrate oligonucleotide to each reaction. Also, the 10X buffer should be designed to optimize reaction conditions for the enzyme being studied. Be sure to calculate the amount of 10X buffer based on the final concentration in the total reaction.
9. The desired reaction time will also depend on the particular enzyme being studied.
10. Again, this step will depend on the enzyme being studied. A stop solution may be necessary in some cases. For example, glycosylases that only remove the damaged base will not cleave the labeled oligonucleotide. In such cases, it is necessary to heat samples to 95°C, which will cause strand breaks at abasic sites, before electrophoresis.
11. Electrophoresis will depend on the length of the damaged oligonucleotide before and after cleavage. As a general rule, the bromophenol blue dye should be one-half to two-thirds down the gel.
12. Film exposure time will depend on the amount of radiolabeled substrate used in each reaction and the amount of the reaction loaded on the gel. We have used exposure times ranging from 1 h to 3 d.
13. The appropriate concentrations of menadione will vary greatly from one cell type to another. For HeLa cells, 300–400 µM menadione induces a level of mtDNA damage that can be readily detected and repaired to some extent within 6 h.

14. Menadione exposure time can be altered. Because menadione is a redox cycler, damage is generated for the entire duration of exposure.
15. Some commercial DNA isolation kits may work just as well.
16. Using normal pipets can shear the DNA. If wide-mouth pipets are not available, you can also cut off the end of a 1-mL micropipet tip (about 0.25 in.).
17. The number of extractions required may depend on cell type. When the aqueous phase is clear and no protein is observed at the interface of the two phases, proceed to the DNA precipitation steps.
18. The gel may remain in the Tris-HCl wash for up to 2 h.
19. Other transfer methods may be used, but vacuum transfer is preferred.

References

1. Zastawny, T. H., Dabrowska, M., Jaskolski, T., Klimarczyk, M., Kulinski, L., Koszela, A., et al. (1998) Comparison of oxidative base damage in mitochondrial and nuclear DNA. *Free Radical Biol. Med.* **24,** 722–725.
2. Yakes, F. M. and Van Houten, B. (1997) Mitochondrial DNA damage is more extensive and persists longer than nuclear DNA damage in human cells following oxidative stress. *Proc. Natl. Acad. Sci. USA* **94,** 514–519.
3. Croteau, D. L. and Bohr, V. A. (1997) Repair of oxidative damage to nuclear and mitochondrial DNA in mammalian cells. *J. Biol. Chem.* **272,** 25,409–25,412.
4. Sawyer, D. E., and Van Houten, B. (1999) Repair of DNA damage in mitochondria. *Mutat. Res.* **434,** 161–176.
5. LeDoux, S. P., Driggers, W. J., Hollensworth, B. S., and Wilson, G. L. (1999) Repair of alkylation and oxidative damage in mitochondrial DNA. *Mutat. Res.* **434,** 149–159.
6. Driggers, W. J., LeDoux, S. P., and Wilson, G. L. (1993) Repair of oxidative damage within the mitochondrial DNA of RINr 38 cells. *J. Biol. Chem.* **268,** 22,042–22,045.
7. Driggers, W. J., Grishko, V. I, LeDoux, S. P., and Wilson, G. L. (1996) Defective repair of oxidative damage in the mitochondrial DNA of a xeroderma pigmentosum group A cell line. *Cancer Res.* **56,** 1262–1266.
8. Druzhyna, N., Nair, R. G., LeDoux, S. P., and Wilson, G. L. (1998) Defective repair of oxidative damage in mitochondrial DNA in Down's syndrome. *Mutat. Res.* **409,** 81–89.
9. Hollensworth, S. B., Shen, C., Sim, J. E., Spitz, D. R., Wilson, G. L., and LeDoux, S. P. (2000) Glial cell type-specific responses to menadione-induced oxidative stress. *Free Radical Biol. Med.* **28,** 1161–1174.
10. Dobson, A. W., Xu, Y., Kelley, M. R., LeDoux, S. P., and Wilson, G. L. (2000) Enhanced mitochondrial DNA repair and cellular survival after oxidative stress by targeting the human 8-oxoguanine glycosylase repair enzyme to mitochondria. *J. Biol. Chem.* **275,** 37,518–37,523.
11. Shimoda-Matsubayashi, S., Matsumine, H., Kobayashi, T., Nakagawa-Hattori, Y., Shimizu, Y., and Mizuno, Y. (1996) Structural dimorphism in the mitochondrial

targeting sequence in the human manganese superoxide dismutase gene. *Biochem. Biophys. Res. Commun.* **226,** 561–565.

12. Tamura, T., McMiken, H. W., Smith, C. V., and Hansen, T. N. (1996) Mitochondrial targeting of glutathione reductase requires a leader sequence. *Biochem. Biophys. Res. Commun.* **222,** 659–663.

13. Thor, H., Smith, M. T., Hartzell, P., Bellomo, G., Jewell, S. A., and Orrenius, S. (1982) The metabolism of menadione (2-methyl-4-naphthoquinone) by isolated hepatocytes. A study of the implications of oxidative stress in intact cells. *J. Biol. Chem.* **257,** 12,419–12,425.

14. Frei, B., Winterhalter, K. H., and Richter, C. (1986) Menadione- (2-methyl-1, 4-naphthoquinone-) dependent enzymatic redox cycling and calcium release by mitochondria. *Biochemistry* **25,** 4438–4443.

25

In Organello Footprinting of mtDNA

Steven C. Ghivizzani, Cort S. Madsen, Chandramohan V. Ammini, and William W. Hauswirth

1. Introduction

Studies of molecular mechanisms coordinating mammalian mitochondrial replicative and transcriptional processes have, for the most part, been limited to in vitro analyses. Although much has been learned from in vitro studies *(1)*, they are often difficult to develop and may not depict biological processes occurring in vivo in a fully accurate manner. We have developed the method of mitochondrial DNA (mtDNA) footprinting as a means to analyze protein–mtDNA interactions within the isolated organelle, in an in vivo-like environment *(2–5)*. This procedure involves the use of dimethyl sulfate (DMS) in a methylation protection assay, similar to that used to analyze protein–nuclear DNA interactions in vivo *(6)*. Dimethyl sulfate is a small molecule that methylates guanine residues at the N-7 position and, to a lesser extent, adenines at the N-3 position, making them sensitive to subsequent cleavage at alkaline pH and elevated temperature. When bound to specific DNA residues, proteins can decrease or intensify purine reactivity to DMS relative to naked DNA. The ability of DMS to readily permeate mitochondrial membranes permits detection of protein–mtDNA interactions that occur within purified organelles. We have employed two strategies (Southern hybridization and primer extension) to visualize and map alterations in mitochondrial DNA methylation resulting from protein binding in organello. Because the entire mitochondrial sequence of numerous species have been determined, it is possible to analyze virtually any mtDNA domain for protein interactions.

From: *Methods in Molecular Biology, vol. 197: Mitochondrial DNA: Methods and Protocols*
Edited by: W. C. Copeland © Humana Press Inc., Totowa, NJ

2. Materials

2.1. Isolation and Purification of Mitochondria from Mammalian Tissue

Solutions should be autoclaved and then prechilled on ice before use.

1. Mannitol–sucrose buffer with $CaCl_2$ (MSB-Ca): 50 mM Tris-HCl, pH 7.5, 210 mM mannitol, 70 mM sucrose, and 3 mM $CaCl_2$.
2. 0.5 M EDTA.
3. Mannitol–sucrose buffer with EDTA (MSB-EDTA): 50 mM Tris-HCl, pH 7.5, 210 mM mannitol, 70 mM sucrose, and 10 mM EDTA.
4. 20% Sucrose buffer: 50 mM Tris-HCl, pH 7.5, 10 mM EDTA, and 20% sucrose (w/v).
5. 1.0 M Sucrose buffer: 10 mM Tris-HCl, pH 7.5, 10 mM EDTA, and 1.0 M sucrose.
6. 1.5 M Sucrose buffer: 10 mM Tris-HCl, pH 7.5, 10 mM EDTA, and 1.5 M sucrose.

2.2. Methylation of Mitochondrial DNA in Organello

1. Phosphate-buffered saline (PBS): 137 mM NaCl, 2.7 mM KCl, 4.3 mM $NaPO_4$, 1.4 mM KPO_4, pH 7.4.
2. Dimethyl sulfate (DMS): Please note that DMS is a potent mutagen and readily permeates living tissues and cellular membranes to methylate genomic DNA. Thus, it is critical that the experimenter perform all operations involving DMS in a chemical fume hood with appropriate protective clothing, including laboratory coat, two pairs of Latex gloves, and face protection. All waste materials containing DMS should be collected and disposed of according to appropriate chemical hazard guidelines.
3. Lysis buffer: 50 mM Tris-HCl, pH 8.0, 25 mM EDTA, 250 mM NaCl.
4. 10% Sodium dodecyl sulfate (SDS): We recommend obtaining SDS from Bio-Rad Laboratories (Hercules, CA), particularly for use in buffer solutions for Southern hybridization (*see* **Subheading 2.7.**).
5. Phenol: Liquid phenol should be saturated with 1 M Tris-HCl, pH 8.0.
6. Phenol–chloroform–isoamyl alcohol: mixed at a ratio of 25:24:1 (v/v/v) and the solution saturated with 1 M Tris-HCl, pH 8.0.
7. Chloroform: saturated with 1 M Tris-HCl, pH 8.0.
8. 95% Ethanol.

2.3. Methylation of Naked Mitochondrial DNA

1. PBS as in **Subheading 2.2.**
2. Lysis buffer as in **Subheading 2.2.**
3. 10% SDS (as in **Subheading 2.2.**) and phenol (as in **Subheading 2.2.**).

4. Phenol–chloroform–isoamyl alcohol as in **Subheading 2.2.**
5. Chloroform as in **Subheading 2.2.**
6. 95% Ethanol.

2.4. Restriction Endonuclease Digestion of mtDNAs

1. Reaction buffer suitable for the restriction endonuclease of choice. This is generally supplied by the manufacturer as a 10X solution.
2. Sterile water: filtered sterilized or autoclaved distilled water.
3. Restriction enzyme.
4. Agarose.
5. 10X TAE buffer: 400 mM Tris, 200 mM glacial acetic acid, and 10 mM EDTA, pH 8.0.

2.5. Methylation of Naked Control mtDNAs

1. 100 mM Sodium cacodylate, pH 7.0.
2. DMS as in **Subheading 2.2.**
3. DMS stop buffer: 1.5 M sodium acetate, pH 7.4, 1 M 2-mercaptoethanol.
4. 10% Piperidine: This reagent should be mixed fresh immediately before use.
5. 7.5 M Ammonium acetate.
6. 95% Ethanol.
7. Tris-EDTA buffer (TE): 10 mM Tris-HCl, pH 8.0, 1 mM EDTA.
8. Formamide loading dye: 10 mL formamide, 10 mg xylene cyanol FF, 10 mg bromophenol blue, and 0.5 M EDTA, pH 8.0.

2.6. Analysis of Mitochondrial DNA Methylation Products by Primer Extension

1. T4 polynucleotide kinase, with suitable reaction buffer (generally supplied by the manufacturer).
2. γ-^{32}P ATP: high specific activity.
3. Oligodeoxynucleotide primer complementary to a specific sequence of mtDNA about 70–200 nucleotides (nts) upstream of the region of interest.
4. Thermostable DNA polymerase with suitable reaction buffer (generally supplied by the manufacturer).
5. Primer extension reaction buffer: 10 mM Tris-HCl, pH 9.0, 50 mM KCl, 0.1% Triton X-100, 2.5 mM MgCl$_2$.
6. 400 µM Each of deoxyribonucleotide, dATP, dGTP, dCTP, dTTP.
7. Formamide loading dye as in **Subheading 2.2.**
8. 6% Polyacrylamide:bisacrylamide (80:1), 7 M urea sequencing gel, at least 15 in in height.
9. 5X Tris-borate EDTA (TBE) electrophoresis buffer: 450 mM Tris-base, 450 mM boric acid, and 10 mM EDTA, pH 8.0.
10. X-Omat x-ray film (Eastman Kodak).

2.7. Southern Hybridization Analysis of mtDNA Methylation Products

1. pBS (+) Phagemid vector (Stratagene, La Jolla, CA). This DNA may be replaced by any plasmid vector that enables generation of a single stranded radiolabeled RNA probe approx 150–200 nt in length. This particular vector was chosen because the multiple cloning site is flanked on either end by T3 or T7 bacteriaphage promoters.
2. Oligodeoxynucleotide primer pair bracketing the region of interest in the mtDNA extending from the reference restriction site approx 200 bp (base pairs) toward the domain of interest.
3. Thermostable DNA polymerase: A suitable reaction buffer will generally be supplied by the manufacturer. Virtually any of the available polymerases will suffice. A high level of replication fidelity is not necessary.
4. 400 μM Each of deoxyribonucleotide, dATP, dGTP, dCTP, dTTP.
5. Molecular biology reagents for subcloning DNA.
6. T7 or T3 RNA polymerase: Reaction buffer and conditions are generally supplied by the manufacturer.
7. Ribonucleotides, ATP, GTP, CTP, UTP: prepared at the concentration recommended for T3/T7 mediated in vitro transcription.
8. α-^{32}P-UTP: high specific activity.
9. Sterile water.
10. 4% Nondenaturing acrylamide gel: 4% polyacrylamide:bisacrylamide (80:1) in TBE.
11. X-Omat x-ray film (Eastman Kodak).
12. 6% Polyacrylamide/7 M urea sequencing gel as in **Subheading 2.6.**
13. Hybond N+ nylon transfer membrane (Amersham, Chicago, IL).
14. Whatman 3MM paper (Whatman Paper Ltd., Clifton, NJ).
15. Na_2HPO_4 buffer: 7.1% (w/v) anhydrous Na_2HPO_4, titrated with phosphoric acid to pH 7.2; this solution is 1 M with respect to sodium ion.
16. Hybridization solution: 1% bovine serum albumin, 1 mM EDTA, 7% SDS, and 0.5 M sodium phosphate buffer.
17. Wash buffer: 1 mM EDTA, 40 mM sodium phosphate buffer, and 1% SDS.

3. Method

3.1. Isolation and Purification of Mitochondria from Mammalian Tissue (See Note 1)

Tissues for the isolation of mitochondria should be obtained fresh from the subject or experimental animal; we have used bovine brain, mouse liver, human placenta, and *Xenopus* eggs and embryos with satisfactory results. Mitochondria are then purified by differential centrifugation and sucrose step-gradient banding. To maintain the integrity of the mitochondria, all procedures should be performed on ice or 4°C.

1. Approximately 200 g of tissue is placed in 500 mL of ice-cold MSB-Ca solution in an electric blender and homogenized on medium speed until the tissue is mostly liquefied.
2. The tissue mixture is then further homogenized by three strokes in a glass homogenizer with a tightly fitting motor-driven pestle. The total volume of homogenate is measured in a graduated cylinder, and 0.5 M EDTA is added to a final concentration of 20 mM.
3. The homogenate is then centrifuged at 1600g for 10 min to pellet the tissue debris. The supernatant is collected and recentrifuged as before. The supernatant from the second spin is collected and centrifuged at 17,000g for 20 min to pellet the crude mitochondria. The supernatant is discarded, and the mitochondrial pellet is washed with 5 volumes of MSB-EDTA and pelleted at 17,000g again.
4. The pellet is then resuspended in 20% sucrose. Each pellet is brought up in 10 mL of 20% sucrose–TE buffer and layered on top of a sucrose step gradient (10 mL each of 1.5 M and 1.0 M sucrose–TE). The banded mitochondria are collected in a minimal volume from the interface between the sucrose layers, diluted with MSB-EDTA and rebanded on this sucrose step gradient.
5. The purified mitochondria are collected, diluted twofold with MSB-EDTA, and pelleted at 23,000g for 10 min. For smaller amounts of beginning tissue, this procedure can be scaled down proportionally with good results.

3.2. In Organello *Methylation of Mitochondrial DNA* (see *Note 2*)

1. After the final centrifugation step, the pelleted mitochondrial (approx 1.7 mL from 200 g of starting tissue) are resuspended in an equal volume of room-temperature PBS, and 200-µL aliquots are placed into sixteen, 1.5-mL Eppendorf tubes. The tubes should then be divided into two equal groups. For 16 tubes, 8 will be used to generate *in organello* methylated mtDNA (tests). For the remaining eight tubes, the mtDNA will be isolated and purified intact, without methylation. After purification, these "naked" DNAs will be treated with DMS to generate comparative controls.
2. To each of the eight test samples, add DMS in twofold increments to final concentrations between 0.50% and 0.062%, immediately mix each by vortexing 2–3 s, and incubate for 3 min at room temperature. The reaction is stopped by diluting the DMS reaction with 1 mL ice-cold PBS. The mitochondria are pelleted by centrifugation for 1 min at 10,000g and washed twice more with 1 mL cold PBS to further dilute the DMS. If desired, mitochondrial pellets can be stored at this point at –80°C for further processing at a more convenient time.
3. To isolate mtDNA from the mitochondrial samples, individual test and control mitochondrial pellets are suspended in 400 µL of lysis buffer, and 25 µL of 10% SDS is added. The solution is vortexed vigorously until it clears. If there is difficulty achieving complete lysis, SDS may be added to a final concentration of 2%. The samples are deproteinized by extracting twice with phenol, twice with phenol–chloroform–isoamyl alcohol (25:24:1, v/v/v), and once with chloroform

and then precipitated with 2.5 volumes of 95% ethanol. The addition of sodium acetate is unnecessary because of the presence of NaCl in the lysis buffer.

3.3. Restriction Endonuclease Digestion of the mtDNAs

1. For Southern hybridization analysis, an equal number of control and test mtDNA samples are centrifuged at $14,000g$ for 15 min in a tabletop microcentrifuge.
2. The ethanol is carefully removed and the mtDNA pellets are resuspended in a 100-µL reaction volume compatible with the restriction enzyme of choice (this step is unnecessary for primer extension analyses). For convenience, double restriction digests can be performed on the same mtDNA sample for the simultaneous analysis of different domains, assuming that the two domains are sufficiently distant from one another. Considerations regarding restriction enzyme site selection are discussed in **Subheading 3.5.**
3. Following digestion, 5 µL of each sample is resolved on a 0.8% agarose gel. Following ethidium bromide staining of the gel, the samples are visualized using ultraviolet light. This procedure verifies complete digestion and provides a means for estimating relative DNA concentrations among samples. Generally, if distinct mtDNA bands are visible on the gel that correspond to expected DNA sizes following digestion, then there is sufficient sample for at least 15 different gel loadings for Southern analysis or at least twice this number of primer extension reactions.

3.4. Methylation of Naked Control mtDNAs

1. Following restriction digestion, all samples are brought to a volume of 200 µL with the addition of 5 µL of water and 100 µL of 100 mM sodium cacodylate. For primer extension analysis, test and control mtDNA pellets are suspended in 200 µL of 50 mM sodium cacodylate.
2. Dimethyl sulfate is then added to the control samples only in twofold increments to final concentrations between 0.5% and 0.062% (any alterations in the methylation conditions that may have been applied to the test samples should likewise be applied to the control samples), and the samples are incubated for 3 min at room temperature. One hundred microliters of cold DMS stop buffer is then added to all of the samples. Three volumes of 95% ethanol are added to all samples, which are then vortexed and centrifuged at $14,000g$ for 15 min.
3. To cleave the mtDNAs at methylated residues, the ethanol is carefully removed (residual ethanol is removed by brief vacuum), and the mtDNA pellets are resuspended in 200 µL of 10% piperidine and incubated 30 min at 90°C. Afterward, the samples are placed on ice and an equal volume of 7.5 M ammonium acetate is added. One milliliter of 95% ethanol is added and the mtDNA precipitated as in **step 2**.
4. To ensure unambiguous resolution of adjacent DNA bands in subsequent sequencing gels, it is critical to remove all traces of piperidine from the mtDNA samples. Generally, three successive ethanol precipitations with ammonium acetate are sufficient. Alternatively, following an initial ethanol precipitation, the samples

may be suspended in 200 µL of water and lyophilized. Samples to be analyzed by Southern hybridization are then suspended in 25 µL of formamide loading dye. Those intended for primer extension are suspended in 50 µL of TE.

3.5. Analysis of Mitochondrial DNA Methylation Patterns by Primer Extension (see Note 3)

For primer extension analysis, 50 pmol of an oligodeoxynucleotide complementary to sequences near the region interest is first end-labeled using T4 polynucleotide kinase. Gel-purified primers are 5′ end-labeled to a high specific activity with α-^{32}P-ATP (ICN, 7000 Ci/mmol) and a T4 polynucleotide kinase kit (USB Corporation–Amersham Life Science) in a standard 50-µL reaction. Free label is removed by two sequential ethanol precipitations in the presence of 2.5 *M* ammonium acetate. One picomole of labeled primer is then mixed in 100-µL reaction volumes containing control and test mtDNAs (approx 1/20 of each sample), reaction buffer, and a thermostable DNA polymerase and incubated for 10 cycles (1 min at 95°C, 2 min at 55°C, and 5 min at 72°C per cycle) in a thermocycler (*see* **Subheading 4.3.**). The samples are then extracted with phenol–chloroform. One-tenth volume of 3 *M* sodium acetate is mixed with the sample, followed by the addition of 2.5 volumes of 95% ethanol. After centrifugation and removal of the ethanol, the radiolabeled samples are resuspended in 25 µL of formamide loading dye. The samples are heated to 95°C for 5 min and 3–4 µL of each sample analyzed by electrophoressis on 6% polyacrylamide, 7 *M* urea sequencing gels at a sufficient voltage and for a sufficient time to read the region of interest at single nucleotide resolution. Following electrophoresis, the plates are separated and the gel is bound to an appropriate size piece of Whatman 3 MM paper and dried using a standard vacuum gel dryer with heat. The dried gel is then exposed to x-ray film. An example primer extension analysis of *in organello* mitochondrial footprinting is shown in **Fig. 1**.

In general, primer extension analysis has proven to be extremely valuable when mtDNA samples are present in limited quantities. However, for *in organello* footprinting of animal mitochondria from large organs, where the starting sample amount is not limiting, Southern analysis is the method of choice because fewer artifacts are seen.

3.6. Southern Hybridization Analysis of mtDNA Methylation Patterns (see Note 4)

1. It is necessary to use mtDNA sequence information to select a restriction site for use as a reference point from which to evaluate patterns of DNA methylation. The mtDNA sequence is also useful for construction of DNAs necessary for the synthesis of strand-specific radiolabeled probes. We have found polymerase chain

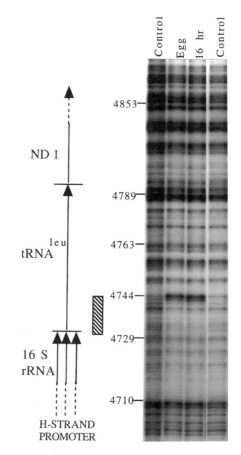

Fig. 1. Primer extension *in organello* footprinting of the mtDNA H-strand mTERF region in *X. laevis* eggs and embryos. The tridecamer mTERF (mitochondrial transcription termination factor) binding site (nucleotides 4730–4743) is indicated by a hatched box. Residues hypersensitive to DMS methylation are evident at nucleotides 4743 and 4744. Control lanes (naked mtDNA methylation products) flank egg and 16 h embryonic *in organello* mtDNA lanes.

reaction (PCR) amplification to be the most efficient method for the generation of specific probes. Oligodeoxynucleotide primer pairs bracketing about 150–200 bp of sequence extending from the restriction site toward the region of interest are synthesized and combined with unmethylated bovine mtDNA in a standard PCR. The amplified mtDNA fragment is then inserted into the pBS(+) phagemid vector (Stratagene, La Jolla, CA) or a similar cloning vector where the inserted DNA fragment is flanked on either side by T7, T3, or other bacterial promoters. Radiolabeled RNA probes complementary to the region and strand of interest are then generated using either the appropriate RNA polymerase or the corresponding promoter site of the appropriate linearized template clone. We usually verify riboprobe synthesis by electrophoresis of the entire reaction in 4% nondenaturing polyacrylamide gels and autoradiography. A small gel slice containing the labeled RNA is excised, crushed, and added to the hybridization solution (*see* **step 3**).

2. Equal amounts of test and control mtDNAs (approx 1/20 of each sample) are loaded onto adjacent lanes of a 6% polyacrylamide, 7 *M* urea sequencing gel and subjected to electrophoresis using the conditions in **step 1**. Following electrophoresis, a precut piece of dry Hybond N+ nylon membrane (Amersham, Chicago, IL) is placed directly on the polyacrylamide gel and allowed to sit for 3–5 min. Because the gel will adhere to the membrane, it can then be easily peeled back and positioned membrane-side down on a single sheet of Whatman 3MM paper (Whatman Paper Ltd, Clifton, NJ). After placing the gel, Whatman paper-side down, onto a standard gel-drying apparatus, it is covered with plastic wrap, and the DNA is transferred to the membrane by applying a vacuum (without heat) for approximately 45 min. This simple technique is fairly easy and yields a sharp, precise transfer of DNA bands.

3. Once the mtDNA has been transferred, it is crosslinked to the membrane by ultraviolet (UV) irradiation. With the gel still attached, the membrane is place gel-side down on a transilluminator at full power for 5 min. Afterward, the polyacryamide is removed by soaking the membrane in a tray of water. While still in the tray, the membrane is rolled up and placed into a rolling hybridization bottle. To prehybridize the membrane, 30 mL of hybridization solution is added and the bottle is placed in a rolling hybridization bottle oven for 30 min at 65°C. Discard the hybridization solution and replace it with 10 mL of fresh hybridization solution, including the crushed polyacrylamide gel slice containing the labeled riboprobe. The DNA blots are hybridized overnight at 65°C. The hybridization solution is decanted (may be saved for later use if desired) and the membrane placed in a container filled with 500 mL of 60°C wash buffer. Perform three 20-min washes at 60°C with gentle agitation. Briefly rinse the membrane once more in wash buffer and expose to x-ray film. An example is shown in **Fig. 2**.

4. Notes

1. Optimally, the footprinting method will be carried from the isolation of the mitochondria through the methylation procedure in a single day. Thus, an early start is suggested. Virtually any tissue can be used for the isolation of mitochondria; however, if the type of tissue is not important to the study, we recommend the use of softer, less fibrous tissues, such as brain or liver. These are fairly rich in mitochondria and are easy to homogenize. It is also suggested that care be taken not to overload the sucrose gradients. Mitochondria of greater purity will be obtained if the homogenate is divided into several tubes per spin.

2. We routinely obtain about 2 mL of purified mitochondria from a single 200-g preparation of mammalian brain, liver, or placenta. Mitochondria from smaller

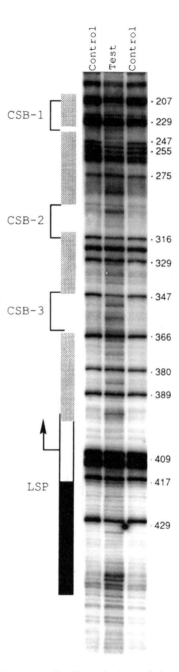

Fig. 2. Southern blot *in organello* footprinting of the mtDNA L-strand D-loop control region in bovine brain. Control lanes (naked mtDNA methylation products) flank an *in organello* mtDNA lane. The locations of LSP (L-strand promoter, filled box) CSB-1, -2, and -3 (conserved sequence blocks 1, 2, and 3, brackets) are indicated. Domains primarily hypersensitive to DMS methylation (shaded boxes) both flank and align with the CSB elements, suggesting a regular, periodic array of protein binding to the mtDNA region.

amounts of starting tissue can be isolated by simply scaling down this protocol. This amount is sufficient for eight test and eight control samples, about the maximum number that can be easily manipulated at one time. Because one of the most difficult steps in a methylation-based footprinting technique is the production of equally intense G ladders in control (naked DNA) and test (DNA–protein) samples, we find that dividing the purified mitochondria into multiple control and test samples allows evaluation of a range of DMS concentrations and incubation times, as well as testing of the reproducibility of results. Because the quality of results will relate directly to the methylation of mtDNA, we suggest that at least two people work together to coordinate the timing of the reactions as accurately and smoothly as possible.

3. In using primer extension analyses, we have occasionally noted the appearance of bands that do not correspond to G or A residues. If such artifactual bands were to occur where G residues are normally found, their presence could lead to incorrect interpretation of methylation reactivities. Many of these types of problems can be alleviated by optimizing reaction conditions empirically. The most important factor for obtaining accurate and reproducible sequences is optimal selection of genomic primers. In theory, a primer can be as close as 10–30 nucleotides to the domain of interest. However, in practice, primers located about 100–200 nucleotides away from the region to be analyzed work best in providing reproducible sequence ladders. About one-tenth of the piperidine-cleaved DNA sample (about 50–100 ng) is sufficient for obtaining a usable primer extension signal. Increasing the primer concentration beyond 1 pmol per reaction does not further enhance signal intensity and leads to higher background. The number of linear amplification cycles necessary to obtain reproducible is another important variable. Amplifying for 10–15 cycles provides adequate signal intensity, whereas nonspecific bands often appear with numbers of cycles greater than this. Nonspecific bands appearing at higher cyles are often one to two nucleotides longer than authentic bands, suggesting that *Taq* DNA polymerase is adding nontemplated nucleotides to the 3′-end of duplex DNA made in earlier cycles. Thermostable polymerases having a 3′–5′ proofreading ability, such as *Vent* DNA polymerase or *Pfu* DNA polymerase, may alleviate this problem. However, neither polymerase performed as well as *Taq* DNA polymerase in head-to-head comparisons over a wide range of dNTP concentrations, annealing temperatures, or ionic strength. For *Taq* DNA polymerase, a final dNTP concentration of 20 μM is best for specificity and reproducibility. Lowering dNTP concentration below this level sacrificed both fidelity and signal intensity. An extension time of 3 min was found to be sufficient for a number of primer–template combinations without compromising fidelity or sensitivity.

Based on these considerations, a typical primer extension reaction consisted of 50–100 ng of piperidine-cleaved mtDNA, 1 pmol of ^{32}P5′-end-labeled primer, 1X *Taq* polymerase reaction buffer (BRL), 2 mM MgCl$_2$, 20 μM each of dATP, TTP, dGTP, dCTP, and 2.5 U of thermostable DNA polymerase (BRL). The reaction is assembled on ice and cycled in a thermal cycler as follows: denaturing at 94°C for 5 min, followed by 10–15 cycles with the profile: denaturing at 94°C for 1 min, annealing at 45–65°C (based on primer T_m) for 1 min, and primer extention at 72°C for 3 min. The tubes are then immediately chilled on ice, 10 µg of carrier tRNA added, and DNA ethanol precipitated in the presence of 2.5 M ammonium acetate. An organic solvent extraction prior to DNA precipitation is usually not necessary. The precipitate is resuspended in 5 µL of loading dye, and 1.5 µL of each reaction resolved in a 6% polyacrylamide, 7 M urea gel as a preliminary test to optimize signal intensities. Using visual inspection or phosphorimager quantitation, the signals in each sample are equalized in a final gel and autoradiographed to obtain the footprint pattern. Autoradiography at room temperature for 12–48 h is usually sufficient to generate adequate signal intensity.

4. After the purified mitochondria are incubated with a limited amount DMS, the mtDNA is then digested with a restriction enzyme at a site near the region of interest. Following breakage of the DNA strands at the methylated positions by alkali and heat, the resulting fragments of DNA are resolved on a denaturing acrylamide gel. Patterns of fragmented DNA are visualized by hybridization to a radiolabeled probe complementary to only one strand of DNA using riboprobes.

A sequence map encompassing the targeted domain is required, first for selection of a restriction enzyme recognition site to be used as a reference point for footprinting and then for construction of a probe for hybridization. Several factors should be considered when selecting the best restriction site for footprint analysis. First, the distance between the restriction site and the domain of interest should be between 70 and 300 bp. DNA fragments smaller than 70 bp are often difficult to detect using standard Southern hybridization conditions and G ladders longer than 300 bp in length may be difficult to resolve. Further, longer G ladders may require empirical optimization of the methylation reaction conditions to generate a uniform distribution of DNA fragments. Second, the restriction enzyme site must recognize only one site near the domain of interest. If a second recognition site is present within a few hundred base pairs on the either side of the domain, then, depending on the size and location of the radiolabeled probe, two G ladders may be superimposed on each other, making an informative analysis impossible. Third, if possible, avoid selecting a reference site separated from the domain of interest by unusually G-rich regions. Homopolymer runs of G residues or G-rich sequences of DNA often require extra manipulation of methylation reaction conditions and are also potential hot spots for heteroplasmic sequences in mitochondrial DNA, which can result in unreadable G ladders.

Careful transfer of the samples from the gel to a solid support is perhaps the most important step for achieving publication-quality results. It may be necessary to perform several blots to attain the proper resolution of the bands as well as equal loadings of DNA. Following electrophoresis, one of the most difficult steps in Southern analysis is the efficient, equal transfer of all DNA fragments to a nylon membrane. When numerous control and test samples are electrophoresed simultaneously, the size and fragility of the gel make handling particularly awkward. This is especially true for electroblotting, where several manipulations of the gel and membrane are required. As handling difficulties increase so do problems with uneven transfer and smearing of the DNA bands. The frequency of poor transfer can be greatly reduced by employing a vacuum blotting technique. As a note of caution, we find that the use of newer, more efficient vacuum pumps may result in poor transfer of higher-molecular-weight fragments (above 200 bp). However, placing an additional sheet of Whatman paper under the membrane often corrects this problem. The extra sheet most likely functions to slow the drying of the polyacrylamide gel, thus allowing more time for DNA transfer.

References

1. Shadel, G. S. and Clayton, D. A. (1997) Mitochondrial DNA maintenance in vertebrates. *Annu. Rev. Biochem.* **66,** 409–435.
2. Ammini, C. V., Ghivizzani, S. C., Madsen, C. S., and Hauswirth, W. W. (1996) Genomic footprinting of mitochondrial DNA: II. In vivo analysis of protein–mitochondrial DNA interactions in *Xenopus laevis* eggs and embryos. *Methods Enzymol.* **264,** 23–36.
3. Ghivizzani, S. C., Madsen, C. S., and Hauswirth, W. W. (1993) *In organello* footprinting. Analysis of protein binding at regulatory regions in bovine mitochondrial DNA. *J. Biol. Chem.* **268,** 8675–8682.
4. Ghivizzani, S. C., Madsen, C. S., Nelen, M. R., Ammini, C. V., and Hauswirth, W. W. (1994) *In organello* footprint analysis of human mitochondrial DNA: human mitochondrial transcription factor A interactions at the origin of replication. *Mol. Cell. Biol.* **14,** 7717–7730.
5. Madsen, C. S., Ghivizzani, S. C., Ammini, C. V., Nelen, M. R., and Hauswirth, W. W. (1996) Genomic footprinting of mitochondrial DNA: I. *In organello* analysis of protein–mitochondrial DNA interactions in bovine mitochondria. *Methods Enzymol.* **264,** 12–22.
6. Nick, H. and Gilbert, W. (1985) Detection in vivo of protein–DNA interactions within the lac operon of *Escherichia coli. Nature* **313,** 795–798.

26

Crosslinking of Proteins to mtDNA

Brett A. Kaufman, Scott M. Newman, Philip S. Perlman, and Ronald A. Butow

1. Introduction

Mitochondrial DNA (mtDNA) nucleoids have been isolated from several organisms, including rat (liver) *(1)*, *Physarum polycephalum (2)*, *Saccharomyces cerevisiae (3)* and *Pichia jadinii (4)*. Most methods for nucleoid isolation have utilized detergent extraction and sucrose gradient centrifugation to separate bulk mitochondrial protein from mtDNA-associated proteins (e.g., **ref. 5**). The disadvantages of these methods include contamination by non-nucleoid proteins and the possibility of losing those nucleoid proteins that are not tightly bound to the complex under the conditions of the fractionation. To circumvent these problems, we have developed an *in organello* formaldehyde crosslinking method for isolating mtDNA nucleoids that stabilizes protein content and reduces the potential for contamination by irrelevant mitochondrial (and cellular) proteins. Additionally, it has long been suggested that mtDNA is associated with the inner membrane. To begin exploring that notion, we have used ultraviolet (UV) crosslinking as a means of detecting the binding of proteins from membrane-enriched fractions to specific sequences of mtDNA.

Formaldehyde crosslinking of proteins to DNA was originally developed to study *Escherichia coli* nucleoid composition *(6)*. This procedure was later used in studies of the assembly and interaction of nucleosomes and transcription complexes in systems such as tissue culture cells, flies, yeast (for a review, *see* **ref. 7**) and trypanosomes *(8)*. Formaldehyde is a highly reactive, membrane-permeable, dipolar reagent that is subject to nucleophilic attack by primary amines (e.g., lysine and cytosine), forming a Schiff base. This intermediate reacts and condenses with a second nucleophile to form a 2-Å bridge, which

From: *Methods in Molecular Biology, vol. 197: Mitochondrial DNA: Methods and Protocols*
Edited by: W. C. Copeland © Humana Press Inc., Totowa, NJ

covalently links the molecules. The strength of this method is that formaldehyde will trap protein–protein, protein–RNA, and protein–DNA interactions without damaging free double-stranded DNA. However, comparison of the relative abundance of proteins within a given preparation is limited, as crosslinking efficiency is dependent on the availability and spatial relationship of nucleophilic amino acids. Because this procedure traps proteins intimately associated with mtDNA, including both direct and indirect interactions, direct DNA binding must be confirmed by electrophoretic mobility shift assay (EMSA) or other methods (e.g., chromatin immunoprecipitation).

We have published a procedure for formaldehyde crosslinking of mitochondrial proteins to yeast mtDNA *(9)* and this chapter discusses the main features of those experiments. Here, we describe both preparative and analytical methods for isolating mtDNA-associated proteins (*see* **Fig. 1**). Our approach begins with highly purified preparations of mitochondria, essentially free of nuclear DNA, followed by formaldehyde treatment to crosslink proteins to the mtDNA in the intact organelles. After quenching the free aldehydes, the mitochondria are lysed and the DNA–protein complexes purified by CsCl density gradient centrifugation. The resulting material may be analyzed using a variety of methods, including direct protein identification by Western blotting or by mass spectrometry, chromatin immunoprecipitation, or comparative analyses of the recovered crosslinked polypeptides derived from strains with various mutations of nuclear and mtDNA.

As an example, we have identified Hsp60p as associated with ρ^+ mtDNA in formaldehyde crosslinked mitochondria *(9)*. Independently, we also identified Hsp60p as one of the proteins that can be UV crosslinked in vitro to mtDNA ori sequences containing BrdU following incubation of the DNA with inverted inner mitochondrial membranes (**Fig. 2**). Subsequent experiments have revealed that Hsp60p is a single-stranded mtDNA-binding protein that binds with specificity to the template strand of ori sequences *(9)*.

2. Materials

2.1. Stock Solutions

Unless otherwise noted, all reagents are from Sigma-Aldrich and are stored at room temperature.

1. 0.1 *M* Phenylmethylsulfonyl fluoride (PMSF) in isopropanol. Stored at –20°C.
2. 1 *M* Spermidine. Stored at 4°C.
3. 2.5 *M* Glycine, pH 7.0.
4. 37% Formaldehyde (Pierce).
5. β-Mercaptoethanol (BME).
6. 3 *M* sodium acetate (NaOAc), pH 5.0.

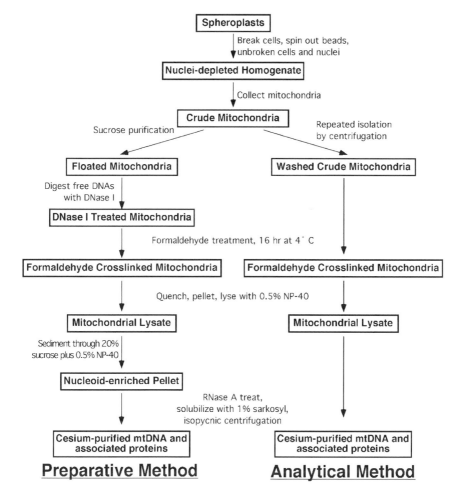

Fig. 1. Flow diagram of preparative and analytic methods for purifying formaldehyde crosslinked, mtDNA-associated proteins. The procedures bifurcate after the isolation of crude mitochondria. On the left is the procedure for generating highly purified mitochondria, formaldehyde crosslinking of those mitochondria, and subsequent steps for isolation of proteins associated with mtDNA. On the right is a streamlined preparation suitable for Western blot analysis.

7. Linear acrylamide (Ambion). Stored at –20°C.
8. 0.5 M Ethylenediaminetetraacetic acid (EDTA), pH 8.0.
9. RNase A (10 mg/mL). Stored in aliquots at –20°C.
10. 0.1 M ZnSO$_4$.
11. 5 M NaCl.
12. 20% Nonidet-P40 (NP-40) in water.

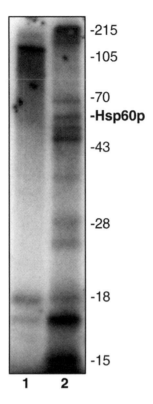

Fig. 2. Autoradiograph of proteins with radioactive label transferred from DNA probes resolved by 12% sodium dodecyl sulfate–polyacrylamide gel electrophoresis (SDS-PAGE). An ori5 DNA probe containing ^{32}P and BrdU was incubated with recombinant Abf2p (lane 1) or inverted ρ^0 mitochondrial membranes (lane 2) and irradiated with UV light. Then, the DNA was absorbed to streptavidin–agarose beads, washed extensively, and digested with MNase I and DNase I to release the label-transferred proteins, which were subsequently resolved by SDS-PAGE and visualized by autoradiography. Polypeptides with apparent molecular weights of 13, 17, 18, 26, 30, 40, 48, 55, 60, 70, approx 150, and approx 200 kDa were visualized. The 60-kDa polypeptide was identified as the mitochondrial chaperone Hsp60p by MALDI-TOF mass spectrometry and Western blot analysis.

13. 20% *N*-Lauryl sarkosine (sarkosyl) in water.
14. 1 *M* Tris-HCl, pH 8.0.
15. 1 *M* KCl.
16. 5 *M* MgCl$_2$.
17. 50% Glycerol.
18. 25X Protease inhibitor cocktail (Roche). Stored at –20°C.
19. Bio-Rad Protein Analysis (Bio-Rad).

20. Bovine serum albumin (BSA) protein standard (Pierce).
21. Yeast extract (Difco).
22. Bacto-peptone (Difco).

2.2. Preparation of Mitochondria

1. SCE: 0.6 M sorbitol, 0.3 M mannitol, 20 mM K$_2$HPO$_4$, 20 mM citric acid, 1 mM EDTA
   pH 5.8 (for Zymolase digestion): filter-sterilize.
   pH 7.0 (for yeast lytic enzyme): Store stock at 4°C. Immediately before preparing cells, add 250 µL PMSF (0.5 mM), 25 µL of BME (7 mM), 50 µL of spermidine (1 mM) in 50 mL of stock solution.
2. 10X Flotation buffer (FB): 100 mM Tricine-KOH, 1 mM EDTA, 50 mM NaCl, pH 7.5.
3. Sucrose solutions: 65%, 55%, 43%, 20%, and 15% (w/w) sucrose in 1X FB. Stored at 4°C.
4. 0.5-mm Glass beads (Biospec Products, Inc.). The beads may be reused after washing with water, hydrolyzing biomaterials with concentrated HCl, rinsing with water, neutralizing with 1 M Tris-HCl, pH 7.4, and oven-drying (approx 180°C).
5. Zymolyase T100 (ICN Biochemicals).
6. Yeast lytic enzyme (ICN Biochemicals).

2.3. Formaldehyde Crosslinking

1. Resuspension buffer (RB): 0.5 M sucrose, 20 mM HEPES, pH 7.4, 2 mM EDTA, 1 mM EGTA. Stored at 4°C.
2. 10X Crosslinking buffer (10X CB): 0.5 M HEPES, pH 7.6, 0.5 M NaCl. Stored at 4°C.
3. 1X CB: RB plus 0.1 volume of 10X CB, 7 mM BME, 0.5 mM PMSF, 1 mM spermidine, and 1X protease inhibitor cocktail. Made fresh before use and kept on ice.
4. Diluted formaldehyde (10%): 0.5 mL 10X CB, 1.35 mL formaldehyde (37%), and 3.15 mL water. Made fresh daily and kept on ice.
5. Dilution Buffer (DB): 0.1 volume CB and 0.5–1% NP-40 in water. Stored at 4°C.
6. Cesium buffer (CS): CsCl (refractive index = 1.365) and 1% sarkosyl in TE.
7. TFA solution: 0.1% trifluoroacetic acid, 60% acetonitrile. Filter-sterilized and stored in glass at room temperature.
8. Acid-wash solution: 1% formic acid, 50% isopropanol. Filter-sterilized and stored in glass at room temperature.
9. DNase I (10 U/µL; Roche Molecular Biochemicals).
10. S1 nuclease (400 U/µL; Roche Molecular Biochemicals).
11. Cesium chloride (Gibco-BRL).
12. 29:1 acrylamide:bis acrylamide (Bio-Rad).
13. 3X Laemmli gel-loading buffer (Bio-Rad).

2.4. UV Crosslinking

1. Sonication buffer (SB): 20 mM HEPES, pH 7.4, and 0.1 M NaCl. Stored at 4°C.
2. PGEM3-ori5 plasmid DNA: 455-bp DraI–NdeI fragment from HS40 (Klenow-treated) was blunt-end ligated and cloned into $Hind$III–EcoRI digested, Klenow-treated pGEM3Zf(+) (Promega). See **ref. *10*** for further details about this segment of mtDNA.
3. DNase I (Amersham-Pharmacia Biotech) in 1 µg/µL, 5 mM NaOAC, pH 5.0, 1 mM CaCl$_2$, and 50% glycerol. Stored in aliquots at –20°C.
4. MNase I (Amersham-Pharmacia Biotech) 1 µg/µL (40 U/µL) in 5 mM NaOAC, pH 5.0, 1 mM CaCl$_2$, and 50% glycerol. Stored in aliquots at –20°C.
5. Binding buffer (BB): 10 mM Tris-HCl, pH 7.4, 4 mM MgCl$_2$, 0.5 mM EDTA, and 50 mM KCl.
6. α-^{32}P-dATP, 800 Ci/mmol (NEN Life Science Products). Stored at –20°C.
7. Materials for PCR experiments:
   a. 10 mM dATP, 10 mM dTTP, 10 mM dGTP, 10 mM dCTP. Stored at 4°C.
   b. 5′ biotinylated, high-performance liquid chromatography (HPLC) purified T7 and Sp6 primers (Gibco-BRL). Stored at –20°C.
   c. 10 mM Bromodeoxyuridine (BrdU) in TE. Stored in amber tubes at –20°C.
   d. *Taq* DNA polymerase (Roche Molecular Biochemicals).
8. Gene Clean (Bio101).
9. Low DNA mass ladder (Gibco-BRL). Stored at –20°C.
10. Amber tubes, 1.6 mL (GeneMate).
11. Streptavidin agarose. Stored at 4°C.

3. Methods

3.1. Preparation of Mitochondria

This section describes a method for generating highly purified mitochondria, free of nuclear DNA. This protocol utilizes differences in sedimentation and buoyancy characteristics to remove other cellular components. Any remaining DNA adhering to the outside of the mitochondria is removed by DNase I digestion (*see* **Fig. 1**).

1. For large-scale preparative experiments, yeast cells are grown in YPG (1% yeast extract, 2% Bacto-peptone, 3% glycerol) or YPGE (1% yeast extract, 2% bacto-peptone, 3% glycerol, 2% ethanol) liquid medium in 11-L batches in 14 L of fermentors to OD of approx 1.6, harvested and washed with water. Slightly higher cell densities in fermentors do not significantly affect the yield of recovered proteins. For small-scale preparations or nonpreparative experiments, 1-L cultures are grown in 4-L flasks with vigorous shaking.
2. To improve digestion of the cell wall, the cells are resuspended in SCE, pH 5.8, (1 mL/g wet weight) with 5 µL/mL BME and shaken lightly (approx 150 rpm) at 30°C for 15 min. During the incubation, suspend Zymolyase in 1 mL of SCE, pH 5.8, (0.5 mg Zymolyase/mL of cell suspension). The cells are pelleted, the

supernatant discarded, and the cells resuspended in the same volume of SCE, pH 5.8, without BME. Next, the 1 mL Zymolyase solution is added to the cell suspension and the mixture is agitated gently at 30°C for approx 35 min (*see* **Note 1**). The spheroplasts are collected by centrifugation at 3000*g* for 5 min at 4°C.

3. To maintain the integrity of the preparation, all subsequent steps are performed on ice or at 4°C unless otherwise indicated. Cells are broken by vigorous shaking in three 1-min intervals with 0.5 pellet volumes of glass beads in 1 mL/g of cells of SCE, pH 7.0, (including BME, spermidine, and PMSF), chilling on ice between intervals.

4. The cytosol is collected as the supernatant after centrifugation at 1600*g*. The pellet from this first centrifugation is homogenized a second time by adding an equal volume of SCE, pH 7.0, and repeating the breakage step in two 1-min intervals. This second homogenate is depleted of unbroken cells, cell debris, and beads by low-speed centrifugation again.

5. After combining the supernatants from both extractions in **step 4**, the lysate is depleted of unbroken cells, nuclei, and glass beads by centrifugation at 2500*g* for 10 min. For preparative experiments, it is critical to remove all pelletable material and up to 12 repetitions of this step may be required.

6. The mitochondria are collected by centrifugation at 27,000*g* for 20 min or 35,000*g* for 15 min. The crude mitochondrial pellet is gradually resuspended (necessary for all following resuspensions) in SCE, pH 7.0, starting with a small volume (approx 200 μL) and increasing until a final volume that fills most of the centrifuge tube, and collect by centrifugation as before in this step (*see* **Note 2**). The supernatant is removed by aspiration and discarded.

7. For gradient purification, the mitochondrial pellet is resuspended in 8–10 mL of 65% (w/w) sucrose. If the initial wet weight of the initial cell pellet is greater than 100 g, the mitochondria are resuspended in a final volume of 18 mL of 65% (w/w) sucrose and equally divided into two prechilled 30-mL tubes. In either case, the 65% (w/w) sucrose is overlayed with 12 mL of 55% sucrose, 8 mL of 43% sucrose, and up to 5 mL of 20% sucrose. Tubes and swinging buckets are balanced with cold 1X FB and the mitochondria are floated at 40,000*g* for 2 h.

8. Floated mitochondria are collected by first removing the material above the 43/55% sucrose interface by aspiration and then collecting the mitochondria at the interface using a Pasteur pipet. The mitochondria are concentrated by centrifugation after diluting the collected fraction to approx 25mL with 15% sucrose in 1X FB. The amount of mitochondrial protein is determined before pelleting the diluted mitochondria using the Bio-Rad Protein Assay (using BSA for calibration) in order to estimate subsequent reaction volumes. The yield is approximately 0.5 mg of purified mitochondria from each gram (wet weight) of cells.

3.2. Formaldehyde Crosslinking Method

This section describes a protocol for using formaldehyde to generate protein–DNA and protein–protein linkages and the subsequent purification of proteins associated with mtDNA. Purified intact mitochondria are treated with formaldehyde and then lysed with detergent to release crosslinked mtDNA–protein complexes, which are then separated from membranes and unassociated proteins in two steps (*see* **Fig. 1**). First, rapidly sedimenting material in the mitochondrial lysate (including crosslinked nucleoids) are sedimented through a sucrose cushion. This step removes approximately half of the total protein and concentrates the protein–DNA complexes in a pellet fraction. Second, the nucleoid-enriched pellet is treated with RNase A, suspended in cesium chloride plus detergent, and subjected to isopycnic centrifugation. The CsCl fractions are analyzed or subjected to further treatments.

1. To further decrease contaminating nuclear DNA and associated proteins in preparative experiments, the mitochondria are resuspended in 15% sucrose in 1X FB at approx 10 mg/mL, adjusted to 40 mM Tris-HCl, pH 8.0, 50 mM KCl, and 2.5 mM MgCl$_2$, and treated with DNase I (5–10 U/50 μg mitochondria) for 1 h on ice. The reaction is terminated by the addition of EDTA, pH 8.0, to a final concentration of 5 mM.

2. To remove the DNase I and to exchange buffers before crosslinking, the mitochondria are pelleted (35,000g for 15 min), rinsed with RB, and resuspended to 2 mg/mL in 1X CB. The crosslinking reaction is initiated by the addition of 0.1 volume of diluted formaldehyde and continued with gentle mixing at 4°C usually for 16 h (*see* **Note 3**). The reaction is quenched by the addition of 125 mM glycine, pH 7.0, and the mitochondria are collected by centrifugation.

3. In preparative experiments (for smaller-scale experiments, *see* **Note 4**), the mtDNA and associated proteins are concentrated by centrifugation. This step is necessary because of rotor volume limitations in the following CsCl gradient centrifugation. First, the mitochondria are resuspended in 1X CB plus 50 mM glycine (approx 8mL), and lysed with 0.5–1% NP-40 and incubated for 5 min on ice. Next, the lysate is diluted threefold with cold DB and layered onto 20% sucrose in DB and the nucleoid-enriched pellet is collected by centrifugation (110,000g, 1 h) at 4°C.

4. The nucleoid-enriched pellet is difficult to resuspend and requires the use of a Dounce homogenizer, using a minimal volume of 1X CB. This suspension is further solubilized by treatment with 1% sarkosyl and RNase A (50 μg/mL) for 1 h at room temperature. For digestion of single-stranded DNA, see **Note 5**.

5. For cesium chloride gradient purification, the suspended crosslinked material is adjusted to a refractive index of 1.365 by adding dry CsCl. This suspension is then diluted with CS buffer until the equivalent of approx 5–15 mg of floated mitochondria can be loaded into each 11-mL crimp cap tubes (*see* **Note 4**). Gradients are formed (260,000g, 16 h, 25°C) in a fixed-angle rotor and gravity fractionated into 1-mL fractions (*see* **Note 6**). Fractions are stored at –20°C.

6. The distribution of mtDNA throughout the gradient is determined by dot blot analysis. Fifty microliters of each fraction is diluted into 250 μL of fresh 100 mM

NaOH and incubated at 37°C for 30 min. Samples are loaded by suction onto a nylon membrane using a dot blot apparatus according to the manufacturer and neutralized with 500 µL of 6X SSC; then, the DNA is ultraviolet (UV) crosslinked to the membrane and the membrane is air-dried. The membrane is probed in Church buffer plus Denhardt's and ssDNA *(11)* with a random-primed mtDNA fragment (*COXII* gene). When DNA is suitably crosslinked, greater than 80% of the DNA sediments in the six highest density fractions as indicated by dot blot analysis (e.g., *see* **ref. 9**). By comparison, in control samples prepared with no formaldehyde treatment all of the mtDNA pellets to the bottom fraction of the CsCl gradient.

7. In order to obtain adequate signals in most types of SDS-PAGE analysis, the resulting material needs to be concentrated. To do this, the material is diluted twofold with TE, adding linear acrylamide as a coprecipitate, adjusted to 0.3 *M* NaOAc, and precipitated with 1 volume of isopropanol. As an example, 40 µL of 3*M* NaOAc, 160 µL of water, and 5 µL of linear acrylamide is added to 200 µL of pooled fractions 1–5. After mixing, 400 µL of isopropanol is added to precipitate the crosslinked DNA. The DNA and associated proteins are pelleted by 5–10 min of centrifugation at maximum speed in a microfuge; the pellet is washed with 70% ethanol and dried. Pellets are boiled in SDS-PAGE loading buffer for 30 min to reverse crosslinks. Depending on the amount of material loaded, it may be necessary to add a few microliters of 0.5 *M* Tris-HCl, pH 6.5, to neutralize the samples (as indicated by the change in color of the indicator dye), as the formaldehyde is converted to formic acid upon heat release of protein. For protein identification, *see* **Note 6**.

3.3. UV Crosslinking

As another means of isolating proteins that directly bind mtDNA, we have employed UV crosslinking using BrdU-labeled mtDNA probes. Our procedure allows the functional identification of membrane-associated proteins with specific mtDNA sequences in vitro, utilizing both the zero crosslink distance and the titratable crosslinking efficiency of BrdU-mediated UV crosslinking. UV crosslinking of proteins to nucleic acids was first described in the 1960s *(12,13)* and later incorporated the use of BrdU-substituted DNA *(14)*. Electrophoresis of UV crosslinked protein–DNA complexes *(15)* led to the description of the binding of adenovirus transcription factor to the major late promoter *(16)*. Our procedure for UV crosslinking of proteins to mtDNA is derived from these reports and from Miyamoto et al. *(17)*.

3.3.1. Isolation of Mitochondrial Membrane Fractions

This procedure uses a published method to purify mitochondrial membranes from ρ^0 yeast strains that lack mtDNA. The membrane fractions contain both right-side out and inverted membranes.

1. To purify mitochondria from ρ^0 cells, Nycodenz step gradients (5–25%, 2 mL/step) are used (*see* **ref. 18**).
2. Membrane fragments are prepared by incubating mitochondria (10–20 mg) in 2 mL SB for 30 min at 4°C, freezing at –80°C, then sonicating (microtip, 70% of the maximal setting) for six 10-s bursts on ice with 30 s between treatments.
3. The membrane fragments are separated from the soluble proteins by centrifugation for 60 min at 226,000*g* and the pellet is resuspended in SB, then diluted with 1 volume of 50% glycerol and stored as aliquots at –20°C.

3.1.2. Generation of ^{32}P-labeled BrdU-Substituted mtDNA Fragments

The high A+T content of mitochondrial DNA requires modification of standard PCR conditions to increase the yield of radiolabeled, biotinylated, BrdU-substituted product.

1. ori5 sequences are amplified by PCR from a recombinant plasmid using biotinylated T7 and SP6 primers in 50-µL reactions containing 0.1 m*M* of dGTP and dCTP, 40 µ*M* of dATP, 10 µL of ^{32}P-dATP, 1 m*M* of BrdU, 2 µ*M* of each primer, and 5U of *Taq* polymerase. To generate adequate labeled DNA product, four 50-µL reactions are used with amplification cycles as follows: 94°C for 3 min, then 30 cycles of 94°C for 1 min, 55°C for 1.5 min, and 72°C for 2 min.
2. The amplified product is gel purified and eluted in 0.2 mL TE (Gene Clean) and then quantitated by EtBr fluorescence using a DNA mass ladder as a standard.

3.1.3. Crosslinking and Resolution of Polypeptides

1. Before crosslinking, DNA (10–20 ng) and protein (30 µg) are preincubated in binding buffer (30-µL volume) for 30 min at room temperature using 1.5-mL amber Eppendorf tubes (*see* **Note 7**).
2. For UV-induced crosslinking, this material is transferred at 4°C to lids from clear 1.5-mL Eppendorf tubes. A long-wavelength UV lamp (Black Ray, UVP) is positioned approx 5 cm from the open lids and the samples illuminated for 30 min. After the mixtures are transferred to Eppendorf tubes, the lids are rinsed with BB (<70 µL). The crosslinked material is solubilized by the addition of 0.5% NP-40. To digest the DNA, each sample is adjusted to 10 m*M* CaCl$_2$, then treated with DNase I (2 U) and MNase (20 U) at 37°C for 1 h and loaded directly onto a 0.75-mm-thick SDS-PAGE gel (*see* **Note 8**). After electrophoresis, the gel was dried and exposed for autoradiography.

4. Notes

1. Digestion of the cell wall is strain and growth phase dependent. The time of digestion is established by assaying sensitivity of the treated cells to dilution with water or SDS. In most experiments 35 min of digestion converts nearly all of the cells to spheroplasts. After breakage of properly digested cells there is

some debris in the supernatant and the low-speed pellet is soft. Alternatively, cells may be digested in SCE, pH 7.0, with yeast lytic enzyme at 1 mg/mL with gentle shaking at 30°C for 45 min.

2. Multiple centrifugation steps may be employed in nonpreparative experiments to decrease membrane contaminants (white halo around dark pellet). The addition of BSA (0.1–1 µg/mL) to the resuspended mitochondria prior to pelleting is optional.

3. The extent of crosslinking of proteins to mtDNA increases with the length of formaldehyde treatment (e.g., *see* **ref. 9**). It may be necessary to determine the optimal time of incubation of mitochondria with formaldehyde, as estimated by the decrease in density of mtDNA in CsCl gradients. Insufficient crosslinking decreases yield, whereas excessive crosslinking leads to a bimodal distribution of the crosslinked mtDNA, with a second peak higher in the gradient. DNA–protein complexes from the two peaks have different levels of Abf2p and Ilv5p, but the less dense peak is otherwise uncharacterized.

4. Although all of the DNA sediments in **step 4**, smaller-scale experiments (i.e., for Western blot analysis) may omit this centrifugation step. Note that when the step gradient is omitted, load less than the equivalent 5 mg of starting mitochondria in each CsCl gradient, as carbohydrates complicate fractionation. Crosslinked material may be further purified by a second CsCl density gradient.

5. To digest single-stranded DNA, **step 5** is modified. The nucleoid-enriched pellet is suspended in 1X CB (including BME, PMSF, and spermidine) in a minimal volume using a Dounce homogenizer. The solution is then adjusted to 150 mM NaCl, 30 mM $ZnSO_4$, RNase A (50 µg/mL), and 0.1% sarkosyl in 6 mL. Then, 400–1000 U of S1 nuclease is added and the samples incubated at room temperature for 1 h.

6. Keratin is a common contaminant in large-scale preparations and may cause interference in some protein identification methods, such as nanospray ion-trap mass spectrometry, but it does not affect Western blot analysis. Because concentrating the crosslinked protein–DNA complexes will also concentrate any contaminating keratin, in preparative applications it is recommended that siliconized, TFA, and acid-washed tubes be used to collect the cesium chloride gradient fractions and that similarly prepared containers be used for all subsequent steps. Filter-sterilization of solutions to remove dust and other solids may decrease keratin contamination. Because linear acrylamide is concentrated with the isopropanol-precipitated mtDNA, the coprecipitate may be extracted with phenol–chloroform to reduce the level of contaminating proteins. For preparative gels, we use commercial reagents to ensure purity (i.e., premixed acrylamide, filter-sterilized running buffers, 3X Laemmli loading buffer).

7. Prior to initiating crosslinking reactions, preliminary EMSA experiments were performed to establish conditions that resulted in efficient binding of the protein preparations to DNA fragments of interest. Recombinant Abf2p was often used as a control for UV crosslinking. These conditions were reproduced during the

initial preincubation of the membrane protein fraction and the ori5 DNA probe prior to UV crosslinking.

8. To prepare proteins for identification, the UV crosslinked, solubilized material was absorbed to BB-equilibrated streptavidin–agarose for 30 min on ice. The protein–DNA adducts were washed 10 times with several volumes of BB plus NP-40. After washing, the bound proteins were released by nuclease digestion (5X increase in DNase I) and resolved by SDS-PAGE. The gel was silver stained, and the bands identified by MALDI-TOF mass spectrometry. The identification of Hsp60p was confirmed by Western blot analysis. It should be noted that using a short-wavelength UV source (Stratagene) produces a similar label transfer pattern.

Acknowledgments

This work was supported by grant GM33510 from the National Institutes of Health and by grants I-6042 and IL-1211 from the Robert A. Welch Foundation.

References

1. Van Tuyle, G. C. and McPherson, M. L. (1979) A compact form of rat liver mito-chondrial DNA stabilized by bound proteins. *J. Biol. Chem.* **254,** 6044–6053.
2. Suzuki, T., Kawano, S., and Kuroiwa, T. (1982) Structure of three-dimensionally rod-shaped mitochondrial nucleoids isolated from the slime mould *Physarum polycephalum. J. Cell Sci.* **58,** 241–261.
3. Miyakawa, I., Sando, N., Kawano, K., Nakamura, S., and Kuroiwa, T. (1987) Isolation of morphologically intact mitochondrial nucleoids from the yeast, *Saccharomyces cerevisiae. J. Cell Sci.* **88,** 431–439.
4. Miyakawa, I., Okazakihigashi, C., Higashi, T., Furutani, Y., and Sando, N. (1996) Isolation and characterization of mitochondrial nucleoids from the yeast Pichia jadinii. *Plant Cell Physiol.* **37,** 816–824.
5. Newman, S. M., Zelenaya-Troitskaya, O., Perlman, P. S., and Butow, R. A. (1996) Analysis of mitochondrial DNA nucleoids in wild-type and a mutant strain of *Saccharomyces cerevisiae* that lacks the mitochondrial HMG-box protein, Abf2p. *Nucleic Acids Res.* **24,** 386–393.
6. Giorno, R., Hecht, R. M., and Pettijohn, D. (1975) Analysis by isopycnic cen-trifugation of isolated nucleoids of *Escherichia coli. Nucleic Acids Res.* **2,** 1559–1567.
7. Orlando, V. (2000) Mapping chromosomal proteins in vivo by formaldehyde-crosslinked-chromatin immunoprecipitation. *Trends Biochem. Sci.* **25,** 99–104.
8. Xu, C. and Ray, D. S. (1993) Isolation of proteins associated with kinetoplast DNA networks in vivo. *Proc. Natl. Acad. Sci. USA* **90,** 1786–1789.
9. Kaufman, B. A., Newman, S. M., Hallberg, R. L., Slaughter, C. A., Perlman, P. S., and Butow, R. A. (2000) *In organello* formaldehyde crosslinking of proteins to mtDNA: identification of bifunctional proteins. *Proc. Natl. Acad. Sci. USA* **97,** 7772–7777.

10. Parikh, V. S., Morgan, M. M., Scott, R., Clements, L. S., and Butow, R. A. (1987) The mitochondrial genotype can influence nuclear gene expression in yeast. *Science* **235,** 576–580.
11. Sambrook, J. and Russell, D. W. (2001) *Molecular Cloning: A Laboratory Manual.* Cold Spring Harbor Press, Cold Spring Harbor, N.Y.
12. Smith, K. C. (1962) Dose dependent decrease in extractability of DNA from bacteria following irradiation with ultraviolet light or with visible light plus dye. *Biochem. Biophys. Res. Commun.* **8,** 157–163.
13. Smith, K. C. (1969) Photochemical addition of amino acids to ^{14}C–uracil. *Biochem. Biophys. Res. Commun.* **34,** 354–357.
14. Lin, S. Y. and Riggs, A. D. (1974) Photochemical attachement of *lac* repressor to bromodeoxyuridine-substituted *lac* operator by ultraviolet radiation. *Proc. Natl. Acad. Sci. USA* **71,** 947–951.
15. Hillel, Z. and Wu, C. W. (1978) Photochemical cross-linking studies on the interaction of Escherichia coli RNA polymerase with T7 DNA. *Biochemistry* **17,** 2954–2961.
16. Chodosh, L. A., Carthew, R. W., and Sharp, P. A. (1986) A single polypeptide possesses the binding and transcription activities of the adenovirus major late transcription factor. *Mol. Cell. Biol.* **6,** 4723–4733.
17. Miyamoto, S., Cauley, K., and Verma, I. M. (1995) Ultraviolet cross-linking of DNA binding proteins. *Methods Enzymol.* **234,** 632–641.
18. Glick, B. S. and Pon, L. A. (1995) Isolation of highly purified mitochondria from Saccharomyces cerevisiae. *Methods Enzymol.* **260,** 213–223.

27

Breeding and Genotyping
of *Tfam* Conditional Knockout Mice

Mats Ekstrand and Nils-Göran Larsson

1. Introduction

Respiratory chain dysfunction is an important contributor to human pathology *(1–3)*. The generation of animal models has much facilitated in-depth studies of pathogenetic mechanisms in mitochondrial disease *(4)*. The function of the respiratory chain is subject to the dual genetic control of both the nuclear and mitochondrial genomes *(1)*.

Mitochondrial DNA (mtDNA) encodes 13 essential subunits of the respiratory chain as well as 22 transfer RNAs and 2 ribosomal RNAs *(5,6)*. Still, the nuclear genome contributes the majority of the subunits constituting the respiratory chain as well as all factors necessary for replication and transcription of mtDNA. One such factor actually necessary for both transcription and replication of mtDNA is mitochondrial transcription factor A (Tfam; previously mtTFA) *(7)*. Tfam is a DNA-binding HMG-box protein that is synthesized in the cytoplasm and is imported into mitochondria. Within mitochondria, it is essential for the initiation of transcription and, subsequently, respiratory chain function (*see* **Fig. 1**). The replication of mtDNA is dependent on an RNA primer generated through transcription *(8)*. This link between an RNA primer and DNA synthesis is at least one of the reasons why also the replication and maintenance of mtDNA is dependent on Tfam. Furthermore, Tfam has a more general DNA-binding ability, raising the question of whether it stabilizes mtDNA in a way like histones do in the context of nuclear DNA.

In order to study the in vivo effects of loss of mtDNA and respiratory chain function we have generated a mouse with a tissue-specific inactivation of the *Tfam* gene *(9)*. To achieve this, we used the bacteriophage P1 recombi-

From: *Methods in Molecular Biology, vol. 197: Mitochondrial DNA: Methods and Protocols*
Edited by: W. C. Copeland © Humana Press Inc., Totowa, NJ

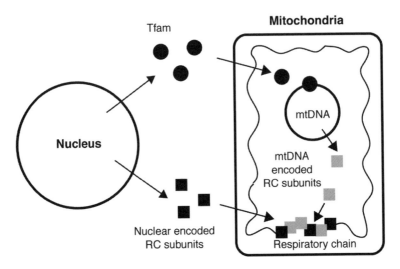

Fig. 1. The biogenesis of the respiratory chain is under dual genetic control. Nuclear genes encode the majority of the respiratory chain subunits (black squares) whereas mtDNA encodes 13 subunits (gray squares). The nuclear and mtDNA-encoded subunits are assembled together to form the respiratory chain. Regulation of mtDNA replication and transcription is performed by nuclear genes. Mitochondrial transcription factor A (Tfam) has a role in both mtDNA transcription and replication. Loss of *Tfam* causes loss of mtDNA and mitochondrial transcripts, which results in a severe respiratory chain deficiency.

nation—or *cre–loxP*—system *(10)*. The strategy is based on the ability of the bacteriophage P1 *cre* (causes recombination) enzyme to recognize short specific sequences (*loxP* sites) and to induce recombination between such sequences *(11)*. The *loxP* site is a 34-bp (base pair) stretch of DNA containing two 13-bp inverted repeats separated by an 8-bp asymmetric spacer determining the direction of the *loxP* site. If the *cre* recombinase finds two *loxP* sites oriented in the same direction, it will induce a recombination event between them, and all DNA intervening the two sites will be deleted. By homologous recombination in embryonic stem (ES) cells, we generated a mouse with exons 6 and 7 as well as the polyA signal of the *Tfam* gene flanked by *loxP* sites (*Tfam^{loxP}*) *(9)*. The insertion of *loxP* sites does not affect the function of the *Tfam* gene because homozygous (*Tfam^{loxP}/Tfam^{loxP}*) animals have a normal life-span as well as normal mtDNA and mtRNA levels *(12)*. These mice can be mated to other transgenic mice expressing *cre* recombinase under the control of different tissue-specific promoters (for a list of available *cre*-expressing animals: http://www.mshri.on.ca/nagy/). This will induce a recombination of the *Tfam^{loxP}* allele only in those tissues where *cre* is expressed, causing a depletion of mtDNA, mtRNA, and respiratory chain function. We have successfully bred *Tfam^{loxP}/Tfam^{loxP}* mice to mice expressing *cre* from heart-, skeletal muscle-, pancreas-, testis-, and brain-specific promoters *(12–14)* and unpublished results.

In this chapter, we describe how to generate mice with a tissue-specific mtDNA depletion by mating mice homozygous for the *loxP*-flanked *Tfam* allele (*Tfam^{loxP}/Tfam^{loxP}*) to mice expressing *cre* recombinase under the control of a Tissue-Specific Promoter (+/*TSP–cre*). Furthermore, we show how to use polymerase chain reaction (PCR) for genotyping the *Tfam* knockout mice and to control for unspecific recombination of the *Tfam^{loxP}* allele. By using a three-primer multiplex PCR, it is possible to distinguish between the normal (*Tfam*) allele, the *loxP*-flanked (*Tfam^{loxP}*) allele, and the recombined (*Tfam^-*) allele. The presence of the *cre* gene is detected with a normal PCR reaction using *cre* specific primers. Finally, we describe how to estimate the recombination frequency by *Pst*I digestion of genomic DNA, Southern blotting, and phosphorimager quantification.

2. Materials

2.1. Solutions

1. 5X TBE: 54 g Tris base, 27.5 g boric acid, 20 mL of 0.5 *M* ethylenediamine tetraacetic acid (EDTA), pH 8.0, and distilled water (dH$_2$O) up to 1 L. Use a 1 : 10 dilution (0.5X TBE).
2. 3X Loading buffer: 0.125% bromphenol blue, 0.125% xylene cyanol FF, and 20% sucrose in dH$_2$O.

2.2. Preparation of Tail DNA

1. Tail lysis solution: 0.5% sodium dodecyl sulfate (SDS), 0.1 *M* NaCl, 50 m*M* Tris-HCl, pH 8.0, 2.5 m*M* EDTA (store at room temperature).
2. Proteinase K: 10 mg/mL in 50% glycerol (store at –20°C).
3. 8 *M* Potassium acetate.
4. Chloroform.
5. TE: 10 m*M* Tris-HCl, pH 8.0, 1 m*M* EDTA, pH 8.0.

2.3. PCR Genotyping of Transgenic Mice

1. 1.25 m*M* dNTP: 12.5 µL each of 100 m*M* dATP, dGTP, dCTP, and dTTP (Amersham Pharmacia Biotech) and 950 µL dH$_2$O.
2. 10X PCR buffer II (Perkin-Elmer).
3. 25 m*M* MgCl$_2$ (Perkin-Elmer).
4. AmpliTaq® DNA polymerase, 5 U/µL (Perkin-Elmer).

2.4. Estimation of Recombination Frequency by Southern Blot

1. NEBuffer 3 (New England BioLabs).
2. *Pst*I (New England BioLabs).

3. 5 M NaCl.
4. Denaturing solution: 1.5 M NaCl, 0.5 M NaOH.
5. Neutralizing solution: 1.5 M NaCl, 1.5 M Tris-HCl (pH 7.5).
6. 20X SSC: 175.3 g NaCl and 88.2 g sodium citrate in 800 mL H_2O. Set pH to 7.0 with NaOH and adjust volume to 1 L.
7. Hybridization solution: 24 g dextrane sulfate, 87 mL dH_2O and 48 mL of 20X SSC. Dissolve by heating. Add 96 mL formamide, 24 mL 100X Denhart's and 1.5 mL of 1 M Tris-HCl, pH 7.5. Filter before use.
8. [α-^{32}P] dCTP Easytides™ (NEN).
9. G-25 Quick Spin Columns (Boehringer Mannheim).
10. Prime-It® RmT Random Primer Labeling Kit (Stratagene).
11. Hybond-C extra nitrocellulose membrane (Amersham).
12. Washing buffer: 1X SSC, 0.1% SDS.

2.5. PCR Primers

Genotyping of the *Tfam* locus: Tfam-A 5′–CTG CCT TCC TCT AGC CCG GG–3′; Tfam-B 5′–GTA ACA GCA GAC AAC TTG TG–3′; Tfam-C 5′–CTC TGA AGC ACA TGG TCA AT–3′

Genotyping of Cre: Cre-F 5′–CAC GAC CAA GTG ACA GCA AT–3′; Cre-R 5′–AGA GAC GGA AAT CCA TCG CT–3′

All primers should be used at a concentration of 10 pmol/μL.

3. Methods
3.1. Mating Strategy of Transgenic Mice

In order to obtain mice that are both homozygous for the *loxP*-flanked *Tfam* allele and carry one copy of the *cre* transgene (*Tfam^loxP/Tfam^loxP*, +/*TSP–cre*), a two-step mating protocol is used. First, a mating consisting of up to three *Tfam^loxP/Tfam^loxP* females and one +/*TSP–cre* male is set up. This cross will give pups with two different genotypes (+/*Tfam^loxP*, +/*TSP–cre*, and +/*Tfam^loxP*) in a 1:1 ratio. When these F1 pups have been genotyped, new matings are set up. The double heterozygous (+/*Tfam^loxP*, +/*TSP–cre*) F1 offspring are back-crossed to homozygous *Tfam^loxP/Tfam^loxP* animals. This will result in four different genotypes (+/*Tfam^loxP*; +/*Tfam^loxP*, +/*TSP–cre*; *Tfam^loxP/Tfam^loxP*, and *Tfam^loxP/Tfam^loxP*, +/*TSP–cre*), each at a frequency of 25% (*see* **Notes 1** and **2**).

3.2. Preparation of Tail DNA

This section describes how to prepare DNA from the small piece of tail that is collected when pups are weaned and marked at about 3 wk of age. It is preferred to collect the tail in a 1.5-mL screw-cap tube because this will minimize the risk of leakage during the lysis process.

1. Add 0.4 mL tail lysis solution and 8 µL proteinase K (10 mg/mL) to the screw-cap microfuge tube. The tail does not have to be minced before digestion. Incubate at 55–60°C on a shaker during very vigorous shaking (approx 250 rpm). Lysis is complete when only a few bone fragments are visible in the tube. This will take between 2 and 12 h, depending on how effective the shaking is.

2. Add 75 µL of 8 *M* potassium acetate and 0.5 mL chloroform to each tube. Vortex and incubate at –70°C for about 15 min. The samples could also be stored at –20°C overnight.

3. Spin at max speed in a benchtop centrifuge for 10 min. Two phases, an upper aqueous phase and a lower chloroform phase, separated by an interface with precipitated proteins will be visible. Transfer the aqueous (upper) phase to a new 1.5-mL tube. Try to get as close to the interface as possible.

4. Add 1 mL of 95% ethanol to the collected aqueous phase at room temperature and invert the tubes several times. At this stage, you should be able to see a white precipitate floating around in the tube.

5. Spin at max speed in a benchtop centrifuge for 10 min. Rinse the pellet with 0.5 mL of 70% ethanol and spin for another 5 min at max speed.

6. Remove as much as possible of residual ethanol and let the pellet air-dry. Dissolve in 100 µL TE. DNA samples can be stored at –20°C for several years.

3.3. PCR Genotyping of Transgenic Mice

This section describes how to use PCR to both detect the presence of the *cre* gene and to distinguish between the wild-type *Tfam*, *loxP*-flanked *Tfam^{loxP}*, and recombined *Tfam⁻* alleles.

3.3.1. Cre PCR

To detect the presence of *cre*, we use the primers Cre-F and Cre-R that are located within the coding part of the *cre* transgene. This ensures that the primers will hybridize with all *cre* transgenes independent of the promoter sequence.

3.3.2. Tfam Multiplex PCR

In order to distinguish between the three different *Tfam* alleles, we use a three-primer PCR approach. The primers Tfam-A and Tfam-B are located around the first *loxP* site in *Tfam^{loxP}* mice. In *Tfam* and *Tfam^{loxP}* mice, they will give products of 404 and 437 bp, respectively, because of the addition of the *loxP* site. In a recombined allele, the Tfam-B primer site will be missing because this is deleted during the recombination event. Instead, the Tfam-C primer, located after the second *loxP* site, will be in proximity to the Tfam-A primer, giving a product of 329 bp (*see* **Fig. 2A**). These PCR products can easily be separated on an agarose gel (*see* **Fig. 2B**).

A

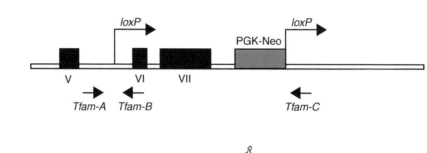

B

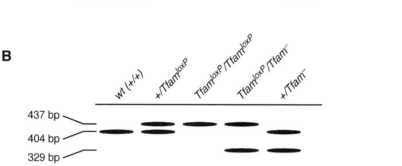

Fig. 2. The *loxP*-flanked *Tfam* locus and PCR genotyping of different *Tfam* alleles: (**A**) the *Tfam*^{loxP} locus and location of different primers used for genotyping; (**B**) schematic drawing showing expected DNA fragments obtained after genotyping of different *Tfam* alleles.

1. Make a PCR mix by mixing the following (for 10 reactions): 32 µL of 1.25 m*M* dNTP, 20 µL of 10X PCR buffer II, 12 µL of 25 m*M* MgCl$_2$, 4 µL each of primers Cre-F and Cre-R or primers Tfam-A, Tfam-B, and Tfam-C (10 pmol/µL), 1 µL AmpliTaq DNA polymerase, and dH$_2$O up to 95 µL.
2. Make 19-µL aliquots of the reaction mix in 0.5-mL PCR tubes and add a drop of mineral oil (if necessary with the thermal cycler used).
3. Add 0.5 µL tail DNA to each PCR tube.
4. Both the *cre* genotyping PCR and the *Tfam* genotyping multiplex PCR are run with the same conditions: 5-min denaturation at 95°C, followed by 35 cycles with 30-s denaturation at 95°C, 30-s annealing at 53°C, and 30-s extension at 72°C, and one final synthesis step of 5 min at 72°C.
5. When thermal cycling is complete, add 10 µL of 3X loading buffer to each tube.
6. Load 15 µL of the samples on a 2% agarose gel (2 g agarose, 100 mL of 0.5X TBE and 2 µL of 10 mg/mL ethidium bromide).
7. Run the gel in 0.5X TBE at 7 V/cm for 45 min.

3.3. Estimation of Recombination Frequency by Southern Blot

This section describes how to estimate the recombination frequency of the *loxP*-flanked *Tfam* allele by Southern blot analysis. When hybridizing a random

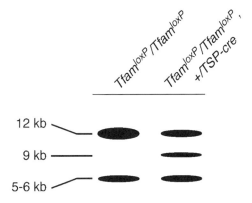

Fig. 3. Schematic drawing showing expected DNA fragments obtained after Southern blot analysis of tissue-specific knockouts. The 12-kb and 5-6kb fragments are obtained after Southern blot analysis of *Pst*I-digested genomic DNA from *Tfam^loxP/Tfam^loxP* animals. The 9-kb fragment results from *cre–loxP*-mediated recombination of the *Tfam^loxP* locus and corresponds to the *Tfam–* allele.

labeled *Tfam* cDNA probe to *Pst*I digested wild-type or *Tfam^loxP/Tfam^loxP* genomic DNA, a large band of 12 kb and a smaller band of approx 5–6 kb will be seen. A recombined *Tfam⁻* allele will give a pattern where the larger 12-kb band is replaced by a shorter 9-kb fragment because of the excision of all DNA intervening the *loxP* sites (*see* **Fig. 3**). In a homogenate of tissue, the relation between this 9-kb "knockout" band and the 12-kb "normal" band can be used to calculate the recombination frequency of the *Tfam^loxP* allele (*see* **Note 3**).

3.3.1. Southern Blot

1. Mix 10–20 µg DNA, 20 µL NEBuffer 3, 10 µL *Pst*I, and dH₂O up to 200 µL. Incubate at 37°C overnight.
2. Precipitate digested DNA by adding 20 µL of 5 *M* NaCl and 500 µL of 95% ethanol to the digestions. Mix and store at –70°C for 30 min. Spin at max speed in a microfuge for 15 min at 4°C, wash with 70% ethanol, and air-dry the pellet. Dissolve in 20 µL dH₂O and 10 µL of 3X loading buffer.
3. Load samples and a size marker on a 0.7% agarose gel (0.7 g agarose, 100 mL of 0.5X TBE, and 2 µL ethidium bromide). Run at 1 V/cm for 16–20 h in 0.5X TBE.
4. Take a picture of the gel and a ruler.
5. Denature the gel by shaking two times for 15 min in denaturation solution. Rinse in dH₂O.
6. Neutralize the gel by shaking two times for 20 min in neutralizing solution. Rinse in dH₂O.

7. Blot over night by sandwiching two long 3MM papers, one short 3MM paper, the gel with edges covered by plastic film, a gel-sized nitrocellulose membrane, a gel-sized 3MM paper, and many paper tissues. The bottom 3MM papers should be presoaked in 20X SSC and the ends of the long 3MM papers should be hanging into a container of 20X SSC. The membrane and top 3MM paper should be presoaked in dH$_2$O. Put a 600-g weight on top of the stack of paper tissues.

8. After blotting, quickly rinse the membrane in 6X SSC and bake it at 80°C under vacuum for 1–2 h.

3.3.2. Preparation of Random Labeled Probe

1. Add 50 ng of *Tfam* cDNA to a single-use reaction tube. Add dH$_2$O to a total volume of 42 μL. Mix, spin down, and transfer to a PCR tube.

2. Boil or incubate at 95°C in a thermal cycler for 10 min. Cool to room temperature.

3. Add 5 μL of [α-^{32}P]dCTP Easytides and 3 μL Magenta polymerase. Mix properly and spin down. Incubate at 37°C for 10 min.

4. Add 2 μL stop solution. Mix, spin down, and put on ice.

5. Spin through a Quick Spin Column to remove unincorporated nucleotides.

6. Boil the probe for 5 min just before adding to hybridization solution.

3.3.3. Hybridization of Probe

1. Prehybridize membrane in a rolling tube with hybridization solution for 4 h at 42°C.

2. Add boiled probe to new hybridization solution preheated to 42°C. Mix and replace prehybridization solution with the new solution containing the probe.

3. Incubate at 42°C overnight in a rolling tube (>16 h).

4. Wash membrane in washing buffer for 15 min at room temperature and two times for 15 min at 65°C.

5. Expose film between 30 min and 2 h, depending on signal strength.

6. Use phosphorimager for quantification.

4. Notes

1. Some *cre* transgenes may be integrated on the same chromosome as the *Tfam* gene. If that is the case, *cre* and *Tfam* will, of course, be linked and the genotypes will not be of the expected Mendelian distribution of 25% of each *(12)*. If they are not very closely linked, however, it will be possible to obtain homozygous mutant animals (*Tfam^{loxP}/Tfam^{loxP}*, +/*TSP–cre*).

2. In a few cases, heterozygous mutant mice (+/*Tfam^{loxP}*, +/*TSP–cre*) give germ cells with a recombined allele (*Tfam$^-$*). This is the result of unspecific expression of the *cre* transgene during oogenesis or spermatogenesis leading to germline disruption of *Tfam*. It should be noted that sperm cells are connected with cytoplasmic bridges during spermatogenesis and this may result in the production of haploid *Tfam$^-$* spermatozoa lacking the *cre* transgene.

3. Quantification from Southern blots can be utilized to calculate the *Tfam* recombination frequency. However, it is important to remember that the hybridization

efficiency of the random labeled probe depends not only on the molar amount of the DNA fragments of interest but also on the length of the fragments (i.e., the number of nucleotides the fragments contain). This means that even if the actual recombination frequency in a given case is 50%, the probe will not bind equally to the recombined 9-kb fragment and the 12-kb normal fragment. This is not an issue if the goal of the experiment is to measure the *relative* recombination frequency for studying changes over time or between different samples. On the other hand, if the goal is to measure the *absolute* recombination frequency of the *Tfam*^{loxP} allele, this discrepancy has to be taken into consideration. One way of doing this is to include genomic DNA from germline heterozygous knockout mice (+/*Tfam*⁻) on the same blot as the samples to be measured. Germline heterozygous knockouts (*see* **Note 2**) have one copy of the recombined *Tfam* allele in every cell (i.e., a recombination frequency of exactly 50% in all tissues). By including DNA from such +/*Tfam*⁻ animals, homozygous *Tfam*^{loxP}/ *Tfam*^{loxP} control animals, as well as two different mixes of these DNAs, it is possible to create a "dilution curve" with, for example, 0%, 20%, 40%, and 50% recombination.

References

1. Larrson, N. and Clayton, D. (1995) Molecular genetic aspects of human mitochondrial disorders. *Annu. Rev. Genet.* **29**, 151–178.
2. Graff, C., Clayton, D. A., and Larrson, N. G. (1999) Mitochondrial medicine— recent advances. *J. Intern. Med.* **246**, 11–23.
3. Larsson, N. G. and Luft, R. (1999) Revolution in mitochondrial medicine. *FEBS Lett.* **455**, 199–202.
4. Wallace, D. C. (1999) Mitochondrial diseases in man and mouse. *Science* **283**, 1482–1488.
5. Anderson, S., Bankier, A. T., Barell, B. G., deBruijn, M. H. L., Coulson, A. R., Drouin, J., et al. (1981) Sequence and organization of the human mitochondrial genome. *Nature* **290**, 457–465.
6. Bibb, M. J., VanEtten, R. A., Wright, C. T., Walberg, M. W., and Clayton, D. A. (1981) Sequence and organization of mouse mitochondrial DNA. *Cell* **26**, 167–180.
7. Parisi, M. A., Xu, B., and Clayton, D. A. (1993) A human mitochondrial transcriptional activator can functionally replace a yeast mitochondrial HMG-box protein both in vivo and in vitro. *Mol. Cell. Biol.* **13**, 1951–1961.
8. Clayton, D. A. (1991) Replication and transcription of vertebrate mitochondrial DNA. *Annu. Rev. Cell Biol.* **7**, 453–478.
9. Larsson, N. G., Wang, J., Wilhelmsson, H., Oldfors, A., Rustin, P., Lewandoski, M., et al. (1998) Mitochondrial transcription factor A is necessary for mtDNA maintenance and embryogenesis in mice. *Nature Genet.* **18**, 231–236.
10. Gu, H., Marth, J. D., Orban, P. C., Mossmann, H., and Rajewsky, K. (1994) Deletion of a DNA polymerase β gene segment in T cells using cell type-specific gene targeting. *Science* **265**, 103–106.

11. Guo, F., Gopaul, D. N., and van Duyne, G. D. (1997) Structure of Cre recombinase complexed with DNA in a site-specific recombination synapse. *Nature* **389,** 40–46.
12. Wang, J., Wilhelmsson, H., Graff, C., Li, H., Oldfors, A., Rustin, P., et al. (1999) Dilated cardiomyopathy and atrioventricular conduction blocks induced by heart-specific inactivation of mitochondrial DNA gene expression. *Nature Genet.* **21,** 133–137.
13. Li, H., Wang, J., Wilhelmsson, H., Hansson, A., Thoren, P., Duffy, J., et al. (2000) Genetic modification of survival in tissue-specific knockout mice with mitochondrial cardiomyopathy. *Proc. Natl. Acad. Sci. USA* **97,** 3467–3472.
14. Silva, J. P., Kohler, M., Graff, C., Oldfors, A., Magnuson, M. A., Berggren, P. O., et al. (2000) Impaired insulin secretion and beta-cell loss in tissue-specific knockout mice with mitochondrial diabetes. *Nature Genet.* **26,** 336–340.

Index

From: *Methods in Molecular Biology, vol. 197, Mitochondrial DNA: Methods and Protocols*
Edited by: W. C. Copeland © Humana Press Inc., Totowa, NJ